Exercises for Introductory Statistics and Probability

K. M. Brown *(with contributions by C. P. Gregory and Y. Feinman)*

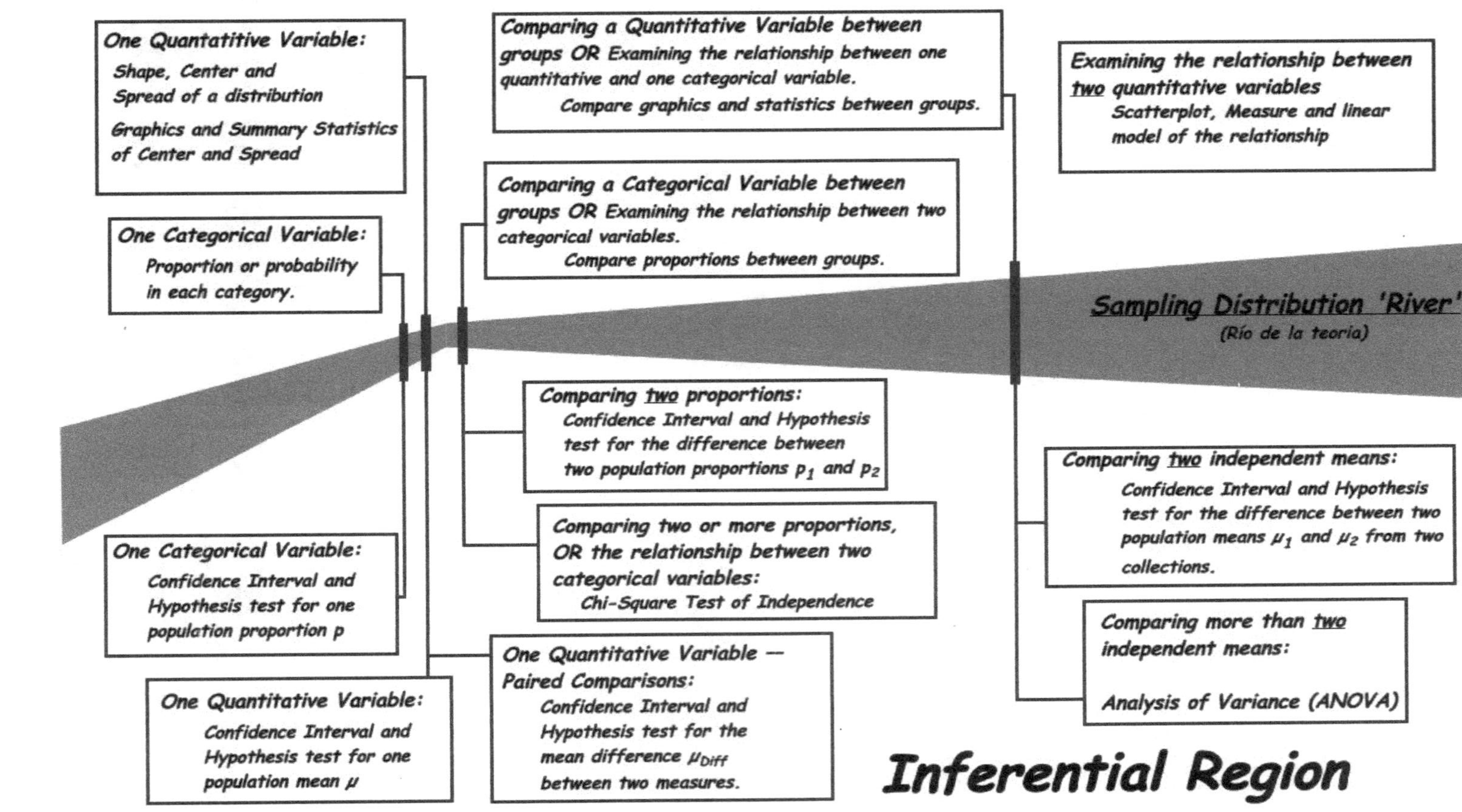

Descriptive Region
One Quantatitive Variable:
Shape, Center and Spread of a distribution
Graphics and Summary Statistics of Center and Spread
One Categorical Variable:
Proportion or probability in each category.
Comparing a Quantitative Variable between groups OR Examining the relationship between one quantitative and one categorical variable.
Compare graphics and statistics between groups.
Comparing a Categorical Variable between groups OR Examining the relationship between two categorical variables.
Compare proportions between groups.
Examining the relationship between two quantitative variables
Scatterplot, Measure and linear model of the relationship
Sampling Distribution 'River'
(Río de la teoria)
Comparing two proportions:
Confidence Interval and Hypothesis test for the difference between two population proportions p_1 and p_2
Comparing two or more proportions, OR the relationship between two categorical variables:
Chi-Square Test of Independence
One Categorical Variable:
Confidence Interval and Hypothesis test for one population proportion p
One Quantitative Variable:
Confidence Interval and Hypothesis test for one population mean μ
One Quantitative Variable — Paired Comparisons:
Confidence Interval and Hypothesis test for the mean difference μ_{Diff} between two measures.
Comparing two independent means:
Confidence Interval and Hypothesis test for the difference between two population means μ_1 and μ_2 from two collections.
Comparing more than two independent means:
Analysis of Variance (ANOVA)
Inferential Region

Table of Contents: Exercises

Unit 1: Describing distributions

Unit 2: Describing relationships between variables

Unit 3: Beyond description: trusting data

Preface: Some Questions about the Course in Statistics

Why is there a sea star on the cover?

Well, it's a nice picture. But it is also because statistics is about everything – sea stars, muscles and anemones included, which is probably why your university system has statistics as a requirement for your program of study. Even without the sea star, think of all of the information (we call it data) that is collected from people when they use social media, shop on-line, even talk. Everywhere you turn, there are data --- masses of data. *The purpose of statistics is to make sense of these data, to see if the data answer questions, to see if the data tell a story.* Tons of data just sitting there do not do it, do not show what patterns or relationships there may be lurking there. Making pictures (a better term is graphics) is a good start. A good graphic may show patterns otherwise hidden, and we now have good software to make the graphics easily. But that is jumping ahead just a bit.

Is statistics hard?

A much better and more useful question is to ask what kinds of challenges you are likely to encounter (how the course is "hard"). Here is a list of them:

- There is a body of technical terminology that must be mastered – a new language, in a sense. Statisticians use what look like common words, but give those words a technical sense connected that deviates somewhat from the "normal" usage. An example is the word "random."
- Statistics uses mathematical symbolism, including some symbols that may be new to you. Having said that, being able to get the answers using formulas is *definitely not* the most important thing, as it may have been in previous mathematics courses. We have software to do calculation; we only need to know what the software is doing, which is harder. Still, the advice is: do not fear symbols; indeed, embrace symbolism!
- What *is* important, and harder is to be able to say what the numbers that are calculated and the graphics that are produced *mean* in the light of questions that we are asking. We call this *interpretation.* This text is focused on interpretation; it gives examples in the ***Notes*** section, and the ***Exercises*** give practice.
- Finally, expect some abstract ideas, and some logic that may take days – perhaps even weeks – to understand. These difficult ideas and logic come after some weeks into the course, and have to do with the thorny question of whether we can actually trust the data that are collected to represent reality faithfully.

What will we be doing?

The core of the course is in the ***Exercises***. They are designed to lead you through how statisticians analyze data (make sense of data) by actually analyzing data. The idea is that one learns by actually using the techniques that statistician use with data. The ***Notes*** provide backup with definitions of technical terms, and explanations. Probably a good strategy is to read through the ***Notes*** for a particular section before doing the exercises, without worrying too much if you do not understand it all at first reading; the questions in the ***Exercises*** will probably address some of the possible confusions.

Practical Requirements for the Course

- ***Online Resources*** It is likely that your college or university, as a part of its course management system, has a site devoted to the course that you are taking; that site should be your first recourse for information. Specifically, there may be online exercises and quizzes. Otherwise, the website for these materials is found at: http://www.cleonestats.blogspot.com/.
- ***Data Sets***: You will be analyzing data – lots of data. The data sets are ***not*** included with the software for carrying out analyses. Rather, they will be found either at the course site at your college or university's course management system, or at http://www.cleonestats.blogspot.com/

For Instructors

§1.1 Exercises on Starting Out

1. ***California Statistics Students*** Here is part of a spreadsheet and the description of variables for data collected in four statistics classes at a college in California in 2009.

Combined ClassData Aut 09

	Gender	AgeYe...	Height	MothersAge	PolticalView	Langua...	NumberHou...	NumberCars	NumberStates	Instructor...	Tattoo
80	M	20	169.0	26	Moderate	2	4	5	2	48	N
81	F	18	160.5	30	Liberal	1	5	3	4	50	N
82	M	19	183.0	27	Moderate	2	5	4	7	52	N
83	F	20	175.0	24	Moderate	1	3	1	6	55	N
84	F	17	173.0	34	Liberal	1	5	4	4	21	N
85	F	25	155.0	34	Liberal	2	3	3	4	62	Y
86	M	21	173.0	23	Liberal	3	2	2	11	55	Y

Variables:	***Description***
Gender:	*Gender of the student: male or female*
Age Years:	*Age of the student (in years)*
Height:	*Height of the student, measured in centimeters*
Mother's Age:	*Age of the student's mother when the student was born*
Political View:	*Self-reported "political leaning" of the student: liberal, moderate, conservative*
Languages:	*Languages spoken by the student*
Number Household:	*Number of people in the household in which the student lives*
Number Cars:	*Number of cars in the household in which the student lives*
Number States:	*Number of states in the USA the student has visited or lived in*
Instructor's Age:	*"Guesstimate" by the student of the age of the instructor*
Tattoo:	*Whether or not the student has a tattoo: yes or no*

Carefully read the description of the variables given above and answer the questions below.

a. List which variables listed in the description are *quantitative* and which variables are *categorical.*

b. What are *possible* values for the variable *Number Household*? (See the ***Notes*** for a definition of *values.* Also, notice that this question asks about *possible* values, not just in the data we have but "absolutely." Is it possible for a household to have fifteen people? Twenty people? Forty people?)

c. A student answers part b by listing: 2, 3, 4, 5 because that is what the student sees in the case table above. Why is ("2, 3, 4, 5") *not* an answer to the question about *possible* values?

d. What values for the variable *Number Household* do you think are likely, and what values do you think are not at all likely? Explain in complete sentences, giving brief reasons for your answer.

e. What are *possible* values for the variable *Number States*?

f. What values for the variable *Number States* do you think are likely and what values not likely?

g. According to the variable description, what are the possible values for the variable *Political View?*

Combined ClassData Aut 09

		Gender M	Gender F	Row Summary
PolticalView	Liberal	35	51	86
	Moderate	38	31	69
	Conservative	5	7	12
	Column Summary	78	89	167

S1 = count( )

h. Calculate the proportion of male students who have "moderate" political views (answer: 48.7%).

i. Calculate the proportion of female students who have "moderate" political views (answer: 34.8%).

j. Use the results of parts h and i to give an answer to the *statistical question:* "Are male or female students in this collection more likely to have 'moderate' political views?"

2. ***Far West Colleges and Universities.*** Here is a spreadsheet showing another collection. This time, we have a collection of colleges and universities from the far West of the United States. (The collection has colleges and universities from the states of Alaska, California, Hawaii, Nevada, Oregon, and Washington.)

FarWestCollegesCollection

	Name	St...	FullTim...	IntlStu...	FTE	PctOverAg...	PctFullTime	Freshman...	Graduation_rate	TuitionF...	Sector
361	Wenatchee Valley College	WA	1952	1	2342	40	63	60	36	2541	Public 2-...
362	West Hills Community College	CA	2158	51	3059	37	45	64	40	624	Public 2-...
363	West Los Angeles College	CA	2121	113	4354	54	24	58	20	810	Public 2-...
364	West Valley College	CA	3314	62	5406	41	35	66	41	708	Public 2-...
365	Western Nevada Community...	NV	954	0	2292	57	19	56	20	1618	Public 2-...
366	Western Oregon University	OR	3800	57	4017	14	88	67	43	4488	Public 4-...

Variables	***Description***
Name	*Name of the college or university*
State	*State in which the college or university is located*
Full-Time Students	*Number of full-time students enrolled*
Int'l Students	*Number of international students enrolled*
FTE	*Full-time equivalent number of students (If a college or university has both full-time and part-time students, a formula is used to express the participation of the part-time students as "full-time equivalent" students. For example, two students each taking six units might equal one full-time twelve-unit student.)*
Pct Over Age 24	*Percentage of the students who are over twenty-four years of age*
Freshman Retention Rate	*Percentage of the first-year students who continue to the second year*
Graduation Rate	*Percentage of all students who graduate*
Tuition Fees	*Total yearly fees for tuition*
Sector	*Whether the college or university is "four-year public," "four-year private," or "two-year public"*
Av Debt	*The average debt that a student has incurred at the end of study*
Pct Grad Debt	*Percentage of graduates who have incurred debt by the end of their study*
Selectivity	*A measure of how selective the institution is for those colleges that area not "open admission." The categories are "minimally," "moderately," or "very selective."*
Pct Hispanic	*Percentage of students who are of Hispanic origin*
Pct White	*Percentage of students who are "White"*
Pct Black	*Percentage of students who are "Black"*
Pct Asian	*Percentage of students who are of Asian or Pacific Island origin*

a. What are the cases for this collection? (See the ***Notes*** for a definition of ***cases.***)

b. State which variables listed above are quantitative and which are categorical.

c. List the possible values for the variable *"State"* for *this* collection.

d. List the possible values for the variable *"Freshman Retention Rate."* (Notice we are asking about "possible" values and not "plausible" or "likely" values.)

e. In the ***Notes,*** there is a section about ***Statistical Questions.*** One of the points made there is that we usually ask *statistical questions* about *collections* and not so much about the individual cases in the collection. Which of the following statistical questions (on the next page) are questions about the collection (Far West Colleges) and which questions focus more on individual cases within the collection? Give a reason for your answer. (You will soon learn to find answers to such statistical questions.) ***PTO***

(i) On average, are tuition fees higher among the "four-year public" colleges and universities or among the "four-year private" colleges and universities?
(ii) Is there more variability in the tuition fees among the "four-year public" colleges and universities or among the "four-year private" colleges and universities?
(iii) Which university has the highest tuition fees?
(iv) Is it true that the higher the tuition fees in a college or university, the higher the average debt will be?

3. ***Potatoes*** You have a collection of potatoes; the cases are potatoes, and you want to study them.
   a. List three (or more) quantitative and two (or more) categorical variables that come to mind for the collection. (See the ***Notes*** for a definition of ***quantitative*** and ***categorical*** variables.)
   b. For each of the variables, list possible values for the variable. (See the ***Notes*** for a definition of ***values.)***
   c. Write three good statistical questions about the collection of potatoes. (See the section in the ***Notes*** about ***statistical questions.)***

4. ***Questionnaire Questions to Variables.*** Often data are collected from people (the cases) using a questionnaire (such as you may have done on the first or second day of class). Here is such a questionnaire item: "With which political party would you say you most closely identify?"
   a. State some possible values for the variable that this question is measuring.
   b. Is the variable categorical or quantitative? Give a reason for your answer.
   c. Give a name for this variable that describes what is being measured. (The name should not be the same as the values; the name should describe what varies.)

5. ***Questionnaire Questions to Variables.*** Here is another questionnaire question: "Estimate the amount of money you spent last month going to movies at theaters."
   a. State *all* the possible values for this variable. (Is $547.17 possible? Is $547.17 likely?)
   b. Is the variable categorical or quantitative? Give a reason for your answer.
   c. Give an appropriate name for the variable (The name may be more than one word.)

6. ***Answering Statistical Questions.*** You have been given the data for the students from several California statistics classes (***CombinedClassDataAut08***, which you will find at the end of this section of exercises). The variables measured in that collection are described in Exercise 1 and on the data sheet you have been given. Here is the statistical question we want to answer from these data.

   Are male students or female students more likely to have a tattoo?

   You will want to look at Example 1 as part of this question depends on that question.
   a. Make a table similar to the one on the right and from the data sheet tally the responses by gender to the question of whether a student has a tattoo.

		Tattoo		
		No	Yes	Row Summary
Gender	Female			
	Male			
Column Summary				

   b. Calculate the percentage of females that have a tattoo. Calculate the percentage of males that have a tattoo. Calculate the overall percentage of students in this collection that have a tattoo.
   c. Using the results of part b, compare the California students with the Penn State students. Use complete sentences in writing your interpretation about any differences you find between the Penn State students and the California students in having a tattoo.

d. In making an interpretation, a student writes, "The differences in the percentages of students having tattoos in Pennsylvania and California are because Easterners are so much more conservative than West Coast students." The student's instructor writes on the paper: "Be careful about unwarranted inferences." What do you think the instructor had in mind when writing this comment?

7. ***Answering Statistical Questions.*** Using the same collection of data on California statistics students (***CombinedClassDataAut08***, at the end of this section of exercises), we want to answer the following question, which happens to be the third statistical question on our list.

   What percentage of students speaks just one language? What percentage speaks two languages? What percentage speaks three languages? Four? Five?

   You will find it convenient to consult Example 2.

   a. From the data sheet, make a table (like the one in Example 2) to tally the number of students who speak one, two, three, etc. languages.
   b. Calculate percentages of those who speak one, two, three, etc. languages.
   c. ***Interpretation*** Look at the percentage of the students that are monolingual (speak just one language). Reflect on whether you think that this number is a relatively big percentage or a relatively small one. Then think of what other information you would like to have to help you decide whether the percentage is relatively big or small.
   d. Write another good statistical question about the variable *Languages* (i.e., the number of languages a student speaks).

8. ***Answering Statistical Questions.*** In our ***Combined Class Data Aut 08*** data set for statistics students (at the end of this section of exercises), we have data on the size of the household, so we can answer the same kinds of questions we did with the Australian student data set in Example 2. *What percentage of students lives in households of two people? What percentage lives in households with three people? What percentage lives in households of four?*

   (NOTE: The number of people in the household includes the student.)

   a. From the data sheet, make a table (like the one in Example 2) to tally the number of students who live in households that have one person, two people, three people, etc. The data are in the variable *Number Household.*
   b. Calculate percentages of those who live in households of one, two, three, etc. people.
   c. ***Interpretation*** Compare your percentages with the ones we found for the Australian secondary students. Are they similar or much different?
   d. You should have found that about 2.4% of the California students live alone and another 9% live in households of size two, whereas only about 2.75% of the Australian secondary school students live in households of size one or two. Does this make sense in light of the fact that one collection is made up of secondary school students and the collection is made up of college students? Explain briefly but well.

9. *Les Écossais.* Here is a spreadsheet for a part of a collection of data from the 1851 Canadian Census for a small part of the province of Québec where (at that date) a large proportion of the population were Highland Scots. Each row in the spreadsheet is a household, and there are just three variables.

1851 Census in Newton and Saint-Télésphore

	NumHH	Age	Place
27	7	50	Newton
28	4	31	Newton
29	5	70	Newton
30	7	45	Newton
31	4	68	Newton
32	6	60	St-Télésphore
33	12	53	St-Télésphore
34	10	60	St-Télésphore
35	7	65	St-Télésphore
36	8	70	St-Télésphore
37	13	50	St-Télésphore

Num HH	*The number of people in the household*
Age	*The age of the head of the household*
Place	*Whether the household was in Newton or in the municipality of Saint-Télésphore*

a. What are the ***cases*** for this collection of data? Be specific.

b. Which variables are ***quantitative*** and which are ***categorical?***

c. What are possible values for the variable *Num HH?*

d. Make up a ***statistical question*** that could be answered with these data.

10. *Les Ozz* Here are more data from the Australian Secondary Student collection. Here we have two new variables: one variable (*Whre Live*) records the state or territory within Australia where the student lived. The states for this collection are:

Qld	*Queensland*
SA	*South Australia*
Vic	*Victoria*
WA	*Western Australia*

Australian Students_2

		MusRegg		Row Summary
		No	Yes	
WhreLive	Qld	64	16	80
	SA	76	14	90
	Vic	99	19	118
	WA	96	14	110
Column Summary		335	63	398

S1 = count ( )

The students were asked about their favorite type of music and were given a list of various types of music. If a student said that reggae was a favorite type of music then "yes" was recorded for the variable *Mus Regg*. If the student did not choose reggae then for that student "no" was recorded for *Mus Regg.*

Our statistical question is:

Is there a difference in the popularity of reggae among students from different parts of Australia?

a. Explain why this statement is not true: "Reggae is more popular for students from the state of Victoria since there are nineteen students who like reggae from that state in our collection and that is more than the numbers liking reggae who are from the other states."

b. The percentage of students from the state of Victoria who list reggae as one of their favorite types of music is 16.10%. Show how this was calculated.

c. Do three other calculations like the one in part b to *calculate* the percentages of students liking reggae for the students from each of the four Australia states.

d. Write an *interpretation* of the calculations you made in part c to answer the statistical question.

Combined Class Data Autumn 08

Case Number	Gender	BirthMonth	BirthYear	Age	Height (cm.)	Mother'sAge	PoliticalView	Languages	NumberHousehold	NumberCars	NumberStates	Instructor'sAge	Tattoo	DominantHand	Class
1	F	Nov	1982	26	181.0	28	Liberal	1	3	3	7	52	Y	Right	AA
2	F	Jun	1985	24	172.0	32	Liberal	1	3	3	12	50	Y	Right	AA
3	F	Nov	1990	18	156.0	26	Liberal	2	9	2	2	53	N	Right	AA
4	F	Jun	1985	24	179.0	28	Liberal	1	4	4	12	55	N	Right	AA
5	F	Nov	1987	21	156.0	24	Liberal	2	4	5	14	55	N	Right	AA
6	F	Jul	1986	23	168.0	23	Liberal	1	5	5	6	47	Y	Right	AA
7	F	Jul	1989	20	166.0	35	Liberal	2	4	4	4	57	N	Right	AA
8	F	Jun	1989	20	160.0	31	Liberal	2	4	4	12	54	N	Ambide	AA
9	F	Feb	1990	18	151.0	35	Moderate	2	5	8	2	53	N	Right	AA
10	F	Aug	1988	21	168.0	30	Moderate	2	4	4	6	52	Y	Right	AA
11	F	Apr	1989	19	174.0	31	Liberal	2	4	2	6	45	Y	Left	AA
12	F	Jan	1989	19	168.0	32	Liberal	3	4	4	3	46	N	Right	AA
13	F	Feb	1989	19	165.0	20	Moderate	4	4	2	2	52	N	Right	AA
14	F	Nov	1978	30	163.0	25	Liberal	3	1	1	4	55	N	Right	AA
15	F	Apr	1991	17	179.0	33	Liberal	3	3	1	12	50	N	Ambide	AA
16	F	Nov	1965	43	173.0	22	Liberal	3	4	4	13	52	N	Right	AA
17	F	Jan	1987	21	168.0	23	Liberal	2	4	3	6	50	Y	Right	AA
18	F	Dec	1987	21	161.0	26	Liberal	3	5	3	2	50	N	Right	AA
19	F	Jun	1989	20	155.0	25	Moderate	2	4	3	8	50	N	Right	AA
20	F	Oct	1986	22	170.0	24	Liberal	1	6	7	3	45	N	Right	AC
21	F	Feb	1976	32	160.0	30	Liberal	2	3	3	4	65	N	Right	AC
22	F	Jan	1990	18	165.0	33	Moderate	2	3	4	8	44	N	Right	AC
23	F	Oct	1957	51	165.0	34	Liberal	2	6	4	5	50	N	Right	AC
24	F	May	1987	21	163.0	28	Conservat	3	2	4	2	55	N	Right	AC
25	F	Nov	1990	18	160.0	31	Moderate	2	2	1	2	55	N	Right	AC
26	F	May	1989	19	161.0	22	Liberal	2	8	4	5	55	N	Right	AC
27	F	Mar	1990	18	167.0	24	Conservat	2	5	4	1	60	N	Right	AC
28	F	Sep	1988	20	170.0	21	Liberal	1	6	3	8	53	N	Right	AC
29	F	Jan	1988	20	156.0	18	Moderate	2	3	3	3	50	N	Right	AC
30	F	Mar	1990	18	155.0	35	Liberal	3	3	3	9	53	N	Right	AC
31	F	Jun	1990	19	166.0	28	Moderate	1	4	4	8	55	N	Right	AC
32	F	May	1990	18	157.0	33	Liberal	1	3	2	5	62	N	Right	AC
33	F	Jun	1989	20	168.0	24	Liberal	1	2	2	7	53	Y	Right	AC
34	F	Mar	1988	20	168.0	23	Liberal	2	4	7	6	62	N	Right	AC
35	F	Feb	1989	19	166.0	30	Moderate	1	5	3	3	55	N	Right	AC
36	F	Feb	1986	22	163.0	22	Liberal	2	6	3	8	54	Y	Right	AC
37	F	Apr	1990	18	170.0	25	Moderate	1	4	3	2	60	N	Right	AC
38	F	Oct	1990	18	177.0	21	Liberal	2	8	7	0	56	N	Right	AC
39	F	Jul	1987	22	167.0	38	Moderate	2	3	2	5	65	N	Right	AC
40	F	Apr	1989	19	157.0	32	Liberal	2	4	9	2	58	N	Right	AC
41	F	Dec	1982	26	161.0	30	Liberal	1	1	1	8	60	Y	Right	AD
42	F	Aug	1988	21	162.0	32	Liberal	1	5	7	8	65	N	Right	AD
43	F	Dec	1988	20	163.0	41	Moderate	2	4	3	4	64	N	Right	AD
44	F	Apr	1987	21	160.0	20	Liberal	2	4	3	2	63	N	Right	AD
45	F	Aug	1985	24	178.0	24	Moderate	2	5	4	10	65	Y	Right	AD
46	F	Jun	1991	18	167.0	30	Moderate	1	5	5	10	70	N	Right	AD
47	F	Jan	1983	25	170.0	23	Liberal	3	2	2	9	55	N	Right	AD
48	F	Sep	1989	19	168.0	28	Liberal	2	4	7	4	68	N	Right	AD
49	F	Mar	1988	20	161.0	35	Liberal	2	2	2	10	60	Y	Right	AD
50	F	Nov	1989	19	164.0	27	Liberal	2	7	6	9	60	N	Right	AD
51	F	Apr	1983	25	169.0	24	Moderate	1	5	3	9	42	Y	Right	AD
52	F	Nov	1989	19	170.0	28	Moderate	2	3	2	6	50	N	Right	AD
53	F	Jul	1987	22	176.0	39	Moderate	2	3	2	5	60	N	Right	AD
54	F	Mar	1988	20	164.0	31	Liberal	2	3	6	5	60	N	Right	AD
55	F	Jan	1990	18	162.0	29	Moderate	3	2	1	10	55	N	Right	AD
56	F	Aug	1990	19	152.0	32	Liberal	1	5	2	18	66	N	Right	AD
57	F	Dec	1981	27	163.5	30	Liberal	1	1	1	37	60	Y	Right	AD
58	F	Jan	1989	19	167.0	27	Moderate	1	4	4	2	58	Y	Right	AD
59	F	Jul	1962	47	165.0	24	Liberal	1	5	2	6	56	N	Right	AD
60	F	Jul	1988	21	155.0	28	Moderate	1	7	4	6	50	Y	Left	AD
61	F	Oct	1990	18	165.0	27	Conservat	1	6	6	8	64	N	Right	AD
62	F	Jan	1984	24	177.0	37	Moderate	1	3	3	19	56	Y	Right	AD
63	F	Jul	1990	19	157.2	27	Moderate	1	5	7	2	55	N	Right	AD
64	M	May	1989	19	167.0	29	Liberal	2	4	2	1	52	N	Right	AA
65	M	Nov	1988	20	167.0	31	Moderate	2	3	2	1	58	N	Right	AA
66	M	Apr	1991	17	169.0	33	Liberal	1	4	2	7	53	N	Right	AA
67	M	Oct	1988	20	154.0	23	Liberal	1	6	5	7	55	N	Right	AA
68	M	Dec	1987	21	178.2	22	Liberal	1	6	4	6	50	Y	Right	AA
69	M	Jun	1987	22	182.0	19	Liberal	2	4	5	3	35	N	Left	AA
70	M	Sep	1986	22	195.0	36	Liberal	1	2	2	8	56	N	Right	AA
71	M	Dec	1988	20	167.0	25	Moderate	1	6	5	1	48	N	Right	AA
72	M	Mar	1988	20	173.0	29	Moderate	2	3	2	8	58	N	Right	AA
73	M	Sep	1989	19	183.0	25	Moderate	1	3	2	6	53	N	Ambide	AA
74	M	Jun	1989	20	173.0	25	Liberal	1	4	4	11	52	Y	Right	AA
75	M	Nov	1987	21	181.0	28	Conservativ	2	4	3	4	52	N	Right	AA
76	M	Feb	1984	24	179.0	39	Moderate	1	3	5	22	48	Y	Right	AA
77	M	Aug	1988	21	185.0	35	Moderate	1	4	3	7	50	N	Right	AA
78	M	Oct	1989	19	190.0	22	Liberal	1	5	4	4	60	N	Right	AA
79	M	Jan	1989	19	172.0	25	Conservativ	1	3	5	5	49	N	Right	AA
80	M	Aug	1989	20	166.0	25	Moderate	2	4	4	5	44	N	Right	AA
81	M	Oct	1989	19	180.0	30	Moderate	1	4	7	5	54	N	Ambide	AA
82	M	May	1982	26	182.0	28	Liberal	2	4	2	4	50	N	Right	AA
83	M	Dec	1986	22	172.0	25	Conservativ	1	4	4	2	52	N	Right	AA
84	M	Oct	1989	19	182.9	32	Conservativ	1	4	2	19	58	N	Right	AC
85	M	Jul	1989	20	167.0	28	Liberal	2	10	6	3	55	N	Right	AC
86	M	Jun	1988	21	180.3	32	Liberal	1	3	3	14	58	N	Right	AC
87	M	Jul	1983	26	167.0	26	Moderate	2	3	5	13	55	Y	Right	AC
88	M	Jun	1989	20	170.0		Conservativ	1	7	1	18	57	N	Right	AC
89	M	Jul	1982	27	182.0	32	Moderate	1	2	2	25	58	Y	Right	AC
90	M	Apr	1990	18	179.0	30	Moderate	2	4	1	4	60	N	Right	AC
91	M	Sep	1989	19	172.0	31	Liberal	1	4	4	3	55	N	Right	AC
92	M	May	1989	19	177.0	24	Moderate	3	4	3	5	55	N	Right	AC
93	M	Jul	1989	20	183.0	34	Liberal	1	4	3	17	52	N	Right	AC
94	M	Oct	1989	19	184.0	31	Liberal	1	5	5	12	53	N	Right	AC
95	M	Feb	1989	19	180.0		Liberal	1	5	3	17	48	N	Right	AC
96	M	May	1990	18	170.0	36	Moderate	2	2	2	2	54	N	Right	AC
97	M	Jun	1985	24	177.0	19	Liberal	2	3	3	3	52	N	Right	AC
98	M	Nov	1988	20	188.0	36	Moderate	2	4	4	17	56	N	Right	AC
99	M	May	1989	19	180.0	28	Moderate	1	4	3	6	61	N	Right	AC
100	M	Dec	1988	20	190.0	35	Conservativ	2	5	3	8	63	N	Right	AC
101	M	Jan	1988	20	173.0	37	Liberal	2	4	6+	10	56	N	Ambide	AC
102	M	Sep	1989	19	124.5		Moderate	2	4	1	7	50	N	Right	AC
103	M	Mar	1988	20	170.0	23	Liberal	1	4	5	5	60	N	Right	AC
104	M	Jan	1986	22	175.0	28	Moderate	2	4	2	2	58	N	Right	AC
105	M	Mar	1983	25	175.0	19	Conservativ	1	3	4	20	56	Y	Ambide	AC
106	M	Jun	1987	22	195.0	35	Liberal	1	2	2	10	60	N	Left	AD
107	M	Apr	1985	23	184.0		Conservativ	2	5	5	1	100	Y	Right	AD
108	M	Aug	1985	24	188.0	27	Moderate	1	4	5	3	60	Y	Right	AD
109	M	Nov	1987	21	167.0	30	Moderate	1	4	3	10	58	N	Right	AD
110	M	Nov	1984	24	169.0	21	Conservativ	2	5	4	4	47	N	Right	AD
111	M	Sep	1989	19	169.0	27	Liberal	1	4	4	7	62	N	Right	AD
112	M	May	1985	23	165.0	28	Conservativ	2	5	4	7	64	N	Right	AD
113	M	Feb	1986	22	181.0	28	Moderate	2	5	1	6	60	N	Right	AD
114	M	Feb	1990	18	180.0	30	Liberal	3	5	4	3	48	N	Right	AD
115	M	May	1987	21	177.0	51	Moderate	2	3	4	2	60	N	Right	AD
116	M	Jun	1988	21	178.0	38	Moderate	1	2	4	12	60	N	Right	AD
117	M	Jan	1990	18	177.0	26	Moderate	2	4	2	2	58	N	Right	AD
118	M	Aug	1987	22	173.0	35	Moderate	1	4	7	16	58	N	Right	AD
119	M	Aug	1990	19	167.0	30	Moderate	2	3	5	4	55	N	Right	AD
120	M	Sep	1989	19	175.0	37	Liberal	1	3	3	8	65	N	Ambide	AD
121	M	Jun	1990	19	180.0	40	Moderate	1	4	4	16	67	N	Right	AD
122	M	Feb	1983	25	183.0	33	Moderate	1	4	4	13	57	N	Right	AD
123	M	Dec	1987	21	176.0	34	Moderate	2	4	6	11	60	N	Right	AD
124	M	Jul	1990	19	186.0	40	Liberal	1	4	3	22	65	N	Right	AD
125	M	Sep	1999	9	189.0	38	Liberal	2	3	1	4	50	N	Right	AD
126	M	Jan	1990	18	173.0	30	Moderate	1	3	3	9	56	N	Right	AD

§1.2 Exercises on Probability

1. ***California College Student Data on Gender and Tattoos:*** In the ***Notes,*** we looked at whether men or women were more likely to have a tattoo; our collection of student data was from Penn State, and the data were collected in the 1990s. This exercise is about students in California in 2008 (***CombinedClassDataAut08***, found in paper form at the end of the exercises to §1.1). Our statistical question is still: "Are male students or are female students more likely to have a tattoo?" That is, which gender is more likely to have a tattoo? You answered a question similar to this one in the previous exercises, but here the answers focus on the calculation and language of probability. Here is the contingency table that you should have.

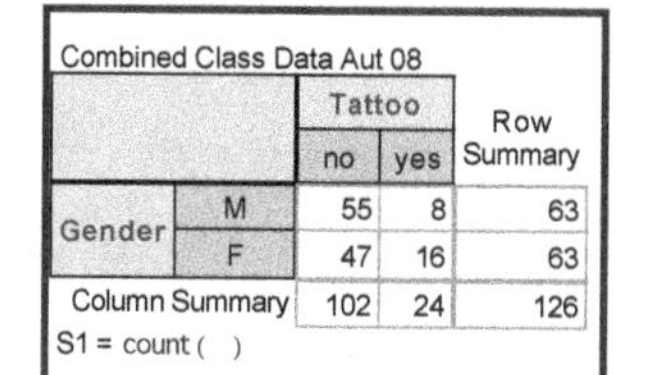

Combined Class Data Aut 08

		Tattoo no	Tattoo yes	Row Summary
Gender	M	55	8	63
	F	47	16	63
Column Summary		102	24	126

S1 = count ( )

- Let ***Y*** be the event that the student has a tattoo;
- Let ***N*** be the event that the student does not have a tattoo;
- Let ***M*** be the event that the student is male;
- Let ***F*** be the event that the student is female.

a. The probability that the person is a female is $P(F)$. Find this probability (by making a fraction) and express the answer as a number between 0 and 1. Round to two decimal places.

b. Find $P(Y)$ using a fraction and then express your answer rounded to two decimal places.

c. Express in English the meaning of $P(Y)$.

d. From the table directly, find the value of the probability $P(F \text{ and } Y)$ showing your calculation and give an interpretation for the number you calculated.

e. Using the notation introduced for conditional probability, the probability that a person has a tattoo given that the person was female is $P(Y \mid F)$. Calculate $P(Y \mid F)$ using the numbers in the two-way table for the California students shown above.

f. Distinguish clearly between the meaning of $P(F \text{ and } Y)$ and the meaning of $P(Y \mid F)$.

g. Using correct notation, find the probability that a person has a tattoo if the person is male.

h. From the analysis you have done for the California students, are females or males more likely to have tattoos? Give a reason.

i. At the right is the two-way table for the Penn State students in 1999. Compare the probabilities you calculated in parts e and g for the California students to the comparable Penn State probabilities. What differences (if any) do you notice?

PennState2

		Tattoo No	Tattoo Yes	Row Summary
Sex	Female	119	18	137
	Male	55	13	68
Column Summary		174	31	205

S1 = count ( )

j. You have calculated $P(Y \mid F)$. Using the California data, calculate $P(F \mid Y)$. (Look at the ***Notes*** at the definition of conditional probability and work out (and write down) a way to remember what number should be in the denominator of the fraction.)

k. Write down in English the difference in meaning between $P(Y \mid F)$ and $P(F \mid Y)$.

l. A student calculates $P(N \mid F) = \frac{47}{102} \approx 0.4608$. Is this student correct? If you think so then give an interpretation of the number. If you think the student is wrong, explain the mistake made.

2. ***Household Sizes*** This exercise uses the variable *NumberHousehold*, the number of people in a student's household using the data in ***CombinedClassDataAut08***. General instructions are shown as the bullets to the right of the vertical line. For detailed instructions, see the software supplement.

   a. Remember that a *sample space* consists of all the possible simplest events for a variable. Even without looking at the actual data, think about the possible responses for *NumberHousehold.* Is "0" possible? Is "5" possible? Is "100" possible for a household? (Notice: the word is "possible," not "plausible.") Then write the sample space for these data using the brace symbols.

   - Open the file ***CombinedClassDataAut08***. Refer to ***Opening Files*** in the software supplement.
   - Use the statistical software to get a ***two-way table*** with the variable *Gender* in the rows, and the variable *NumberHousehold* in the columns. Refer to the software supplement.
   - Since we are comparing males and females, and since *Gender* is the row variable, get row proportions or percentages.

   b. For all of the students (male and female together), calculate $P(X=4)$ and give an interpretation of the number that you found.

   c. For all of the students, calculate the probability $P(x=4 \text{ or } X=5)$. [See the section on the definition of the probability of the union (or "or" probability) of two events in the ***Notes.***]

   d. Give an interpretation of $P(x=4 \text{ or } X=5)$ in the context of "randomly choosing a student."

   e. Calculate the probability $P(X<4)$. In the ***Notes §1.2*** (in the section *'Events Can Also Refer to Numbers')* there is a similar calculation but for Australian high school students. Compare what you have found for this collection of California college students to the calculation of $P(X<4)$ for the Australian high school students, shown in this table.

Number Household	1	2	3	4	5	6	7	8	9	30	Total
Counts:	1	10	63	142	112	49	12	6	2	1	398
Percent of cases:	0.25%	2.51%	15.83%	35.68%	28.14%	12.31%	3.02%	1.51%	0.50%	0.25%	100.00%

   Does the difference in the probabilities make sense, knowing that the Australian students are high school students?

   f. Again, for all of the *California* students, calculate $P(X \geq 6)$. Compare what you get to $P(X \geq 6)$ for the *Australian* high school students shown and interpret in the context of the comparison.

   g. For the *California* students, calculate and give the meanings in English of the conditional probabilities $P(X<4 \mid M)$ and $P(X<4 \mid F)$. Your answer should begin: " $P(X<4 \mid M)= \ldots$ shows the probability...".

   h. Compare the probabilities $P(X<4 \mid M)$ and $P(X<4 \mid F)$. Is there a difference in these probabilities? If so, do you think that the difference is big? Or small?

   i. For the data as they were collected, the events $P(X=4)$ and $P(X=5)$ are *mutually exclusive* since it is understood that the number in the household is the biggest number of people who normally live there. What should be the value of $P(X=4 \text{ and } X=5)$?

j. Compare $P(X<4 \mid F)$ with $P(X<4)$ and determine whether the events $X<4$ and F are independent or not independent. (Refer in the ***Notes*** to the section on "Independence.")

k. Calculate the probability $1-P(X \geq 6)$ and write the probability using just one $P(\quad)$ statement, without the "1 – . . ."

3. ***Political View and Gender*** This exercise is to be done using software using the data in the file ***CombinedClassDataAut10***. The general software instructions are shown below to the right of the vertical line. One of the variables measured is the *Political View* of students. Using the categories of that variable and the categories of *gender,* we can define these events:

L = Liberal, ***Mod*** = Moderate, ***C*** = Conservative, ***M*** = Male, ***F*** = Female

- Open the file ***CombinedClassDataAut10***. Refer to ***Opening Files*** in the software supplement.
- Use the statistical software to get a ***two-way table*** with the variable *Gender* in the rows, and the variable *PoliticalView* in the columns. Refer to the software supplement.
- Since we are comparing males and females and since *Gender* is the row variable, get row proportions or percentages.

The two-way table produced should have the same count entries as the one shown here, although the rows and columns may be in a different order, and will also have row proportions or percentages.

CombinedClassDataAut10

		PolticalView			Row Summary
		Liberal	Moderate	Conservative	
Gender	M	43	39	7	89
	F	43	44	14	101
Column Summary		86	83	21	190

S1 = count ( )

Statistical question:

Are male students or female students more likely to have "liberal" political views?

We will use this contingency table to answer these questions.

a. Confused Conrad (whom you will meet often in these exercises) says: "The answer is easy: since there are equal numbers of male and female students who have liberal political views, males and females have the same probability of having liberal political views." What is wrong with CC's thinking? Be very specific.

b. Calculate the probability $P(L \mid F)$ and give an interpretation in the context of the data.

c. Calculate the probability $P(L \mid M)$ and give an interpretation in the context of the data. Check the answers to parts b and c using the output from the statistical software.

d. If a student's answer is 0.5 for both part b and part c, what conditional probabilities has this student actually calculated: $P(L \mid F)$ and $P(L \mid M)$ or $P(F|L)$ and $P(M|L)$? Give a reason for your answer.

e. In the notation $P(L \mid F)$, what is the "given"? In the calculation, what total is used for the denominator?

f. Suppose a student's answer for both parts *b* and *c* is 0.2263; this is not correct. Your job is, by doing some calculations, to decide what mistake the student made. Then use the correct notation to show the probabilities that the student actually (but wrongly) calculated.

g. Use the correct probabilities and write an answer (in English) to the statistical question in italics next to the contingency table above:

Are male students or female students more likely to have "liberal" political views?

h. Are the events L and F *mutually exclusive events* or *not mutually exclusive*? Give a reason for your answer.

i. Are the events L and F *independent events* or *not independent*? Give a reason for your answer. [Compare $P(L \mid F)$ and $P(L)$ to determine whether the events L and F are independent or *not*.]

j. Check the ***Notes*** for the addition rule and use that rule to calculate $P(M \text{ } or \text{ } C)$. There should be three terms in your calculation. (Notice, we have asked for M "male" and not Mod "Moderate.")

k. Calculate $P(not \text{ } L)$ using the formula for calculating "*not*" probabilities given in the ***Notes***. Show your work completely. In terms of political views, what probability has the calculation of $P(not \text{ } L)$ given you? Think in the context of the variable.

l. To the question "Find the probability that a student has conservative political views *given* that the student is a male," Confused Conrad writes: $P\left(\frac{7}{21}\right) = 0.33\overline{3}$. Find *all* the mistakes CC has made (and there is more than one), explain what they are, and give the correct answer to the question, using the correct notation (which CC has not done).

m. To the question: "Find the probability that a student both has *moderate* political views *and* also is a male student," a student writes: $P(Mod) + P(M) = \frac{83}{190} + \frac{89}{190} = \frac{172}{190} \approx 0.9053$. Find and explain *all* the mistakes and do the correct calculation.

4. *Number of States Visited* This exercise will also use the collection ***CombinedClassDataAut10.*** The variable we will analyze (*Number States*) is about the number of states California students have visited. Some students have never left California whereas others are very well-travelled.

a. Without looking at the data, write the sample space for the variable *Number States* if the categories are to be regarded as events. Your answer should have $S = \{ \cdots \}$ and then the appropriate numbers inside the braces. Notice that you will want to use ellipses inside braces {. . .} but notice also that you can be a bit more specific than for the *Number Household* problem.

b. Are the events listed in your sample space *mutually exclusive*? Give a reason for your answer.

c. Is the variable *Number States* quantitative or categorical?

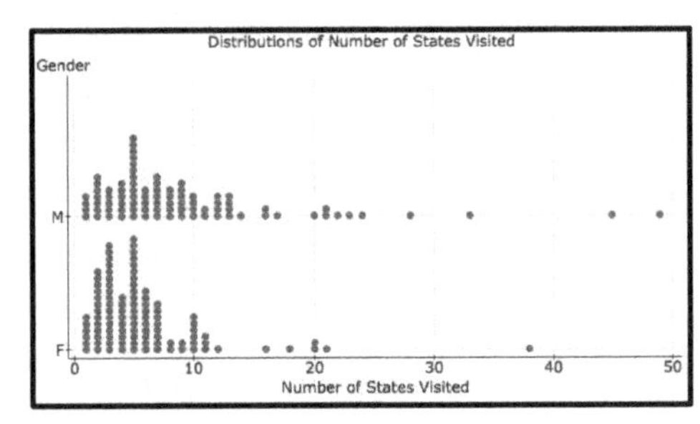

- Open the file ***CombinedClassDataAut10***. Refer to ***Opening Files*** in the software supplement.
- Use the statistical software to get a ***dot plot*** of the variable *NumberStates* grouped by the variable *Gender*. Refer to the software supplement.
- If possible, get vertical grid lines on the plots.

The plot should be similar to the one on the right.

For these data, the number of males is $n_M = 86$ and the number of females is $n_F = 101$.

d. By counting the dots on the dot plot and using the information above, calculate the probabilities $P(X>10|M)$, $P(X>10 | F)$ and $P(X>10)$. Notice that you only have to count the dots where $X > 10$, and that the total number of dots for the males and for the females was given.

e. Interpret the probabilities you have just calculated to answer the question: "Are males or females more likely to have visited more than ten states?"

f. [Optional, but helpful] On the dot plot, either by hand or using software, indicate the dots for students who had visited $X > 10$ states for the males and females. The software being used may allow this to be done easily, but some software does not, and it must be done by hand.

To answer the questions below and to check the answers to the questions above, use the two-way table shown here. Software may be used to get this two-way table as well.

CombinedClassDataAut10

		MoreThan10States		Row Summary
		No	Yes	
Gender	M	62	24	86
	F	91	10	101
Column Summary		153	34	187

S1 = count ( )

g. Calculate and give interpretations of $P(X>10 \text{ and } M)$, $P(X>10 \text{ and } F)$, $P(M)$ and $P(F)$.

h. Apply the formula $P(A | B) = \frac{P(A \text{ and } B)}{P(B)}$ to get $P(X>10 | M)$ and $P(X>10 | F)$. In the ***Notes*** see the subsection on ***Conditional Probability ("If").***

i. Find the probability you will get if you use the rule for ***"Not"*** and calculate $1 - P(X>10)$. Express this probability in symbols without the "1" and give the meaning in words.

j. Find the probability that a student has visited ten or fewer states if that student is male and express this probability using the correct notation.

k. Find the probability that a student is male and that student has visited ten or fewer states. Use the correct notation.

l. Find the probability that a student is male if the student has visited ten or fewer states. Use the correct notation.

5. ***Review and Political Views and Tattoos.*** This exercise reviews the material on probability, the analyses you made between gender and having a tattoo, and between gender and political views, and applies the ideas to the relationship between a student's political views and whether or not the student has a tattoo.

Combined Student Data Spr 09

		Tattoo		Row Summary
		n	y	
Gender	f	54	32	86
	m	50	14	64
Column Summary		104	46	150

S1 = count ( )

- Use the file ***CombinedClassDataSpr09*** and software to get the two-way table relating the variables *Gender* and *Tattoo* for the data collected in Spring 09.

a. Calculate $P(Y)$ for Spring 09 using the two way table and compare the results to what was found for ***CombinedClassDataAut08*** and for Penn State in 1999, discussed in the ***Notes.*** The tables for Autumn 08 and PennState are shown here. Comment on the comparison.

Combined Class Data Aut 08

		Tattoo		Row Summary
		no	yes	
Gender	M	55	8	63
	F	47	16	63
Column Summary		102	24	126

S1 = count ( )

PennState2

		Tattoo		Row Summary
		No	Yes	
Sex	Female	119	18	137
	Male	55	13	68
Column Summary		174	31	205

S1 = count ( )

b. In answer to our statistical question "Are males or females more likely to have a tattoo?" we got different results for the Penn State collection and the California collection. Use the correct notation and analyze these Spring '09 data to answer the same question. Do the calculation and make the interpretation in the context of the question.

c. ***Gender and Political View revisited*** In Exercise 3, you also looked at the relationship between political view and gender for the Autumn '10 sample, asking the question, "Are females or males more likely to have liberal political views?" Look at this same relationship for Spring '09. Follow the directions below and then use the correct notation and analyze these Spring '09 data to answer the question about gender and political view.

- Use the file ***CombinedClassDataSpr09*** and software to get the two-way table relating the variables *Gender* and *PoliticalView* for the data collected in Spring 09.

d. Do you think that having liberal political views makes it more (or less) likely that a student has a tattoo? Or do you think it works the other way: having a tattoo means liberal political views are more (or less) likely? If you think "political views affect having a tattoo" (that is, using an arrow: "political views -> tattoo") then you will use $P(Y \mid L)$ compared with $P(Y \mid Mod)$ and $P(Y \mid Con)$. If you think "having a tattoo affects political views," ("tattoo -> political views") you will use $P(L \mid Y)$ compared with $P(L \mid N)$, $P(Mod \mid Y)$ compared with $P(Mod \mid N)$, and $P(Con \mid Y)$ compared with $P(Con \mid N)$ to analyze the data.

- Use the file ***CombinedClassDataSpr09*** and software to get the two-way table relating the variables *PoliticalView* and *Tattoo* for the data collected in Spring 09.

e. Choose one of the options ("political views -> tattoo" or "tattoo -> political views, do the calculations (or get software to do the calculations) and give an interpretation for the relationship between political views and having a tattoo among students in the samples analyzed.

6. ***Real Estate Data: Style of House and Region in a County*** The data for this exercise concerns houses that were sold in 2007–2008 in San Mateo County, California. There were a total of 3,947 houses sold in the period from June 2007 to June 2008, but the data for this exercise uses just $n = 400$ selected randomly from the 3,947. Part of the spreadsheet is shown here. Each row is a different house sold.

RE San Mateo 0708

	Caseno	Beds	Baths	SqFt	Age	Style_2	Region	List_Price	Sale_Price	Diff_Sal...	Time_in...
7	2037	3	2	1060	60	Other	South	475000	362980	-112020	118 d
8	2204	2	2	1160	82	Ranch	North	685000	640000	-45000	101 d
9	2836	2	2	1100	62	Other	South	998000	965000	-33000	22 d
10	2257	3	2	1950	42	Other	Central	949000	953000	4000	0 d
11	1754	3	2	1530	57	Other	South	719000	720000	1000	0 d
12	1036	3	2	2250	33	Traditional	Central	1399000	1600000	201000	11 d
[illegible]	[illegible]	[illegible]	[illegible]	[illegible]	[illegible]	[illegible]	[illegible]	[illegible]	[illegible]	[illegible]	[illegible]

For this exercise, we will be only interested in the variables *Style3* and *Region*. *Style3* records for each house the style of the house in one of three categories: Traditional, Ranch or Other. All of the houses are in San Mateo County, but that county can be divided into four regions, namely: Central, Coast, North, and South. For this exercise, we give these letters to the categories:

Region: **C** = Central, **CST** = Coast, **N** = North, **S** = South.
Style3: **Con** = Contemporary, **T** = Traditional, **R** = Ranch, **O** = Other.

- Use the file ***San Mateo RE Sample Y0708*** and software to get the two-way table relating the variables *Region* and *Style3* for the data for houses in San Mateo County

a. What are the cases for these real estate data?

b. Are the variables *Style3* and *Region* quantitative or categorical?
The two-way table should be equivalent to the one shown here.

c. Using the numbers in the table, calculate each of the following probabilities and for each give an interpretation.

San Mateo RE Sample 0708

		Region				Row Summary
		Central	Coast	North	South	
Style3	Contemporary	16	15	22	36	89
	Other	56	17	30	57	160
	Ranch	36	6	15	36	93
	Traditional	20	4	11	23	58
	Column Summary	128	42	78	152	400

S1 = count ( )

(i.) $P(Con)$ (ii.) $P(N)$ (iii.) $P(Con \text{ and } N)$ (iv.) $P(Con \mid N)$ (v.) $P(N \mid Con)$

d. Using the probabilities just calculated, you can determine whether the events **Con** and **N** are independent events or not independent events. Show your work and come to a conclusion.

e. Is it true that $P(Con \text{ and } N)$ is equal to $P(Con) + P(N)$? Show evidence for your answer.

f. Is it true that $P(Con \text{ or } N)$ is equal to $P(Con) + P(N)$? Show evidence for your answer.

g. Use correct probability notation to show that the proportion of Contemporary houses sold on the Coast is greater than the proportion of Contemporary houses sold in the North region.

h. To answer part g, a student calculates and puts down as an answer $P(15/400) \approx 0.0375$. There are many things wrong with what is written, both in the calculation that is done and in the use of the probability notation. Find and correct all of the errors.

7. *Size of Household in Different Places* This exercise repeats and extends the analysis done in Exercise 2 but uses different collections. We think of the values of the variable—the household sizes—as events whose number is designated by X, so that $X = 5$ means a household of size five. We begin with number of students who reported living in households of size, 1, 2, 3, . . . etc. for the Combined Class Data for Spring 09, which is shown here.

Combined Class Data Spr 09

	NumberHousehold									Row Summary
	1	2	3	4	5	6	7	10	14	
	6	15	35	45	32	8	3	1	1	146

S1 = count( )

a. Is the variable ***Number Household*** a categorical variable or a quantitative variable?

Below are three similar tables, one each for students in South Africa, for students in Australia, and for students in New Zealand. All of the data for these three countries are for high school students, and not for college students.

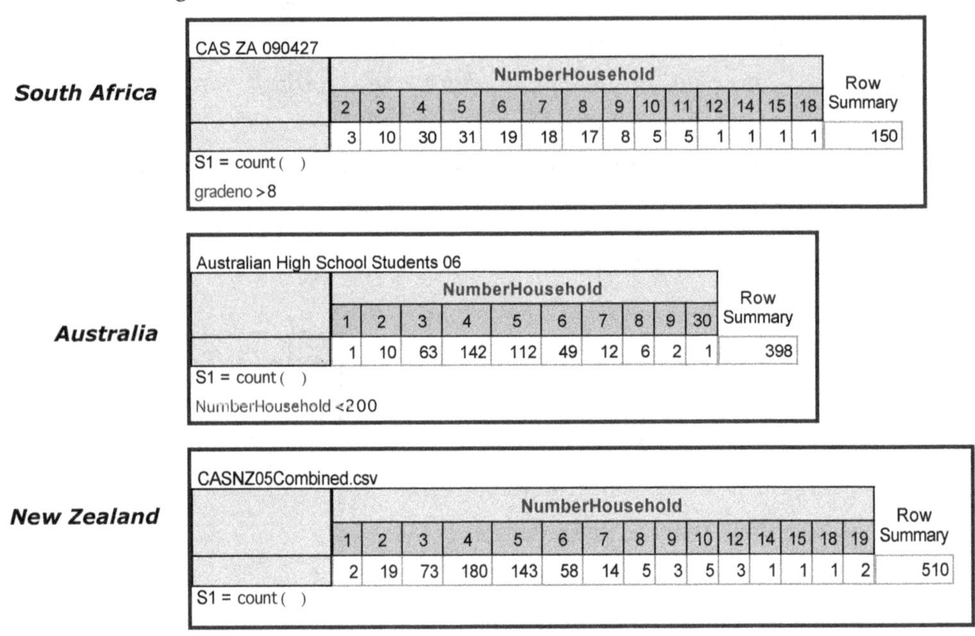

South Africa

CAS ZA 090427

	NumberHousehold														Row Summary
	2	3	4	5	6	7	8	9	10	11	12	14	15	18	
	3	10	30	31	19	18	17	8	5	5	1	1	1	1	150

S1 = count( )

gradeno >8

Australia

Australian High School Students 06

	NumberHousehold										Row Summary
	1	2	3	4	5	6	7	8	9	30	
	1	10	63	142	112	49	12	6	2	1	398

S1 = count( )

NumberHousehold <200

New Zealand

CASNZ05Combined.csv

	NumberHousehold															Row Summary
	1	2	3	4	5	6	7	8	9	10	12	14	15	18	19	
	2	19	73	180	143	58	14	5	3	5	3	1	1	1	2	510

S1 = count( )

Statistical Question: *Do the proportions of students who live in "large households" differ among these places?* We will define "large households" as households of five or more people.

b. What is the appropriate type of probability calculation for this comparison: conditional ("if") probability, intersection ("and") probability, or union ("or") probability? Give a reason for your answer.

c. Calculate the probabilities and display them using the correct notation.

d. Using the results of your calculations, write a clear, coherent answer to the statistical question stated just above part b (in italics). (You should be able to do this in one to three sentences.)

e. Suppose someone told you that the calculations you have done are not valid because there are different numbers of students surveyed in the different countries ($n_{Califronia} = 146$, $n_{South\ Africa} = 150$, $n_{Australia} = 398$ and $n_{New\ Zealand} = 510$). What would you say to this person? Would you agree? Make a case.

§1.3 Exercises on Distributions

1. ***California Statistics Students*** Here is a table showing the counts for the number of states visited (*NStatesVisited)* by students in the collection of ***CombinedClassDataSpr08.*** The table shows that there were just six students who had only visited or lived in one state, thirteen students who had visited two states, fourteen who had visited three states, etc.

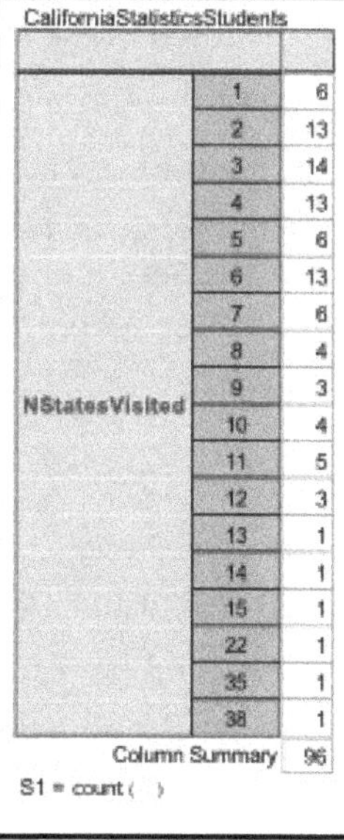

CaliforniaStatisticsStudents

NStatesVisited	1	6
	2	13
	3	14
	4	13
	5	6
	6	13
	7	6
	8	4
	9	3
	10	4
	11	5
	12	3
	13	1
	14	1
	15	1
	22	1
	35	1
	38	1
	Column Summary	96

S1 = count ( )

   a. What are the cases for this distribution? Are the cases students, or are they states, or are they visits, or something else?
   b. The definition of a distribution is given in the ***Notes*** (in a box). Write it is about the table shown here that convinces you that what is shown here fits the definition of a distribution.
   c. *Review*: Calculate the percentage of California students that have visited (or lived in) fewer than four states, using the probability notation of §1.2.
   d. *Review:* Calculate the percentage of California students that have visited (or lived in) four or more states, using the probability notation of §1.2.
   e. *Review:* If you add the probability in part c to the probability that you get in part d, you should get 1 (or something very near to it). Give a reason why you should get "1."
   f. In symbols, the answer to part e should be $P(X<4)+P(X\geq 4)=1$. Which probability rule of §1.2 ("and," "or," "if," or "not") does this illustrate? Explain your answer.
   g. ***Bad Dot Plots and Good*** The first dot plot shows an error sometimes made. The scale for the variable *NStatesVisited* must have equal intervals, unlike the bad example shown here. The scale shown here is terribly wrong, even though it is neat. The space between 15 and 22 should not be the same as between 14 and 15. By hand, make a dot plot of the distribution of *NStatesVisited.* The second dot plot shows what the scale should be, but the dot plot is not finished.

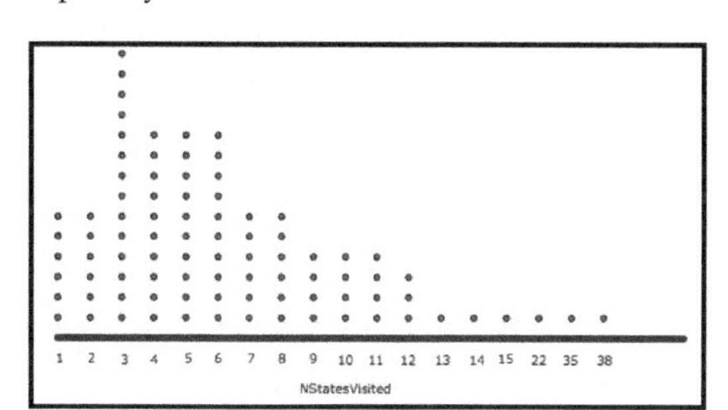

- Using the file ***CombinedClassDataSpr08*** use software to get a dot plot for the variable *NStatesVisited.*
- Using the file ***CombinedClassDataSpr08*** use software to get a histogram for the variable *NStatesVisited.* See the ***Software Supplement*** for ***Histogram.***

   h. Determine the bin width of the histogram that software has produced.
   i. Copy the histogram (a printed copy is good idea) and shade in the bars that correspond to the answer to part c above.

2. *Students' Grades in Costa Rica* The distribution of letter grades for some students in Costa Rica was shown in the ***Notes.*** Here is a part of the spreadsheet for the entire collection.

Grades Stats

	Semester	Sex	Exam1	Exam2	Exam3	Course...	LetterG...
108	2001-1	M	70.5	70.5	58	57	F
109	2001-1	M	89.4	84.5	82.5	83.8685	B
110	2001-1	F	79.5	56	81	68.25	D
111	2001-1	M	93.9	88	71	80.06	B
112	2001-1	M	73.5	78.5	72	77.0333	C
113	2001-1	M	94.8	84.5	77.5	84.17	B

Variables: ***Description***

Semester *Year and semester of the course*
Sex Gender of the student: male or female
Exam1 *Score on Exam 1* *(The scores are out of 100.)*
Exam2 *Score on Exam 2* *(The scores are out of 100.)*
Exam3 *Score on Exam 3* *(The scores are out of 100.)*
Course_Grade: *Overall course grade, expressed as a percentage* *(The grade included homework and other assignments as well as the tests.)*
LetterGrade: Letter grade for the course: A, B, C, D, F

a. For this collection, what are the cases?

b. Which variables are quantitative, and which variables are categorical?

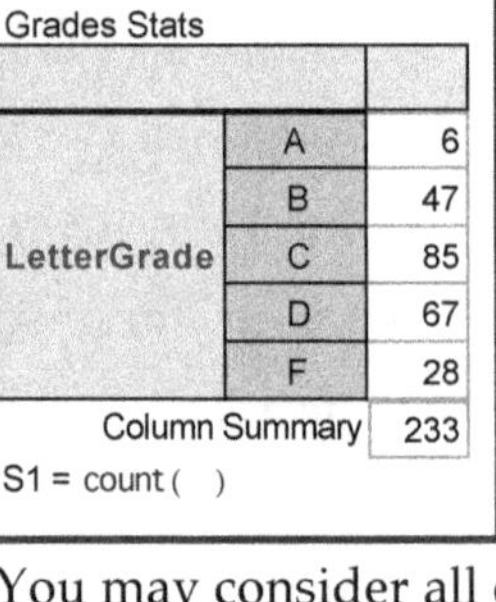
Grades Stats

LetterGrade	A	6
	B	47
	C	85
	D	67
	F	28
Column Summary		233

S1 = count ( )

c. Calculate the percentage of students who earned each of the grades A, B, C, D, and F. You may wish to make a small table to organize the answer.

d. Use the probability notation of §1.2 correctly to show the probability of getting an A.

e. Make a statistical question that can be answered with just these data. You may consider all of the variables listed and not just what you see in the *LetterGrade* two-way table.

3. *Household Sizes for California Students and Western Australian Secondary Students*

A natural statistical question is:

Are there differences in the distributions of household size comparing students in California and Western Australia?

CA and WA Comparison

		Place: California	Place: Western Australia	Row Summary
NPeopleHH	1	4 0.0625	0 0	4 0.031496063
	2	9 0.140625	2 0.031746032	11 0.086614173
	3	11 0.171875	12 0.19047619	23 0.18110236
	4	21 0.328125	27 0.42857143	48 0.37795276
	5	13 0.203125	16 0.25396825	29 0.22834646
	6	2 0.03125	3 0.047619048	5 0.039370079
	7	3 0.046875	2 0.031746032	5 0.039370079
	8	0 0	1 0.015873016	1 0.0078740157
	14	1 0.015625	0 0	1 0.0078740157
Column Summary		64 1	63 1	127 1

S1 = count ( )
S2 = columnProportion

a. Inspect the percentages and the counts in the table of the two distributions. Find one difference between the distributions and write what that difference is (using complete sentences, etc.).

b. Compare the proportions of students in the two places who live in households having five or more people. Use the probability notation of §1.2 correctly and interpret your results in the context of the data.

c. These two distributions have almost the same total count (64 for California and 63 for Western Australia). Would we be able to compare the two distributions by calculating proportions or percentages if the total counts were very different? (For example, suppose we had 180 students for Western Australia but still just 64 for California.) Give a reason for your answer.

4. *Household Sizes for California Students*

Here are the distributions of the household sizes (*NPeopleHH*) for male and female California statistics students.

a. By hand, make a double dot plot similar to the one shown in the ***Notes*** that shows the two distributions of household size for the male and the female students.

b. In your judgment, are there differences in the two distributions or are they essentially very similar?

Califnornia Statistics Students 1999

		Gender		Row Summary
		Female	Male	
NPeopleHH	1	1	3	4
	2	8	1	9
	3	6	5	11
	4	9	12	21
	5	6	7	13
	6	1	1	2
	7	2	1	3
	14	0	1	1
Column Summary		33	31	64

S1 = count( )

5. *Mothers' Ages of California Students: Stemplots* (Use the ***CombinedClassDataAut08*** sheet.)

a. By hand, make an ordered stemplot of the distribution of the variable *MothersAge* for the female students. Use ***stems of two*** (see the ***Notes*** for an explanation of what stems of two are).

b. By hand, make an ordered stemplot of the distribution of the variable *MothersAge* for the male students. Use stems of two as well.

- The stem plots made by hand may be able to be checked using software. Consult ***Stem Plots*** in the ***Software Supplement***.

c. For each of the distributions shown in the stemplots (the male and the female) calculate the percentage of students whose mothers were age 30 or more when the student was born, using the conditional probability notation of §1.2 to write your results, using $X \geq 30$ in your notation. Not all the students gave information; use as denominators only the total for which you have information.

d. Interpretation: Would you say that the percentages show that the mothers of male and female students are basically similar or would you say that there is an important difference?

e. Use the stemplot for the female students you made to answer part a to create a dot plot (by hand) for the data on *MothersAge* for the female students.

f. Use the stemplot or the dot plot you made (in part e) for the female students to make a histogram (by hand) with bin width of two years for the data on *MothersAge* for the female students.

g. Is it possible from your stemplot to calculate the percentage of female students whose mothers were less than twenty-seven27 years old? If so, make the calculation. If it is not possible, say why it is not possible.

h. Is it possible from your histogram to calculate the percentage of female students whose mothers were younger than 27 years old? If so, make the calculation. If it is not possible, say why it is not possible.

i. Is it possible from your dot plot to calculate the percentage of female students whose mothers were younger than 27 years old? If so, make the calculation. If it is not possible, say why it is not possible.

- Open the file ***CombinedClassDataAut08.***
- Use software to get a histogram of the variable *MothersAge* grouped by *Gender* (that is, one histogram for males and another for females.)
- Make the histograms with *binwidth* equal to 2, and start the first bin so that the bins represent ages $18 \le MothersAge < 20$, $20 \le MothersAge < 22$ etc. so that the bins are those that you used with the stems-of -two stem plot.

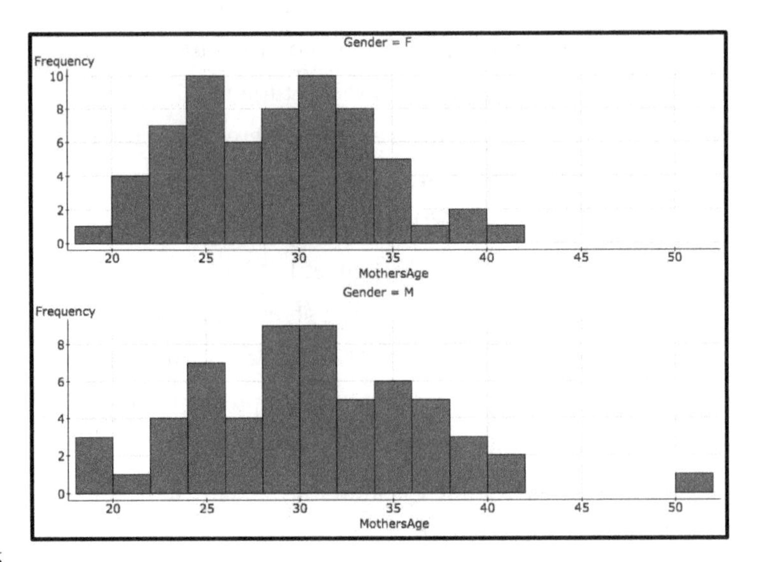

1. On your hand-drawn histogram for the female students, shade in the proportion of students whose mothers were 30 or older:

   $P(x \ge 30 \mid F) = 27/63 \approx 0.4286$

7. *SF Espresso and Café Ratings* There is an interesting website (*http://www.coffeeratings.com/*) that comes from the attempt by one man to "find, taste, and review every noteworthy espresso" he could find in the city of San Francisco. For each espresso and each café he uses a scale of 10 (10 being the best), and then for the overall score he averages the "espresso" score and the "café" score.

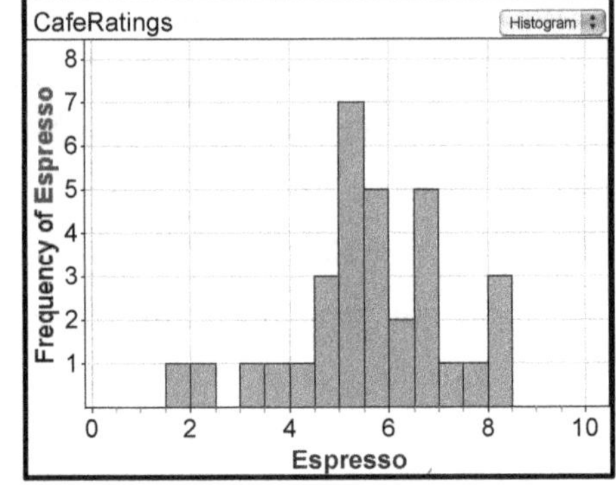

a. Here is a histogram of the n = 32 **Espresso** ratings for just the *Embarcadero* neighborhood. What is the *bin width* for this histogram?

- Using software, open the file ***CoffeeRatingsSFEmbarcadero.***
- Make the histogram shown above, whose bars start at Espresso rating = 1.5, have bin width = 2, and where the vertical axis shows frequencies. Get horizontal grid lines.

b. Use the histogram you have made to find the proportion of cafés rated at 7 or more and express your answer in probability notation.

c. If you add up the numbers for all of the bars in the histogram, what should the sum be? Why?

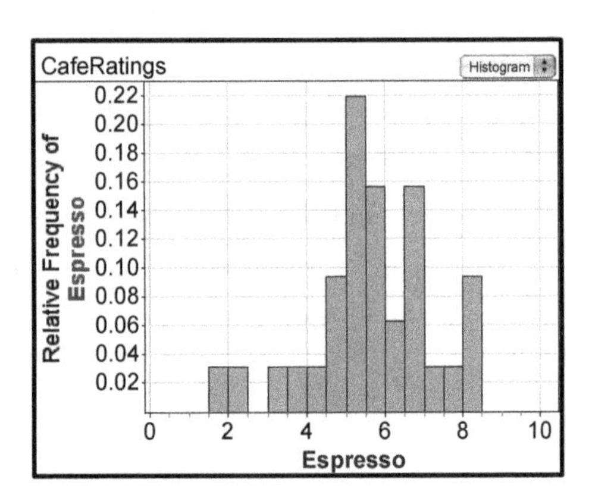

- Using software, change the vertical axis in the histogram from frequencies to relative frequencies.

d. Use this relative frequency histogram to estimate $P(X < 4)$. Show a sum of proportions.

e. If you add up the proportions for all of the bars in the histogram, what should the sum be? Why? (Note: With n = 32 cases, using the "frequency" rather than the "relative frequency" histogram is more accurate. However, the goal here is to get accustomed to relative frequency histograms, which have proportions on the vertical scale.)

8. ***SF Espresso and Café Ratings (continued)*** Shown below is a bar graph that is not a histogram. It is for the same data as the histogram in Exercise 7. Some spreadsheet software will by default produce this kind of bar graph. To see the difference between a histogram and this kind of bar graph, answer these questions.

a. What variable is on the horizontal axis in the histogram above? And what variable is on the horizontal axis in the bar graph shown here?

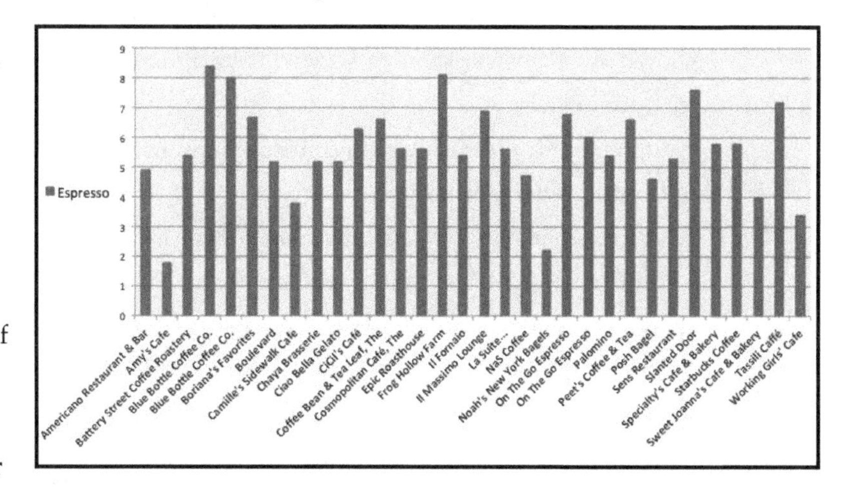

b. What is on the vertical axis in the histogram? And what is on the vertical axis in the bar graph shown here?

c. What does the height of a bar indicate for a histogram? And what does the height of a bar in the bar graph indicate. (To answer this question you may want to focus in on a single bar for both graphics.)

d. Put into your own words the difference between a histogram and a bar graph such as the one here.

9. ***Remember when you were fourteen*** In Example 3 in the ***Notes*** for §1.3 we looked at the proportions of male and female Australian secondary students who were in the height interval 160 cm ≤ *Height* < 180 cm, and we found that the proportions in this "mid-height" range were quite similar. What happens if you look at fourteen-year-olds instead of thirteen-year-olds? Here are the histograms.

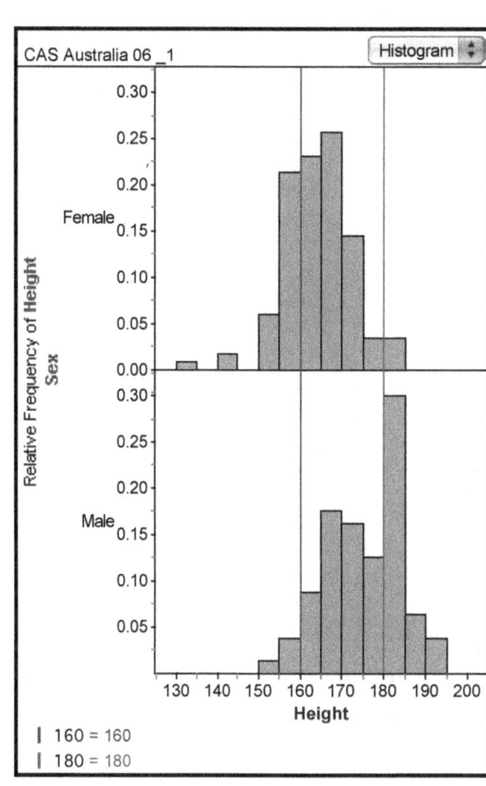

a. From the histograms, calculate the proportion of male students and the proportions of female students who are in the height interval 160 cm ≤ *Height* < 180 cm.

b. Write an interpretive sentence comparing what was found in Example 3 with what you have found here from part a.

c. Express your answer to part a in probability notation.

d. If you chose a fourteen-year-old male Australian secondary student from the collection, what is the probability that the student is 180 centimeters or taller? Use probability notation.

e. If you chose a fourteen-year-old female Australian secondary student from the collection, what is the probability that the student is 180 centimeters or taller? Use probability notation.

f. Interpret (i.e., put in a sensible English sentence or two) what you have found in parts d and e.

10. *Time taken to travel to school in South and Western Australia* The variable *TimeSchool* measures the time it takes a student (in Australia) to get from home to school.

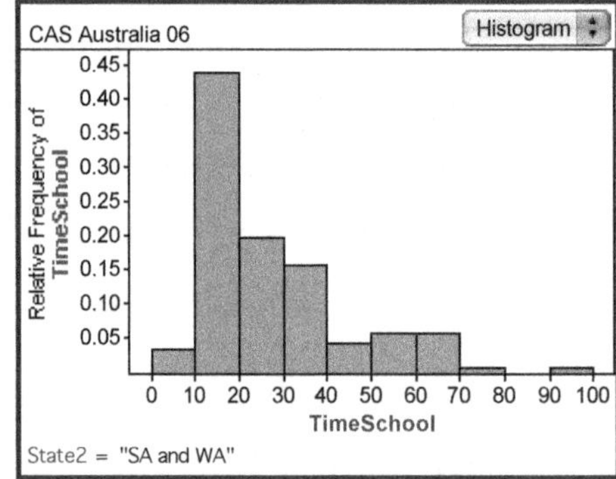

a. If you add up the relative frequencies for all the bars shown, what answer should you get? Give a reason for your answer.

b. Add the relative frequencies to confirm your answer to the question in part a.

c. Find the probability that a student takes 40 or more minutes to get to school if that student is chosen at random from the South and Western Australian collection. Use probability notation.

d. By doing as little work as possible, find the probability that a student takes less than 40 minutes to get to school. Show what you did.

11. *Les Écossais* Here are data showing the distributions of household size (*NumHH*) for households in two communities in Quebec in two census years in the nineteenth century. The cases are households; for the 1831 census there were $n_{1831} = 93$ households, and for the 1851 census, $n_{1851} = 97$. The variable is the number of people per household or *NumHH*. (Some background: Gaelic-speaking Highland Scots immigrants inhabited most of the households. The immigration had begun in the latter part of the eighteenth century but had continued during the nineteenth century.) Our statistical question is: *Is there evidence that the distribution of sizes of households for this community changed between 1831 and 1851?* The questions are on the next page.

Household Census of 1831 | ***Household Census of 1851***

Scots in Quebec (Histogram) — Annee = "1831" | Scots in Quebec (Dot Plot) — Annee = "1851"

These distributions can be pictured using a histogram (1831) and a dot plot (1851). You now have a chance to compare the usefulness of each. (Actually you can easily make a dot plot, by hand, from the histogram or a histogram from the dot plot.) ***PTO***

a. Make any kind of calculation (or calculations) you wish to make using the graphics you have been given in order to answer the statistical question that we have posed (in italics above) about whether there are any differences in the distributions of the sizes of households for 1851 and 1831.

b. Write a kind of mini-report giving your answer to the statistical question stated above about whether there are any differences in the distributions of the sizes of households for 1851 and 1831.

c. Which of the graphics did you find more useful: the histogram or the dot plot? Give a reason for your answer. Note: A good way to do this exercise is to use teamwork. Find a partner and together decide what kinds of calculations it would be good to have. Then one of you should do the calculations for the 1831 data and the other the calculations for the 1851 data. Then bring your calculations together and discuss what the two of you think they mean. One completely valid answer is that there is very little difference between the 1831 and the 1851 distributions of the number of people in households.

12. *The Midge Question* In 1981, two new varieties of a tiny biting insect called a midge were discovered by biologists W. L. Grogan and W. W. Wirth in the jungles of Brazil. They dubbed one kind of midge an Apf midge and the other an Af midge. The biologists found out that the Apf midge is a carrier of a debilitating disease that causes swelling of the brain when a human is bitten by an infected midge. Although the disease is rarely fatal, the disability caused by the swelling may be permanent. The other form of the midge, the Af, is quite harmless and a valuable pollinator. In an effort to distinguish the two varieties, the biologist took measurements on the midges they caught. The two measurements taken were of wing length (WL) and antennae length (AL), both measured in centimeters.

- Using software, open the file entitled ***Midges*** and use the data to answer the question below in a short joint report. Decide what to do and then do it with software.

Question: Is it possible to distinguish an Af midge from an Apf midge based on wing and antenna length? Write a report that describes to a naturalist in the field how to classify a midge he or she has just captured. (This question came from Daniel J. Teague.)

[Reference: Grogan, William L., Jr. and Willis Wirth. 1981. "A new American genus of predaceous midges related to Palpomyia and Bessia (Diptera: Ceratopogonidae)." *Proceedings of the Biological Society of Washington* **94** (4): 1279-1305.]

13. ***Real Estate Data*** The data for this exercise concerns houses that were sold in 2007–2008 in San Mateo County, California. There were a total of 3,947 houses sold in the period from June 2007 to June 2008, but the data for this exercise uses just n = 400 selected randomly from the 3947. Here is a part of the spreadsheet for these data, where each row is a different house sold. For this exercise, we will be only interested in the following variables.

RE San Mateo 0708

	Caseno	Beds	Baths	SqFt	Age	Style_2	Region	List_Price	Sale_Price	Diff_Sal...	Time_in...
7	2037	3	2	1060	60	Other	South	475000	362980	-112020	118 d
8	2204	2	2	1160	82	Ranch	North	685000	640000	-45000	101 d
9	2836	2	2	1100	62	Other	South	998000	965000	-33000	22 d
10	2257	3	2	1950	42	Other	Central	949000	953000	4000	0 d
11	1754	3	2	1530	57	Other	South	719000	720000	1000	0 d
12	1036	3	2	2250	33	Traditional	Central	1399000	1600000	201000	11 d

---

Time_in_Market *records the time it took in days for the house to sell*

ListSale *is a categorical variable with three categories:*

"Over" if the sale price of the house was more than the list price.
"Same" if the sale price of the house was the same as the list price
"Under" if the sale price of the house was under the list price

TMlessthan20 *is a categorical variable with three categories:*

"On market less than 20 days"
"On market 20 days or more"

Region *indicates the region of San Mateo County in which the house was located.*

---

a. What are the cases for these real estate data?

b. Is the variable *Time_in_Market* a quantitative variable or a categorical variable?

c. Is the variable *ListSale* a quantitative variable or a categorical variable?

- Using software, open the file ***San Mateo RE Sample Y0708***.
- Get relative frequency histograms of the variable *Time_in_Market* grouped by the categories of the variable *ListSale* . Use bars with *binwidth* = 20 starting at *Time_in_Market* = 0 days.

The graphic should resemble the one here.

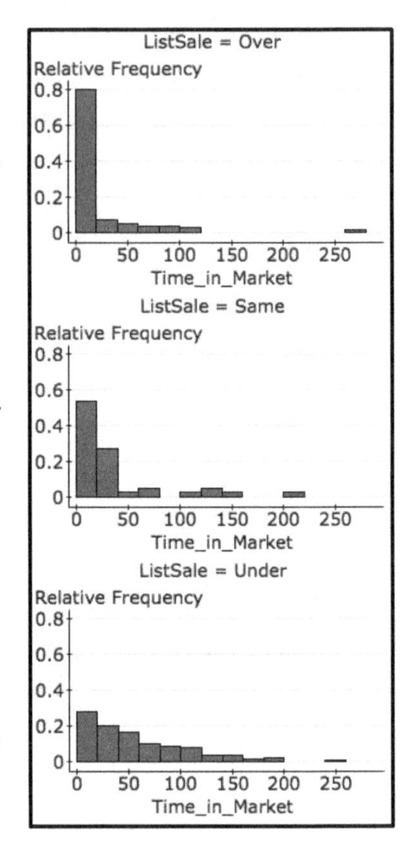

d. For each of the three categories of the variable *ListSale* ("Over" = O, "Same" = S, and "Under" = U) use the histogram to estimate the proportion of houses in the sample that spent less than 20 days on the market before selling. Express the results using good probability notation; i.e., for the "over" category, you should have a value for $P(X < 20 \mid O)$ and likewise for the other two categories.

e. Copy the output into a word-processing document and shade in the bar or bars on the plot that represent your answers to part d.

Follow the directions on the next page to check your answers.

- Using software with the file ***San Mateo RE Sample Y0708*** make a two-way table with the variables *ListSale* and *TMlessthan20* and get percentages or proportions in the correct direction to confirm the results made from inspecting the histograms.
    f. *Interpretation.* Try to express in English (or some other language—but make it elegant) what the comparison of the three proportions means in the context of whether the houses sold for over, under, or the same as the list price.
    g. *Review:* Are the events $X < 20$ and $X \geq 20$ mutually exclusive? Give a reason for your answer.
    h. *Review:* Are the events $X < 20$ and $X \geq 20$ independent? Give a reason for your answer.

- With software, get either histograms (as shown below) or two-way tables to investigate the differences in *Time_in_Market* by the regions of the San Mateo County. For the histogram, use the same bin width as in the histogram above.
    i. Use either the histograms or the two-way tables to explain in which regions the houses appear to sell fastest and in which regions the houses sell at a slower pace. Give evidence for your conclusions.

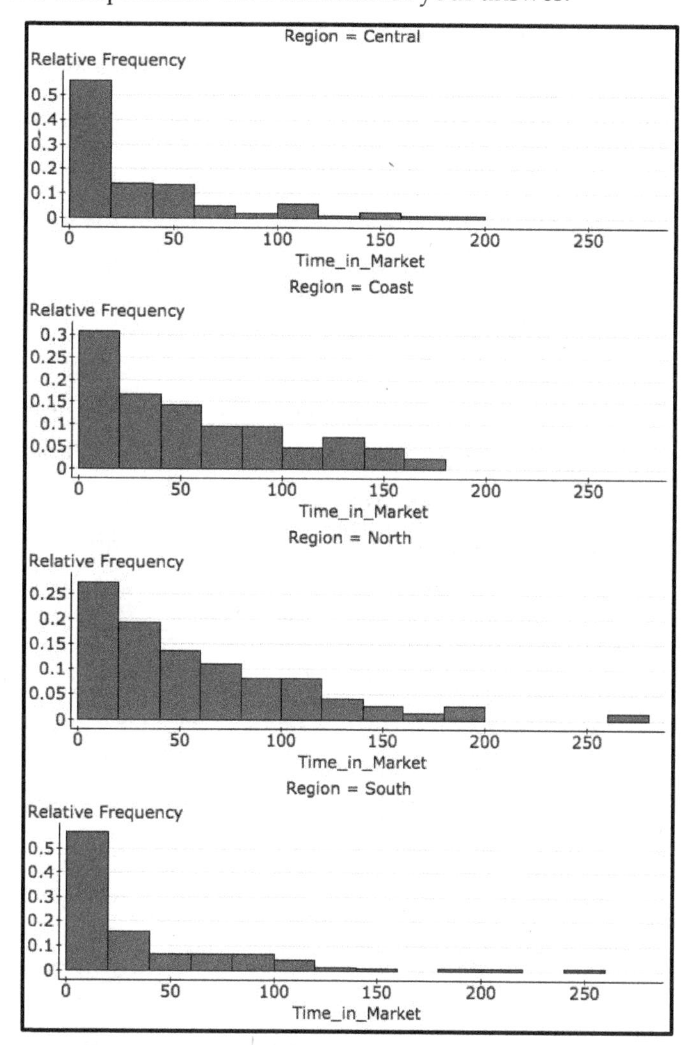

San Mateo RE Sample 0708

		Region				Row Summary
		Central	Coast	North	South	
TMlessthan20	On market 20 days for more	56 0.4375	29 0.690476	53 0.726027	66 0.434211	204 0.516456
	On market less than 20 days	72 0.5625	13 0.309524	20 0.273973	86 0.565789	191 0.483544
	Column Summary	128 1	42 1	73 1	152 1	395 1

S1 = count ( )
S2 = columnProportion

§1.4 Exercises: Shape, Center, and Spread

1. ***Review and the Shape of a Distribution.*** Here is a graphic of the distribution of the number of full-time students (the variable *FullTimeStudents*) for just the two-year colleges.

   a. *Review:* What are the cases for these data?

   b. *Review:* Is the variable *FullTimeStudents* a quantitative variable or a categorical variable?

   c. *Review:* What is the *bin width* for the histogram?

   d. *Review:* About what proportion of two-year colleges have fewer than 2000 students? You may express your answer as $P(X < 2000) =$

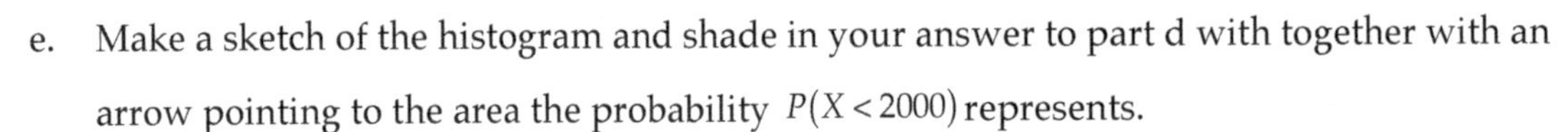

   e. Make a sketch of the histogram and shade in your answer to part d with together with an arrow pointing to the area the probability $P(X < 2000)$ represents.

   f. What name should we give to the shape of this distribution: *right-skewed, left-skewed,* or *symmetric?* Give a reason for your answer.

   g. At about how many *FullTimeStudents* does the center of this distribution appear to be? (The answer does not require any calculation at this point; just judge roughly where the center of the distribution is. You may well give your answer as a range of values.)

   h. What is the *range* of values for this distribution?

   i. In the ***Notes***, there are several graphs showing the same variable *FullTimeStudents* but for the public four-year universities. Which distribution has the wider spread, this one or the one for the four-year universities? Give a reason for your answer.

2. ***Selectivity and Tuition Fees.*** (Note: this exercise continues on the next page.) The distributions shown are from the Far West colleges collection for just the private schools. We can compare the tuition fees for colleges and universities that are "very selective" in their admission policies with the tuition fees for institutions that are "minimally or moderately selective." Some numbers: there are $n_{vs} = 23$ for the "very selective" schools, and their average tuition fee is $27,609; there are $n_{mms} = 52$ for the "minimally or moderately selective" schools, and their average fee for tuition is $20,905.

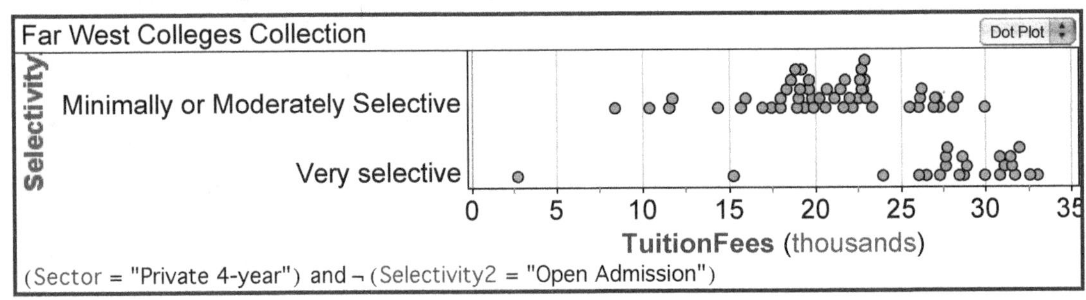

   a. What are the cases for these data?

   b. Judging from the dot plots and averages, which distribution appears to have the higher center?

c. On average, judging by the plots and the numbers given, which type of college or university, the "very selective" or the "minimally or moderately selective," typically has the higher tuition fees? Give a reason (or reasons) for your answer.

d. Is the distribution of tuition fees for the "minimally or moderately selective" institutions *left-skewed, right-skewed, symmetric, bimodal,* none, or a mix of these? Give a reason for your answer, and if you answer "none" or a "mix," state which type of shape is in the mix.

e. For this and the next question (but not for the questions following that), disregard the two outliers for the "very selective" group. Disregarding these two outliers, which group has a distribution of tuition fees with the greater spread? Give a reason for your answer.

f. A common confusion is to think that if the center of a distribution is large (far to the right) then it must necessarily be that the spread of that distribution is big as well. Explain how part e shows that this confusion is not correct.

g. *Review:* Using the dot plots (counting dots), the notation "*TF*" for the variable *TuitionFees,* the $n_{vs} = 23$ ("very selective") and $n_{mms} = 52$ ("minimally or moderately selective"), calculate $P(TF \geq \$25000 \mid MMS)$, and $P(TF \geq \$25000 \mid VS)$. [*Hint:* There is a very easy way to get $P(TF \geq \$25000 \mid VS)$ using one of the probability rules.]

h. Express in English what $P(TF \geq \$25000 \mid MMS)$ and $P(TF \geq \$25000 \mid VS)$ tell you.

i. Explain how your answer to part h agrees with your answer to parts b and c (which are really the same question).

3. *Ages of People and Dogs.* The idea of this exercise is to be acquainted with the ideas of *shape, center,* and *spread* by thinking about where on a scale a distribution is, what shape the distribution has, and how spread out is a distribution that you can imagine. The variable is *age*, and in most of the parts of the exercise, the cases will be people. What you are to do is to sketch a rough ***density curve*** on a scale of ages in answer to each scenario. Here is an example: "Sketch the distribution of the ages of the students in your class." A good sketch would be like this one.

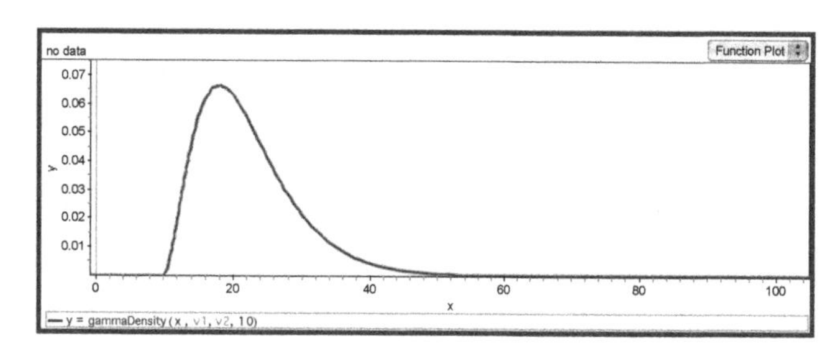

Draw a scale for each of the scenarios (a through f). Here are the scenarios:

a. The ages of the children in a fifth-grade class. (Think: what are the ages of the children likely to be?)

b. The ages of the people in a fifty-year reunion of a fifth-grade class.

c. The ages of people attending a performance of the latest hot rock band.

d. The ages of the residents of a retirement home.

e. The ages of the fans at a San Francisco 49ers game.

f. The ages of all the dogs in your county.

g. The age of death for classical music composers of the 1700s and 1800s.

4. ***Ages of People and Dogs*** Look at the sketches of the distributions that you made for the scenarios a through f in the exercise on "Ages of People and Dogs." In answering the questions below, also include the example distribution on the ages of the students in your class. For each of the questions below, give a reason for your answer in a complete sentence.
   a. Which of the distributions of *Age* is farthest to the right (and thus has the greatest center)?
   b. Which of the distributions of *Age* is farthest to the left (and thus has the lowest center)?
   c. Which of the distributions of *Age* has the greatest spread in the variable?
   d. Which of the distributions of *Age* has the least spread in the variable?
   e. Which of the distributions (if any) are likely to be right-skewed?
   f. Which of the distributions (if any) are likely to be left-skewed?
5. ***Tuition Fees and Type of College*** The data are from the Far West colleges collection and show the distributions of tuition fees for the three types of colleges and universities that we have been examining.

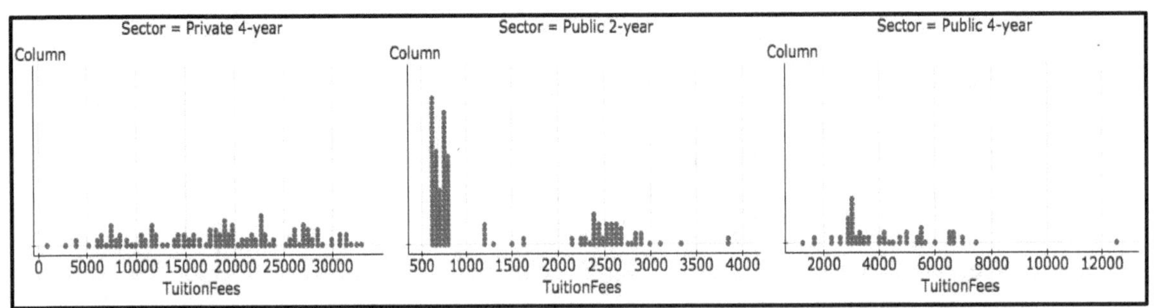

   a. What are the cases for these data?
   b. In the graphic shown above, notice that the scales differ greatly: for the private colleges, the fees range up to $30,000, whereas for the Public 2 year, the maximum is $4000. Put the three distributions (the three sectors) of tuition fees in order of smallest center (or location), middle center (or location) and biggest center (or location). Say how the ranking you have made in terms of center makes sense from what you know otherwise about the fees for different types of educational institutions. Notice that the scales for Tuition Fees differ greatly.
   c. Put the three distributions (the three sectors) of tuition fees in order of smallest spread, middle spread, and greatest spread. Say how the ranking you have made in terms of *spread* makes sense (or does *not* make sense) from what you know otherwise about the fees for different types of educational institutions.
   d. Do any of the distributions appear *bimodal*? Which?
6. ***Real Estate Data*** (Note: This exercise continues on the next page.) The data for this exercise concerns houses that were sold in 2007–2008 in San Mateo County, California. There were a total of 3,947 houses sold in the period from June 2007 to June 208, but the data for this exercise uses a sample of $n = 400$ selected randomly from the 3,947. For this exercise, the variables of interest are:

SqFt	*records the size of the living area of the house in square feet.*
List_Price	*the price that the seller first asked for when the house was put on the market.*
Region	*"regions" of San Mateo County: "Central," "Coast," "North," and "South."*

a. Characterize the three variables we will analyze as either quantitative or categorical.

b. What are the cases for our data?

- Open the file ***San Mateo RE Sample Y0708.***
- Make histograms of the variable *SqFt* grouped by the variable *Region.*
- Use a *bin width* of 500 square feet, and start the bars at 0 square feet.
- Use relative frequencies. It will be helpful to get grid lines, as shown here.

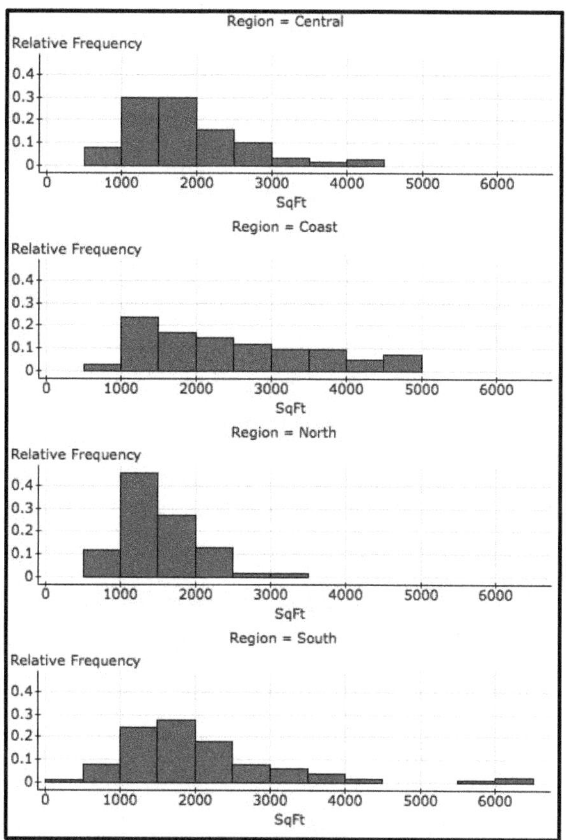

c. What is the *bin width* of these histograms?

d. What are the most appropriate words to characterize the shape of these distributions? (See ***Shape*** in the ***Notes.***)

e. Which region appears to have the greatest variability in the sizes of houses (in of *SqFt)?*

f. Judging from the histograms, are there differences by region in the average size of houses? Give a reason for your answer.

- Make relative frequency histograms of the variable *List_Price;* start the bars at 0, and use *bin width* equal to 250 000, as shown here.

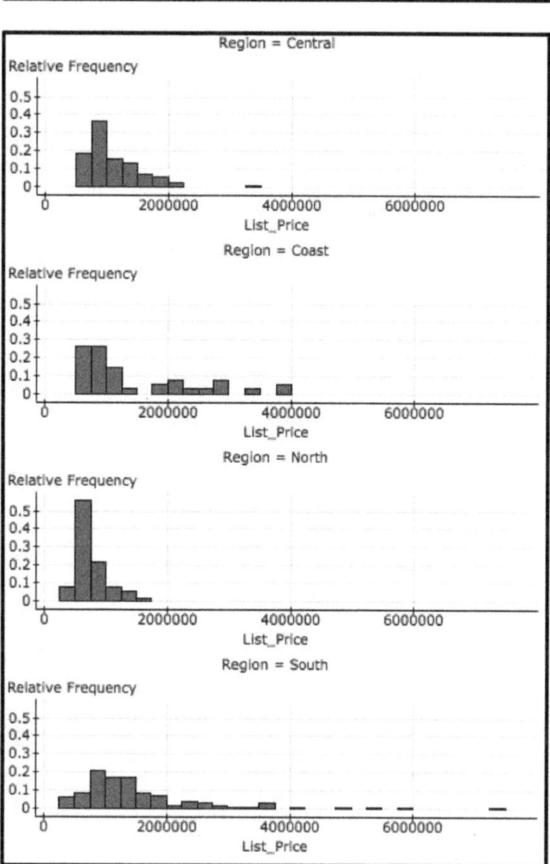

g. What are the most appropriate words to characterize the shape of these distributions? (See ***Shape*** in the ***Notes.***) Give a reason for your answer.

h. Which region appears to have the greatest spread or variability in the sizes of houses (in the distributions of *List_Price*)?

i. Can you say which region has the distribution of *List_Price* with the highest center or lowest center, or do the locations of the distributions appear similar? Explain.

j. Using the correct conditional probability notation and using the histograms, calculate the proportions of houses in each of the regions whose *List_Price* was less than $750,000. (You will have to look carefully at the histograms.)

§1.5 Exercises on Measures of Center/Location

1. According to the ***Notes***, there is a relationship between skewness and whether the mean or the median is larger for a distribution. Express in words what this relationship is.

2. ***Mammals*** These are data from the collection ***Mammals2***. The data come from a study that related various characteristics of mammals to the number of hours the mammals typically sleep. The authors rated each mammal as to how likely the mammal was to be the prey of other animals. Here are the life spans in years of life (it appears to be maximum life span) of the thirteen mammals most likely to be preyed upon and the fourteen mammals least likely to preyed upon. So, n = 13 for the "most-preyed upon mammals" and n = 14 for the "least-preyed upon mammals." The numbers for life span are rounded to nearest whole year of life, although in the data the numbers are given to tenths of a year.

Mammals Most Likely to be Prey

Number	Species	Life Span
1	Chinchilla	7
2	Cow	30
3	Donkey	40
4	Giraffe	28
5	Goat	20
6	Ground squirrel	9
7	Guinea pig	8
8	Horse	46
9	Lesser short-tailed shrew	3
10	Okapi	24
11	Rabbit	18
12	Roe deer	17
13	Sheep	20

Mammals Least Likely to be Prey

Number	Species	Life Span
1	Arctic fox	14
2	Big brown bat	19
3	Cat	28
4	Chimpanzee	50
5	Eastern American mole	4
6	Genet	34
7	Giant armadillo	7
8	Gorilla	39
9	Gray seal	41
10	Gray wolf	16
11	Human	100
12	Jaguar	22
13	Little brown bat	24
14	Red fox	10

   a. Do you think that on average the least-preyed upon mammals should have a longer life span, shorter life span, or pretty much an equal life span to the most-preyed upon mammals?

   b. Find the mean life span of the most likely to be prey and the mean life span of the least likely to be prey. Show your calculations in an organized way so that someone reading your answer will know what you have done. Do the results fit your expectations as expressed in part a?

   c. Make ordered stemplots for the life spans of the most likely to be prey and the life spans of the least likely to be prey and from these calculate the medians.

   d. For the least likely preyed upon, the human appears to be an outlier. Now, suppose humans lived to be only fifty-two years old rather than one hundred. If that were so, would the mean life span change for the distribution? Would the median life span change? Explain your answer. You can base your explanation on calculations, but you can also give an explanation based upon what is said in the ***Notes***.

   e. Find the five-number summary for the life spans for the most-preyed upon mammals.

   f. Find the five-number summary for the life spans for the least-preyed upon mammals.

   g. Use your results from parts e and f to make a pair of box plots to compare life spans of the most- and least-preyed. (Your graphic should resemble the example in the ***Notes*** that compares the *PctOverAge24* in the two-year and four-year institutions.)

PTO

h. In the collection of data, "Predation Level" runs from Level 1 (least-preyed upon) to Level 5 (most-preyed upon). Here are box plots for three levels of predation—the two you have been working on, in addition to all the levels between. The dots indicate outliers; your box plot for the least-preyed upon should have the whisker extending to the human, with a life span of one hundred years. In the next section, you will learn how such box plots are made. Here is the question: do the box plots show evidence that life span in mammals is affected by their predation level, or not?

3. *Gestation* "Gestation" means the typical length of pregnancy for a mammal. Here the length of time is expressed in days. The histogram shown shows the distribution of Gestation for the mammals in our data.

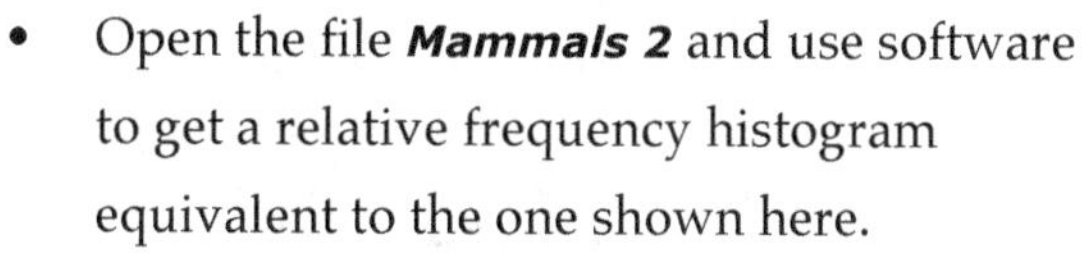

- Open the file ***Mammals 2*** and use software to get a relative frequency histogram equivalent to the one shown here.

a. What are the cases for these data?

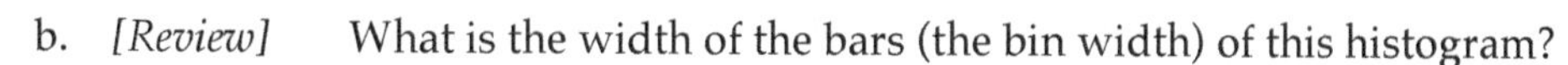

b. *[Review]* What is the width of the bars (the bin width) of this histogram?

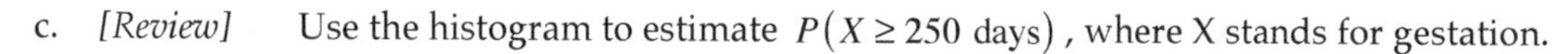

c. *[Review]* Use the histogram to estimate $P(X \geq 250 \text{ days})$, where X stands for gestation.

d. *[Review]* Use the histogram to estimate the proportion of mammals that have lengths of pregnancies (gestation) less than one hundred days. Use the correct probability notation.

e. *[Review]* Either make a sketch of the histogram from software or make a copy of the histogram and shade in and label the answers to parts c and d with the probability notation pointing to the shading.

f. For this distribution of gestation, which measure of center will be bigger, the mean or the median? Give a reason for your answer.

- Using software, get a table of summary statistics showing the mean and the five number summary (Minimum, Q_1, Median, Q_3 Maximum) for the variable *Gestation*. Consult the software supplement for ***Summary Statistics***.

Summary statistics:

Column	n	Mean	Min	Q1	Median	Q3	Max
Gestation	58	142.35345	12	35	79	210	645

g. Confirm that your answer to f is correct. Write the mean and the median using the correct symbols.

h. Use the five-number summary to make a box plot by hand.

- Using software, get a box plot using the variable *Gestation*. Consult ***Box Plots*** in the ***Software Supplement***. The box plot that software produces for the variable *Gestation* may well show two outliers. The formulas for determining outliers in box plots are presented in §1.6. Some software gives the choice of showing or not showing outliers.

4. ***A Wrong Idea*** Is it true that the more cases there are in a collection, the bigger the mean or the median will be? No! It is not true. The mean and median do not get bigger because the number of cases n (the count) is bigger. Notice that in the formula for the mean $\overline{x} = \frac{\sum_{i=1}^{n} x_i}{n}$ we divide by the number of cases (or count) n.

Here are more data on mammals. There are two collections of species of mammals. One collection includes mammals that are more exposed to danger from other animals (because they are bigger?) and there are n = 9 of these mammals. The second collection consists of mammals that are less exposed to danger from other animals. There are n = 12 of these mammals for which we have data. For both collections, we have the gestation period (length of pregnancy) in days recorded.

Species More Exposed to danger from Other Animals	Gestation (days)
Cow	281
Donkey	365
Giraffe	400
Goat	148
Horse	336
Okapi	440
Rabbit	31
Roe deer	150
Sheep	151

Species Less Exposed to Danger from Other Animals	Gestation (days)
Desert hedge hog	
Echidna	28
European hedge hog	42
Galago	120
Golden hamster	16
Owl monkey	140
Phanlanger	17
Raccoon	63
Rhesus monkey	164
Rock hyrax	225
Slow loris	90
Star nosed mole	
Tenrec	60
Tree shrew	46

a. Make (by hand or using software) comparative dot plots for the variable *Gestation.*

b. For each of the two collections, calculate (by hand) the mean and median for the variable *Gestation*.

c. Is it true or false that the collection with the bigger count (the number of cases) has the bigger mean? Explain your answer using your calculations.

5. ***Danger, Length of Pregnancy, and How to Read Summary Tables***.

- Open the file **Mammals 2** and use software to get a table of summary statistics equivalent to the one shown here. See the software supplement for ***Summary Statistics.***

Summary statistics for Gestation:
Group by: Danger_Level

Danger_Level	n	Mean	Min	Q1	Median	Q3	Max
High Danger	19	222.5	21.5	112	170	365	624
Intermediate Danger	22	104.13636	16	28	45.5	140	645
Low Danger	17	102.23529	12	38	60	120	310

a. Looking at the means and the medians for the three danger levels, can you say that mammals in greater danger tend to have a longer or shorter length of pregnancy? Defend your answer with numbers.

b. Think about the definition of quartiles. For the low danger mammals, what number should replace the question marks in the probability statement $P(X \geq ???) = 0.25$ if X stands for gestation (days of length of pregnancy)? Give a reason for choosing the number.

c. Judging from the Five number summary, can you predict which *Danger Level* box plot for *Gestation* will have the box farthest to the right if the box plots are shown horizontally. Give reasons for the answer.

- Using the file **Mammals 2** get software to make comparative boxplots of *Gestation* grouped by *Danger Level* to check the answer to part c.

6. *The Fellowship of the Ring* In J. R. R. Tolkien's classic, the heroes of the story are the hobbits. Soon after starting their journey from the Shire the hobbits came to a place named Bree. This town was in an area where both Men and Hobbits lived. The Big Folk and the Little Folk (as they called one another) were on friendly terms, minding their own affairs in their own ways, but rightly regarding themselves as necessary parts of the Bree-folk. Nowhere else in the world was this peculiar but excellent arrangement to be found.

(J. R. R. Tolkien, The Fellowship of the Ring, Ballantine Books Edition, p. 206)

We have managed to get a sample of the heights (measured in cm) of Bree-folk who happened to be in The Prancing Pony, the inn in Bree. There are $n = 26$ Bree-folk represented, where $n_{Little} = 12$ and $n_{Big} = 14$. Here are the data.

Height (cm)
108
107
109
102
164
107
165
170
106
172
100
115
106
179
172
181
166
181
111
167

a. Make a dot plot of the data and comment on what you see.

b. Calculate the mean and the median for all of the data. Do the mean and the median make any sense in the context of these data? Explain.

c. Here is the box plot for the data. Compare the box plot to your dot plot. Comment on the usefulness of the box plot for these data.

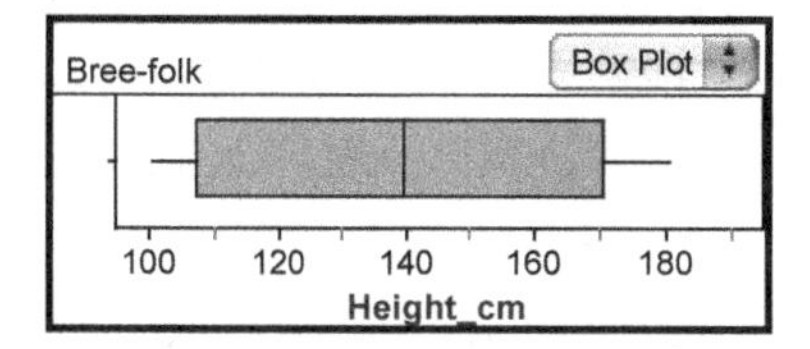

d. Calculate means and medians that do make sense to the Bree folk and explain their sense.

7. *Real Estate Data: Size of Houses* This exercise concerns a sample of data on houses that were sold in San Mateo County in the years 2007 to 2008. The collection you will use does not include all the houses that were sold; it is just a sample of the entire population of all houses that were sold. You will look at the variable SqFt. This variable records the amount of square feet in the living area of the house, so it is a measure of the size of the house.

- Using ***San Mateo RE Sample Y0708,*** get dot plots for the variable *SqFt* grouped by *Region.*

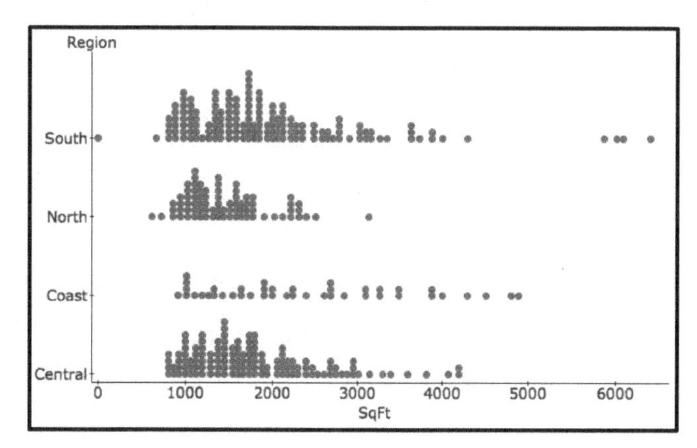

a. What are the cases for this collection?

b. What is the most appropriate name for the shape of the distributions?

c. Which measure of center/location should give a more accurate notion of the location of the distributions, the mean or the median? Give a reason.

d. Judging from the plots, which region would you guess has the largest houses on average?

- Using ***San Mateo RE Sample Y0708,*** get ***Summary Statistics*** for the variable *SqFt* grouped by *Region.* Get the table shown on the next page with mean and the Five Number Summary, and "filter" for *SqFt > 0* to exclude the "house" with zero area. See ***Filtering*** in the software supplement.

e. Use the mean and the median to rank order the four regions in average size of house.

Summary statistics for SqFt:
Group by: Region

Region	n	Mean	Min	Q1	Median	Q3	Max
Central	128	1832.5469	800	1260	1675	2220	4200
Coast	42	2427.6429	920	1430	2210	3270	4890
North	78	1477.5897	610	1140	1390	1710	3130
South	152	1989.8553	0	1350	1756.5	2310	6400

f. A student (your friend) persists in thinking that a higher number of cases will lead to a higher mean value. Write a short message to your friend using this summary table to show your friend that it is not so that a higher count leads to a higher mean or median.

8. *Great Lakes Colleges and Universities I* The same variables are measured as in the Far West collection. We will look at the variable Freshman Retention Rate, which is the percentage (it is given as a percentage) of first-year students who proceeded to the second year of study. Our statistical question will be how the Sector of institution (public two-year, public four-year, or private four-year) is related to Freshman Retention Rate.

- Open the file ***GreatLakesCollegesUniversities*** and get a graphic showing box plots of the variable *Freshman Retention Rate* grouped by the variable *Sector,* the variable that shows the type of institution.

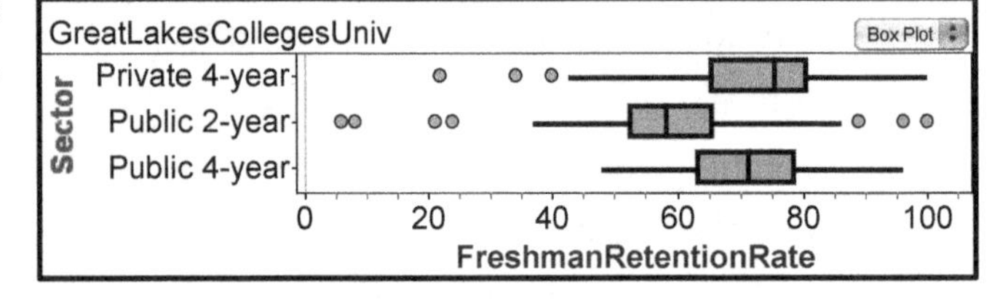

- With the same file, get ***Summary Statistics*** that include the mean and the median. The output should be similar to what is here.

Summary statistics for FreshmanRetentionRate:
Group by: Sector

Sector	n	Mean	Median
Private 4-year	209	72.770335	75
Public 2-year	149	58.214765	58
Public 4-year	81	70.802469	71

a. What are the cases for these data? (That is, what is it that we are talking about?)

b. Interpret what the means and the medians as well as the appearance of the box plots tell you about the differences in the average Freshman Retention Rate in the different types of institutions. Your interpretation should use the numbers but must say something about the type of institution and what the variable Freshman Retention Rate measures.

c. What would be the most appropriate word for the shapes of the three Freshman Retention Rate distributions: right-skewed, left-skewed, or symmetrical?

d. Explain how the relative sizes of the means and medians support your conclusion in part c.

Special Exercise A: Mammals, Means, Medians, and Resistance

The ***Notes*** have introduced two measures of the location of a distribution—the mean and the median. The median is resistant to skewness and outliers but the median is not. Here we investigate and show that the median is not affected by (that is: *is resistant to*) outliers, but the mean is affected by outliers (that is: *is not resistant* to outliers.)

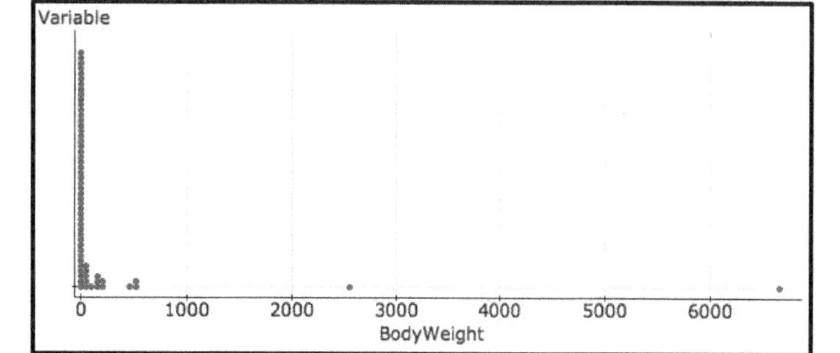

- Using the file ***Mammals 2*** get a dot plot for the variable *BodyWeight.* The plot show resemble this one.

1. Record the names and the body weights of the two outliers in this collection. If you are not accustomed to metric measures, know that the conversion factor is 2.2 lbs/kg. You may be able to highlight the dots on the graph using software so that those observations are also highlighted in the data spreadsheet.

- Using software, get ***Summary Statistics*** showing the mean, median, and the count.

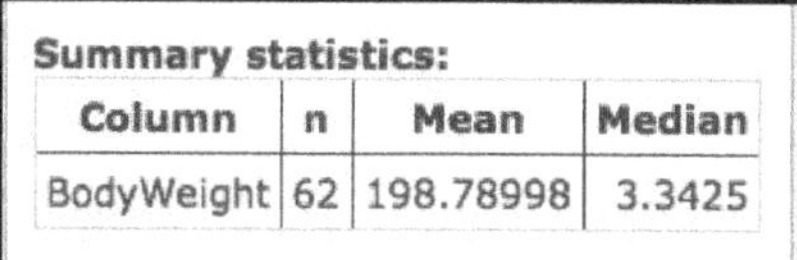

Summary statistics:

Column	n	Mean	Median
BodyWeight	62	198.78998	3.3425

Notice the huge difference between the values of the mean and the median.

•2. ***Elephants slimming***. Some statistical software have dynamic features that allow a dot (that is, an observed value) to be changed by dragging. The purpose of this dynamism is to answer "what if" questions. If the software you are using has these nice features, drag the dot for the African elephant so that he loses weight until he is just a bit heavier than the Asian elephant (the next outlier). The idea here is to see the effect of outliers on the mean and the median. To determine whether your software has these features, consult the software supplement under ***Cool Stuff***. If your software lacks these features, move on to the next software instruction before Question 3.

   a. Record the body weight of the mammals now that the African elephant has shrunk down. Compare it with the mean weight before the heavy African elephant slimmed.

   b. Has the slimming program affected the median? Why is this?

• c. Continue to shrink the body weights of both elephants so that they are both under 1000 kilograms. Record the mean body weight and the median body weight. Has the median changed? Has the mean changed?

- Using software, filter the data so that the data set includes only animals whose body weight is less than 1000 kg. See the software supplement on ***Filtering***.

3. a. Describe what has happened to the mean and median for the filtered sample for the mammals that weigh less than 1000 kg. compared with the full sample of mammals.

   b. Is the distribution of body weights without the elephants still right-skewed? Compare the mean and median.

Special Exercise S: Summation or Sigma Notation

What sigma: $\sum$ means:

- $\sum_{i=1}^{6} x_i$ means the same as $x_1 + x_2 + x_3 + x_4 + x_5 + x_6$. That is, $\sum_{i=1}^{6} x_i = x_1 + x_2 + x_3 + x_4 + x_5 + x_6$
- In English, $\sum_{i=1}^{6} x_i$ means: add the x's, starting with the first one x_1 and ending with the last one x_6 .
- The x_i in the formula is just a "general term" for the values of the variable we are using, showing that the "i" can be 1, 2, 3, etc., up to the last one.

Example: Some very young children are just learning about geometry and measurement. Each of the children is given a small stick of a different length. Here are the measurements and the children's names.

Num	Name	Length of Stick (cm)
1	Adam	4
2	Bashir	7
3	Cindy	6
4	David	3
5	Eben	2
6	Fiona	8

We are interested in the mean length of the sticks. The mean length of the sticks can be written as $\bar{x} = \frac{\sum_{i=1}^{n} x_i}{n}$. In our example, since n = 6, this can be expanded as:

$$\bar{x} = \frac{\sum_{i=1}^{n} x_i}{n} = \frac{\sum_{i=1}^{6} x_i}{6} = \frac{x_1 + x_2 + \cdots + x_6}{6} = \frac{4+7+6+3+2+8}{6} = \frac{30}{6} = 5\ cm$$

Facts about sigma for those seeing it for the first time:

- Sigma notation $\sum_{i=1}^{n} x_i$ does not mean $\sum$ multiplied by x.
- $\sum$ is a symbol that commands: "Add the x's."

Questions to answer:

1. Two more children are added to the group, and all the children are given new sticks. For this new set of data, find the mean length of the sticks and express your work showing both the sigma notation correctly and also the expansion of the notation as shown above.

Num	Name	Length of Stick (cm)
1	Adam	6
2	Bashir	8
3	Cindy	5
4	David	7
5	Eben	7
6	Fiona	8
7	Gina	9
8	Huw	6

2. A student writes the answer to question 1 as follows.

$$\bar{x} = \frac{\sum_{i=1}^{n} x_i}{n} = \frac{\sum_{i=1}^{8} x_i}{8} = \frac{\sum 6+8+5+7+7+8+9+6}{8} = \frac{56}{8} = 7\ cm$$

However, there is one thing wrong with the answer, beside which the instructor writes, "Not needed." Determine what the error is and write for your own use (and to answer this question) what is wrong.

Order of Operations

In elementary algebra, you reviewed that the order of operations for arithmetic is: "Parentheses (or grouping symbols), Exponentiation, Multiplication/Division, Addition/Subtraction." Addition and subtraction come last. When sigma notation is used, the addition implied by the notation is often the last thing to be done. Here is an example; one of the formulas in the next section is called

the ***variance***, $s^2 = \frac{\sum_{i=1}^{n} (x_i - \bar{x})^2}{n-1}$. (We will study the meaning of this in detail in §1.6.)

Expanded, everything to the right of the sigma $\sum$ is done for each of the cases then the results are added. The subtraction inside the parentheses and squaring come first. So, for our example, since $\bar{x} = 5$ the expansion is:

$$\begin{aligned} s^2 &= \frac{\sum_{i=1}^{n} (x_i - \bar{x})^2}{n-1} \\ &= \frac{(x_1 - \bar{x})^2 + (x_2 - \bar{x})^2 + (x_3 - \bar{x})^2 + \cdots + (x_n - \bar{x})^2}{n-1} \\ &= \frac{(4-5)^2 + (7-5)^2 + (6-5)^2 + (3-5)^2 + (2-5)^2 + (8-5)^2}{6-1} \\ &= \frac{(-1)^2 + (2)^2 + (1)^2 + (-2)^2 + (-3)^2 + (3)^2}{5} \\ &= \frac{1+4+1+4+9+9}{5} = \frac{28}{5} = 5.6 \end{aligned}$$

3. For the sticks for eight children where $\bar{x} = 7\ cm$ calculate s^2 using

$$s^2 = \frac{\sum_{i=1}^{n} (x_i - \bar{x})^2}{n-1}$$ showing the expansion as above.

(The new data have been shown here once again.)

Num	Name	Length of Stick (cm)
1	Adam	6
2	Bashir	8
3	Cindy	5
4	David	7
5	Eben	7
6	Fiona	8
7	Gina	9
8	Huw	6

4. The ***standard deviation*** s is the square root of the variance, so $s = \sqrt{s^2} = \sqrt{\frac{\sum_{i=1}^{n} (x_i - \bar{x})^2}{n-1}}$. We will often want this rather than the variance. Notice here that since the square root applies to the entire sum, in the order of operations it comes last (everything inside the square root must be worked out first). For our example $s = \sqrt{5.6} \approx 2.37$, calculate s from s^2 for the eight children data.

5. Here are data for the number of bedrooms for n = 10 houses sold in 2007 to 2008 in one town in San Mateo County.

Address	Beds	Baths
384 ELM AV	2	2
2781 OAKMONT DR	4	2
2170 FLEETWOOD DR	3	2
35 TANFORAN AV	2	2
2990 SAINT CLOUD DR	3	2
2421 FLEETWOOD DR	4	2
546 5TH AV	3	2
2421 FLEETWOOD DR	4	2
1625 JUNIPER AV	3	2
2400 LEXINGTON WY	3	2

   a. Express the formula for the mean number of bedrooms for these houses using sigma notation but inserting the correct number for the size of the collection and then expand the formula (as you did above in the sticks problem) and work out $\overline{x}$. (The numerical answer is 3.1.)

   b. Express the formula for s using sigma notation; insert the correct number for the size of the collection and the mean, and then expand the formula to calculate s. (The numerical answer is 0.738.)

   c. Without really calculating, but by looking at the formulas and data, you should be able to state the values of $\overline{x}$ and s for the number of bathrooms for these data. Explain how you determined your answers.

6. ***Common Student Errors.*** Here are two errors that are made using the formula for the variance; the errors are illustrated using the number of bedrooms data. For each of the errors, put into words what the error is. (The answer has to be more specific than "the student did not use the formula correctly.")

   a. To calculate the variance using the formula $s^2 = \frac{\sum_{i=1}^{n}(x_i - \overline{x})^2}{n-1}$ a student writes what is written just below; what mistake is the student making about the formula?

$$s^2 = \frac{(2-7)^2}{10-1} = \frac{(-5)^2}{9} = \frac{25}{9} \approx 2.78$$

   b. The mistake in the calculation below may be more difficult to spot, but it has to do with order of operations. What is the mistake?

$$\begin{aligned} s^2 &= \frac{\sum_{i=1}^{10}(x_i - 7)^2}{10-1} \\ &= \frac{(2+4+3+2+3+4+3+4+3+3)^2 - (7)^2}{10-1} \\ &= \frac{(31)^2 - (7)^2}{9} \\ &= \frac{961-49}{9} \\ &= \frac{912}{9} \\ &= 101.3\overline{3} \end{aligned}$$

§1.6 Exercises on Variation: Measurement and Interpretation

1. ***Explaining variability.*** Read the first paragraph of the ***Notes*** for this section. It is clear that the tuition fees higher among the private colleges and universities, but it is also clear that there is much more spread or variability among what is charged by the private colleges.

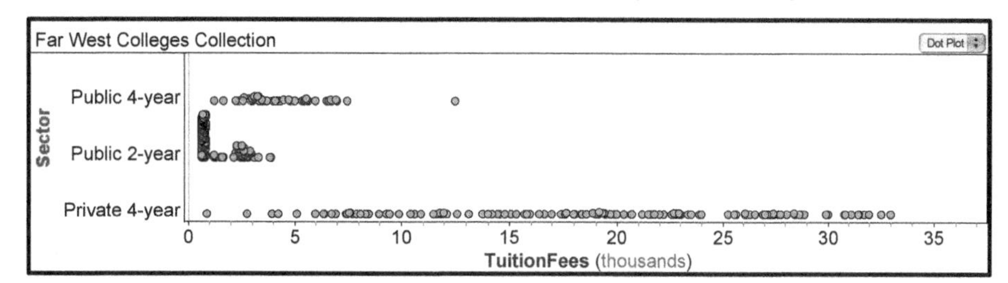

   a. Imagine that you are explaining to a friend (perhaps by e-mail) what this graphic tells you about tuition fees in public and private colleges. You can attach the graphic to your e-mail. You need to tell your friend what the dot plot shows about the center (or location) of the distributions but *also* about the differences in the *spreads* of the distributions between the private and the public institutions.

   b. Now assume that it is *not* possible to attach the graphic; explain the features of the graphic without your friend being able to see it. How will you explain the fact that the distribution of tuition fees at private universities is much more spread out or varied than are the fees at the public institutions?

2. ***Calculating variance by hand.***

   a. There is a worked example on page 82 of the calculation of the variance. In that example, two of the terms (the things that get added) were left out so that the calculation would actually fit on the page. Your task (the "answer" to this question) is to work out and write down the two missing terms.

   b. Some people find it easier to calculate the variance by hand if the data are arranged in a chart. Here are the data for *percentage over age 24* for the two-year schools, showing in schematic form the calculation of the sums in the calculation of the variance. Fill in the missing entries. (Use the small Roman numerals to refer to the cells.)

i		x_i	$(x_i - \text{x-bar})$	$(x_i - \text{x-bar})^2$
1	Blue Mountain Community College	40	-2.41	5.8081
2	Central Oregon Community College	36	-6.41	41.0881
3	Chemeketa Community College	45	(i)	6.7081
4	Clackamas Community College	44	1.59	(ii)
5	Clatsop Community College	51	8.59	(iii)
6	Columbia Gorge Community College	55	12.59	158.5081
7	Klamath Community College	31	-11.41	130.1881
8	Lane Community College	44	(iv)	(v)
9	Linn-Benton Community College	37	(vi)	29.2681
10	Mt Hood Community College	32	-10.41	108.3681
11	Oregon Coast Community College	44	1.59	2.5281
12	Portland Community College	53	10.59	112.1481
13	Rogue Community College	48	5.59	31.2481
14	Southwestern Oregon Community College	41	(vii)	(viii)
15	Tillamook Bay Community College	39	(ix)	(x)
16	Treasure Valley Community College	40	-2.41	5.8081
17	Umpqua Community College	41	-1.41	1.9881
	SUMS	721	0.03	726.1177
	x-bar	42.41		

   c. Use the sums given to calculate the variance s^2.

   d. Calculate the standard deviation s. ***PTO***

e. One of the columns shows $(x_i - \bar{x})$. The sum is shown at the foot of that column. What should the value of $\sum_{i=1}^{n}(x_i - \bar{x})$ be? The value shown here is *not* exactly what the value should be because we have introduced some rounding error. (See the section in the ***Notes: Looking at the variance formula closely and thinking graphically.***)

3. *Interpreting Measures of Variation*

a. Confused Conrad (who has not been reading the ***Notes***) says, "If the mean is bigger then the standard deviation must also be bigger." Explain to CC how the summary statistics for *PctOverAge24* for colleges grouped by sector show his idea to be wrong.

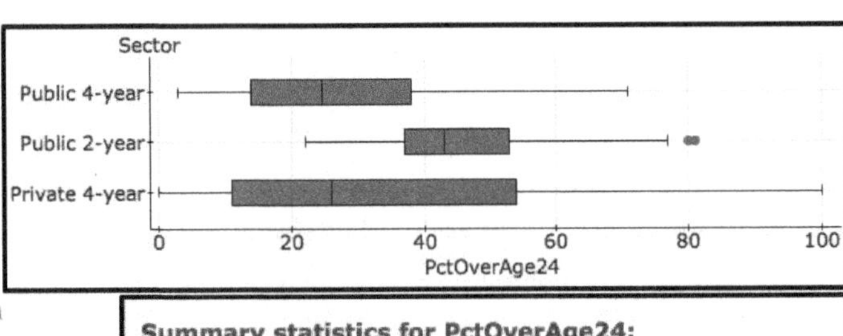

Summary statistics for PctOverAge24:
Group by: Sector

Sector	Mean	Median	Std. dev.	IQR	n
Private 4-year	34.539007	26	28.948604	43	141
Public 2-year	45.468571	43	11.343026	16	175
Public 4-year	27.032258	24.5	17.309116	24	62

b. Confused Conrad also says, "For the private colleges the s (the standard deviation) is 28.9, and that tells me that about 28.9% of the students in private colleges are over age twenty-four, and this is higher than the percentage in the state four-year schools." This is not a correct interpretation. What mistake has CC made? (A correct interpretation compares the standard deviations for the three categories to determine which category has the biggest *variability* and not the center or location of the distributions.)

Summary statistics for TuitionFees:
Group by: Sector

Sector	Mean	Median	Std. dev.	IQR	n
Private 4-year	18501.612	19083	7816.0387	13424	139
Public 2-year	1340.6686	788	895.16486	1711	172
Public 4-year	4285.2581	3656	1856.2444	2466	62

c. Here is a summary table showing means, medians, standard deviations, and *IQR*s for the variable *TuitionFees* for the public two- and four-year schools and the private four-year schools. Interpret the measures of center/location in the context of the data (in context means: give an interpretation that says something about how tuition fees are different in the three types of colleges).

d. Give an interpretation of the two measures of variation (Standard deviation and IQR) in the context of the data.

e. The graphic below shows exactly the same distributions as the dot plot in Exercise 1 (Explaining Variability). Which graphic (for you) gives the clearer picture of the differences between tuition fees amongst the colleges in the different sectors? Are there features of the distributions that are obscured by either one of the graphics choices?

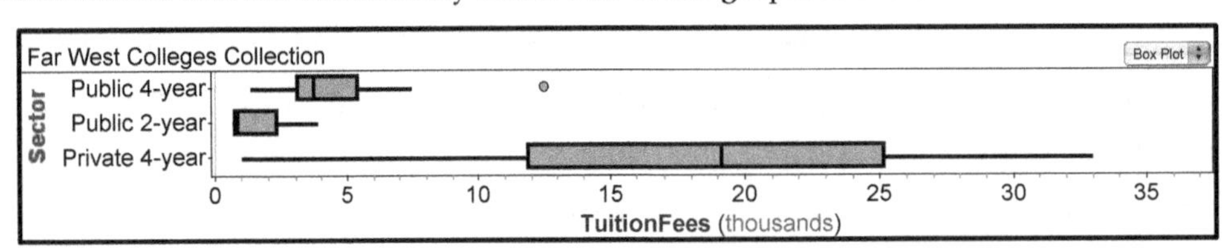

4. ***Students' Grades.*** A professor at the Instituto Tecnológico de Costa Rica kept careful records of students' scores for a calculus course that he taught semester after semester. There were three tests (all marked out of 100) and then the final course grade that depended upon the tests but also upon quizzes and homework. These records are for several semesters, but the tests given across the semesters are comparable.

Summary statistics:

Column	Mean	Median	Std. dev.	IQR	n
Exam1	67.305993	68.5	15.183399	20.5	267
Exam2	77.820974	80.7	14.041802	18.2	267
Exam3	73.546442	76.8	17.806656	25.9	267
Course_grade	71.583521	73	11.872266	13.7	267

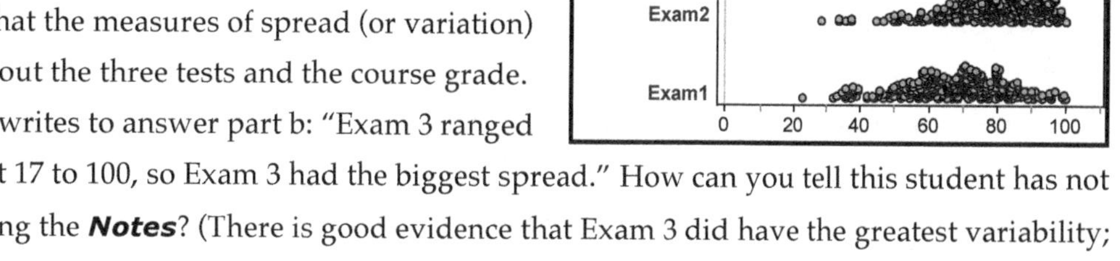

a. Explain what the measures of center tell you about the three tests and the final course grade. Use the numbers.

b. Explain what the measures of spread (or variation) tell you about the three tests and the course grade.

c. A student writes to answer part b: "Exam 3 ranged from about 17 to 100, so Exam 3 had the biggest spread." How can you tell this student has not been reading the ***Notes***? (There is good evidence that Exam 3 did have the greatest variability; what is that evidence?)

5. ***Great Lakes Colleges and Universities.*** We will look at the variable *Percentage Over Age 24* and its relationship with the type of college or university measured by the variable *Sector.* This relationship was examined in the ***Notes*** for the colleges in the Far West region. Our question is whether the relationships are similar for colleges and universities in the Great Lakes region.

- Using the file ***GreatLakesCollegesUniversities*** get box plots for the variable *PctOverAge24* grouped by the variable *Sector.* Also get the summary statistics mean, median, standard deviation and IQR for the *PctOverAge24* grouped by *Sector,* as shown here.

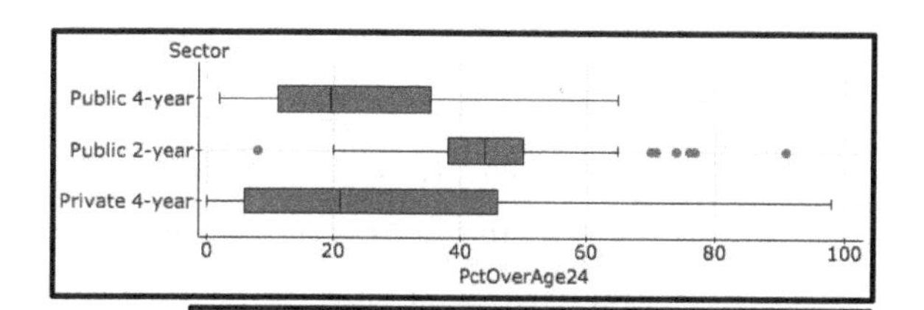

Summary statistics for PctOverAge24:
Group by: Sector

Sector	Mean	Median	Std. dev.	IQR	n
Private 4-year	28.991304	21	25.827948	40	230
Public 2-year	45.107383	44	10.939725	12	149
Public 4-year	22.54878	19.5	14.19455	24	82

a. Which type of college is on average the highest *PctOverAge24?* Refer to the summary statistics.

b. Which type of college has the *smallest* amount of variability in the distribution of the variable *PctOverAge24?* Refer to the summary statistics.

c. Which type of college has the *largest* amount of variability in the distribution of the variable *PctOverAge24?* Refer to the summary statistics.

- d. Use software to get the five-number summary for the public four-year colleges and then use the results to calculate the ***Lower*** and ***Upper Fences***. Use the results of the calculations to explain why no outliers are shown for the variable *PctOverAge24* for public four-year colleges.

e. Use the five-number summary for the public two-year colleges to calculate the ***Lower*** and ***Upper Fences*** for this type of college. Notice that there are data beyond the fences.

- f. To see the influence of outliers on the standard deviation, we will filter out the outliers. Filter using the expression *(Sector = "Public 2-year")* and *(PctOverAge24 > 20)* and *(PctOverAge24 < 68)* or one equivalent to this. (Refer to the software supplement for ***Filtering***.) Compare the standard deviation *s* and *IQR* with the values that include the outliers. Which has changed the most, *s* or *IQR*? Cite what feature of the measures of variation mentioned in the ***Notes*** this comparison illustrates.

g. True or false, and explain: "In 25% of the two-year colleges in the Great Lakes region (with the outliers *included*), half of the students are over age twenty-four."

6. ***Real Estate Data: Time in the Market*** In this exercise you will look a sample of data on houses that were sold in San Mateo County in the years 2007 to 2008. The collection you will use does not include *all* the houses that were sold; it is a *sample* of the entire *population* of all houses that were sold. Our statistical question is whether there are differences among the regions of San Mateo in the length of time houses stay in the market before they are sold. The variable that measures the interval between the date the house was put on the market and the date the house was sold is called *TimeInMarket.*

- Using the file ***San Mateo RE Sample 0708*** get box plots for the variable *TimeInMarket* grouped by the variable *Region*. Also get the summary statistics mean, median, standard deviation and IQR for the *TimeInMarket* grouped by *Region*.

Summary statistics for TimeInMarket:
Group by: Region

Region	Mean	Median	Std. dev.	IQR	n
Central	33.429688	15.5	39.404761	36.5	128
Coast	54.857143	41	47.123383	73	42
North	59.890411	45	52.745343	68	73
South	34.467105	15	42.288182	41.5	152

a. Judging from the shapes of the distributions, which measures of center and spread would be best to use? Give a reason for your answer based upon the ***Notes.***

b. Use the measures that you chose in part a to explain in the context of the data any differences or similarities in *center* in the variable *TimeInMarket* among the four regions for the sample of houses that were sold in 2007–2008. In which regions or regions do houses typically stay on the market longer? Refer to the numerical summaries for the explanations.

c. Use the measures that you chose in part a to explain in the context of the data any differences or similarities in *spread* in the variable *TimeInMarket* among the four regions for the sample of houses that were sold in 2007–2008. In which regions or regions do houses typically have the greatest variety of times in the market?

- Using the file ***San Mateo RE Sample 0708*** get box plots for the variable *TimeInMarket* grouped by the variable *Region*. Also get the summary statistics mean, median, standard deviation and IQR for the *TimeInMarket* grouped by *Region*. (This output is not shown here).

d. Answer question b for the variable *SalePrice* instead of *TimeInMarket.*

e. Answer question c for the variable *SalePrice* instead of *TimeInMarket.*

Exercises 7–10: Measures of Spread and Colleges and Universities

Colleges and Universities. The next four exercises use the data on colleges and universities also used in the ***Notes.*** The data shown in the case table are for the Far West. Some of the variables measured for each college or university are:

Collection 1

	Institution_Name	Sector	State	Fees	FRR	PctGrad	Av_Debt
48	California State University, San Marcos	Public 4-year	CA	2776	75	35	13112
49	California State University, Stanislaus	Public 4-year	CA	2807	80	52	16200
50	Canada College	Public 2-year	CA	652	68	42	
51	Cascadia Community College	Public 2-year	WA	2384	61	34	
52	Central Oregon Community College	Public 2-year	OR	3489	50	13	
53	Central Washington University	Public 4-year	WA	4278	80	52	27025

Sector	*Categories: "public two-year," "public four-year," or "private four-year" institution*
State	*The state in which the college or university is located: AK, CA, HI, NV, OR, WA*
Fees	*Total tuition fees for one year*
FRR	*Freshman retention rate: the percentage of freshmen in the college or university who continue*
PctGrad	*The percentage of students in the college or university who get a degree*
AvDebt	*The average debt of students at the college or university (measured for four-year institutions)*

7. Interpreting Data on Tuition Fees for Four-Year Colleges

Our statistical question is:

How do fees differ between public and private institutions?

Summary statistics for TuitionFees:
Where: "Sector" != "Public 2-year"
Group by: Sector

Sector	Mean	Median	Std. dev.	IQR	n
Private 4-year	18501.612	19083	7816.0387	13424	139
Public 4-year	4285.2581	3656	1856.2444	2466	62

a. What are the cases for these data?

b. Before looking at the data, first write down some ideas about how you think the distributions of the variable *Tuition Fees* will differ between the private and public institutions.

c. Compare the measures of center shown for the variable Fees and write what the measures tell you about the private institutions compared with the public institutions. A good start to what you write is: "On average the fees for private colleges and universities are…" In your sentence, you may wish to cite either the mean or the median or both.

d. In general, what does a small value for a standard deviation tell you about a distribution? What does a big value tell you?

e. Now look at the measures of spread for the two types of institutions. For the private colleges, both the standard deviation and the iqr are much bigger than they are for public colleges and universities. What does that tell you? Good words to use in your interpretation are: "variable" or "diverse" or even "more different"—although that is slightly clumsy English.

f. For these data, the measures of center (mean and median) and also the measures of spread (standard deviation and IQR) are both higher for the private colleges. Does that mean, in general, that if the measures of center are higher, the measures of spread must also be higher? Why or why not?

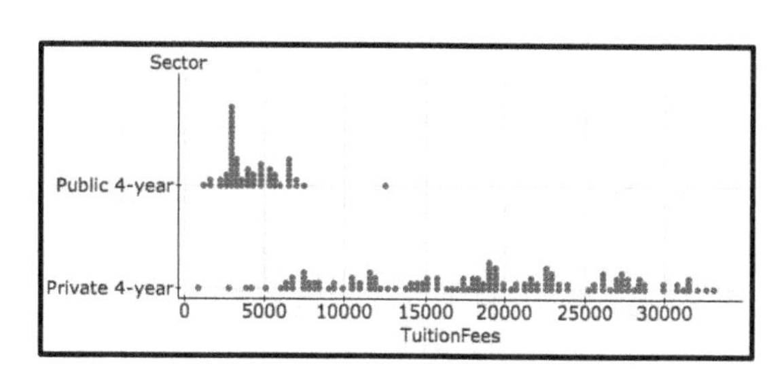

g. Explain why this statement is false: "The reason that the mean *Tuition Fees* is bigger for the private colleges is because the count $n = 139$ whereas for the public colleges it is just $n = 62$."

h. Explain in ordinary English that would be understandable to a fellow student not taking statistics what it would mean to a student looking for a college or university that the diversity or variability of fees is greater for the private colleges.

8. *Calculations on community college fees in Hawaii* There are just seven community colleges in Hawaii. Here are the data for the fees that they charge for a year.

a. Find the five-number summary and from that the Inter-Quartile Range (IQR) for these data.

b. Determine whether there are outliers for these data by calculating the ***Lower and Upper Fences***.

c. Construct a box plot for the distribution of fees.

d. Find the mean annual fees charged for the community colleges in Hawaii. (Answer: $1175.86)

Name	Fees ($/year)
Hawaii Community College	1240
Honolulu Community College	1158
Kapiolani Community College	1188
Kauai Community College	1158
Leeward Community College	1153
Maui Community College	1166
Windward Community College	1168

e. By hand, calculate, showing all of the steps, the standard deviation for these data using the formula for the standard deviation. (Answer: $30.50.)

- Use software with the file ***HawaiianCommCollegeFees*** to check the answers. If the software calculation of Q1 and Q3 *includes* rather than *excludes* the median (as in the ***Notes***) in the calculations of the quartiles then the results will be slightly different.

9. *How do tuition fees in community colleges differ in different states?*

a. A common mistake is to put too much importance on the number of cases in different categories. Generally, the *counts* do not tell you much about a distribution. However, here, the numbers corresponding to *count (.)* do tell you something. What do they say about community colleges in the three states?

b. Look at the measures of center (mean and median). Write what these tell you about the distributions of fees in the three states.

c. Look at the measures of spread. Write what these tell you about the distributions of fees in the three states. (Good words to use are "variability" or "diversity.")

Summary statistics for TuitionFees:
Where: "Sector" = "Public 2-year" and (State="CA" or State="OR" or State="WA")
Group by: State

State	Mean	Median	Std. dev.	IQR	n
CA	726.08182	722	99.212138	128	110
OR	2711.5294	2696	415.08706	485	17
WA	2577.0571	2541	295.07735	215	35

d. Suppose that you have a friend who has been your fellow student at a community college in California. Now this friend is planning to move to the Pacific Northwest (either Oregon or Washington) and after gaining residency there, plans to continue studying in a community college. What can you tell your friend based upon these data? (Do not forget to speak to your friend about diversity in fees.)

e. Based upon what you did in question 8, compare the distribution of fees for the community colleges in Hawaii with the distributions for fees on the Pacific Coast.

10. *Percent graduating differ for private and public four-year colleges*

Refer to the Summary Statistics and the box plots below.

a. What do the measures of center tell you about the percent graduating in private and public institutions?

b. What do the measures of spread tell you about the percent graduating in private and public institutions?

c. Comment on this assertion in light of the numbers and the box plots below: "If you attend a private school, you definitely have a better chance of graduating."

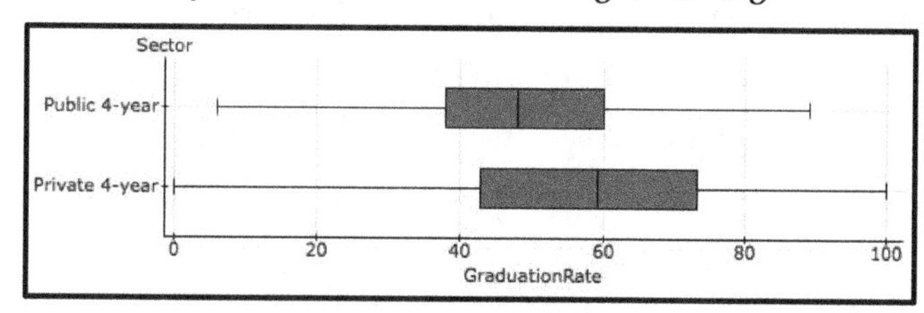

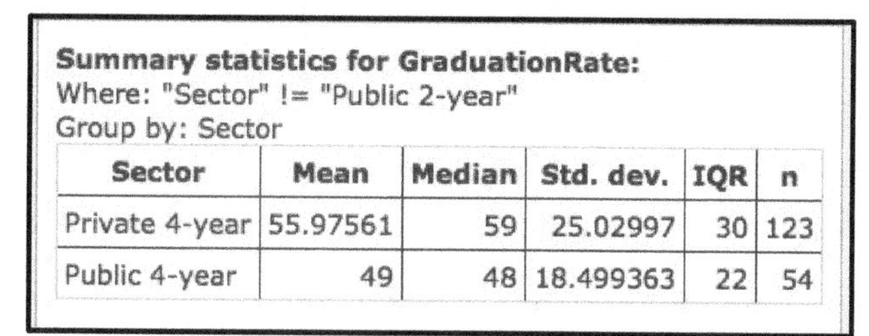

Summary statistics for GraduationRate:
Where: "Sector" != "Public 2-year"
Group by: Sector

Sector	Mean	Median	Std. dev.	IQR	n
Private 4-year	55.97561	59	25.02997	30	123
Public 4-year	49	48	18.499363	22	54

§1.7 Exercises: Models for Distributions

1. ***Heights of Male California Students: Preparing to use the Normal Model.***

We will look at the distribution of heights of male students in statistics classes in a college in Northern California. This first exercise just sets the stage, using a histogram of the distribution. Exercises 2 and 3 follow on to actually employ the Normal Model.

Summary statistics:
Where: Gender="m"

Column	Mean	Std. dev.	n
Height	175.71286	8.0766055	210

a. ***Using the histogram.*** The histogram shown has the vertical axis in frequencies (counts). Use the histogram to estimate the proportion of male California students in this collection that were shorter than 170 centimeters. (To help: there were three students with heights less than 155 cm and there are n = 210 students. You can estimate the number of students between 160 cm and up to 165 cm and also the number of students between 165 cm and up to 170 cm by inspecting the graphic closely.)

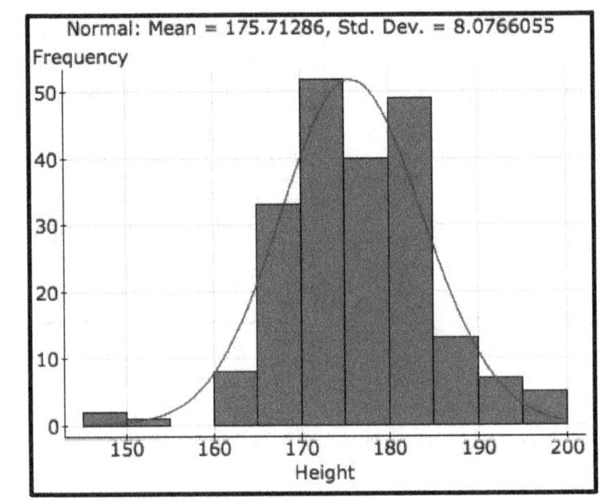

b. Express your answer to part a in probability notation, using X to represent height.

c. ***Using the histogram.*** Use the histogram to estimate the probability $P(X \geq 185)$. We will compare the result we get using the Normal Model with this estimate. Notice that the Normal Model curve leaves "gaps" so that for this small sample, it is not a perfect fit.

d. ***Sketching the Normal model.*** Using $\mu = 175.7$ cm. and $\sigma = 8.1\text{ cm}$, work out the values of $\mu - 3\sigma$, $\mu - 2\sigma$, $\mu - \sigma, \mu$, $\mu + \sigma$, $\mu + 2\sigma$, $\mu + 3\sigma$. Then, make a horizontal axis, and place the values just calculated, and draw a reasonable looking Normal Distribution Model, using the fact that for a Normal distribution one standard deviation is at the inflection point of the curve. Make another scale just below the scale for $\mu - 3\sigma$, $\mu - 2\sigma$, $\mu - \sigma$, μ, $\mu + \sigma$, $\mu + 2\sigma$, $\mu + 3\sigma$ and mark -3 for $\mu - 3\sigma$, then -2 for $\mu - 2\sigma$ and so on until you have 3 for the value for $\mu + 3\sigma$. This second scale indicates the number of standard deviation units a height is away from the mean height 175.7. It is the units used for the ***Standard Normal Distribution,*** which has mean $\mu = 0$ and $\sigma = 1.$, and is the distribution used on the Normal Chart.

e. On the sketch that drawn for part d, shade in the proportion of students that are shorter than 170 cm.

f On the sketch that drawn, shade in $P(X \geq 185)$.

g. Perhaps using a different kind of "shading," show on your picture $P(170 \leq X < 185)$.

- Check the answers to parts e, f, and g with software with a ***Distribution Calculator*** feature.

2. ***Heights of Male California Students: Calculating the z score***. This exercise follows on from Exercise 1 and concentrates on the next steps in using the Normal model: calculating the z score. We are using the Normal model with $\mu = 175.7$ cm and $\sigma = 8.1$ cm.

   a. Find the z score for the height x = 170.

   b. If Exercise 1 was done, use the sketch done in part d. Otherwise, do Exercise 1d. The scale that runs from – 3, – 2, . . .3 is the scale for the ***Standard Normal Distribution*** (the Normal distribution that has $\mu = 0$ and $\sigma = 1$). On the sketch locate the z score calculated in part a.

   c. Shade in on the sketch, $P(z < -0.704)$ [Recognize z =-0.704 as the answer to part a].

   d. Find the z score for the height x = 185.

   e. On the sketch made, show by shading the proportion of the Normal model distribution that is greater than the z score that you got as the answer for part d.

- Check the shading on the sketch in e using software with a ***Distribution Calculator*** feature. The sketch using the Standard Normal Distribution should look like the sketch shown in Exercise 3b.

   f. Without really calculating, find the z score for the mean height of 175.7 cm. Explain briefly why your answer makes sense.

3. ***Heights of Male California Students: Using the Normal Probability Chart.***

   a. Use the ***Normal Probability Chart*** to find $P(z < -0.70)$, which is the z score that you should have from Exercise 2a. You should have z = - 0.704, but the chart can only handle two decimal places. This is the z score for the Normal model with $\mu = 175.7$ cm and $\sigma = 8.1$ cm for the height of 170 cm.

   b. Your answer to Exercise 2b should resemble this drawing of the Standard Normal Distribution. On your sketch, indicate both the z score, z = - 0.704 and also the answer to part a for $P(z < -0.70)$. (Hint: One of these quantities should be on the horizontal axis, and the other as an "arrow" pointing to the shaded area. Get the two correct.)

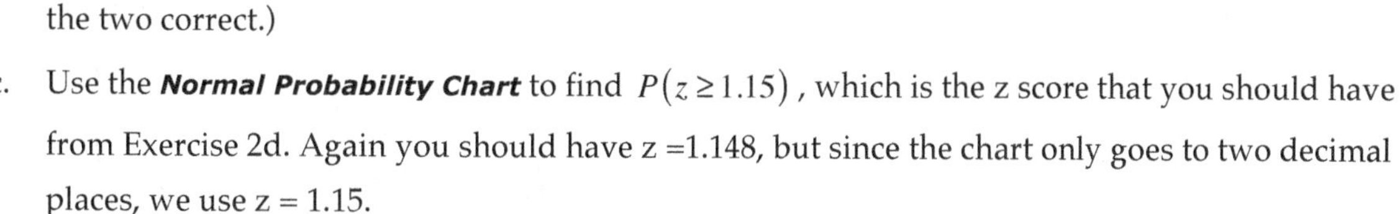

   c. Use the ***Normal Probability Chart*** to find $P(z \geq 1.15)$, which is the z score that you should have from Exercise 2d. Again you should have z =1.148, but since the chart only goes to two decimal places, we use z = 1.15.

   d. A student writes as the answer to part c the number 0.8749, and that shows that the student is reading the chart correctly. But can this be the answer to the question, "What is the probability that z ≥ 1.15?" or "find $P(z \geq 1.15)$?" Here is the picture of the ***Standard Normal Distribution*** that you should have from Exercise 2e.
   Explain (perhaps with the aid of the sketch) what is wrong with the wrong "answer" 0.8749.

   e. On your sketch (that resembles the one shown here) show both the z score of 1.15 and also the answer to $P(z \geq 1.15)$.

f. Use the ***Normal Probability Chart*** (or your work above) to find $P(z<1.15)$ and explain what this number means in terms of the heights of male California students.

g. Express the answer to part f in probability notation using X to indicate height.

h. Make a new ***Standard Normal Distribution*** sketch and shade in $P(-0.70<z<1.15)$ [The "blank" sketch, without shading is shown here, and the shaded answer can be checked using software having a ***Distribution Calculator*** feature.]

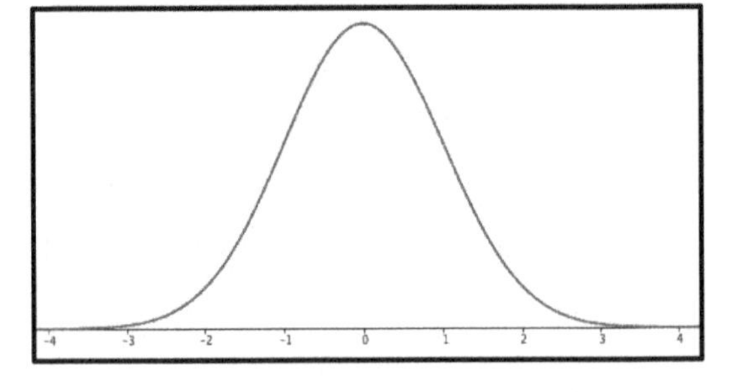

i. Use the ***Normal Probability Chart*** (or your work above) to find the numerical value of $P(-0.70<z<1.15)$. (Hint: You may find it helpful to perform a subtraction.)

j. Express the meaning for the heights of male California students of the value you found in part I, and the shading you made in part h. You may find it helpful to use the probability notation using X to indicate height. (Again, you can check it with software.)

k. ***Jack the Giant***. Jack is 210 cm. tall. Find Jack's z score using the model μ = 175.7 cm and σ=8.1 cm.

l. ***Jack the Giant*** Locate Jack on one of your sketches. (You should have found a z score bigger than any shown on the ***Standard Normal Distribution*** drawings above.

m. ***Jack the Giant*** Use the ***Normal Probability Chart*** (or by thinking) to find the probability that a male California student is as tall or taller than Jack.

n. ***Jack the Giant*** A calculator gives $P(X\geq 210)\approx 0.00001146$ using the Normal model μ = 175.7 cm and σ=8.1 cm. Does this agree with what your sketch and the ***Normal Probability Chart*** tell you? Explain.

4. *Heights of Australian High School Students: What about outliers?* This exercise uses data from the Census at School data collected from Australian high school students.

- Using the file ***CASAustralia08A*** get the boxplots for male and female heights shown here.

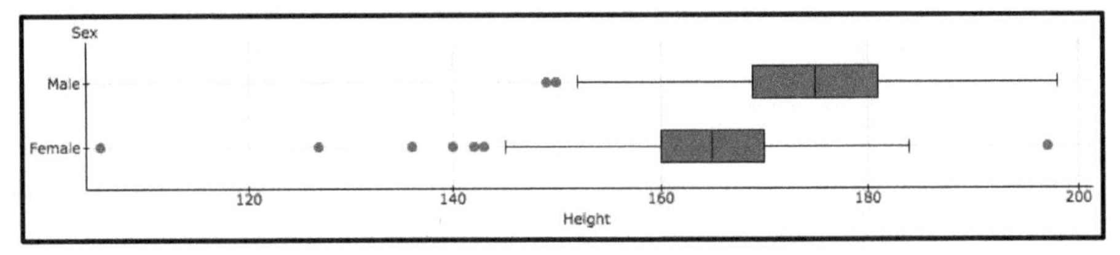

a. *[Review]* Are the heights of the males or the heights of the females more variable?

   – First, use the ***Range = Max – Min*** to answer the question about the differences in variability in height between males and female students.

   – Then use the ***IQR***s to answer the question about the difference in spread in height between male and females. You can estimate the IQRs from the graph or get it using software.

Then say why the two measures give you different answers (*Hint:* Think *resistance*).

- Using the file ***CASAustralia08A*** get the summary statistics for male and female heights, including mean, median, standard deviation and IQR.

Summary statistics for Height:
Group by: Sex

Sex	Mean	Median	Std. dev.	IQR	n
Female	164.391	165	9.0759367	10	289
Male	174.70336	175	9.4750148	12	268

b. *[Review]* Judging from the standard deviations, which height distribution has the greater spread, the heights of the male students or the heights of the female students? Cite the correct measures.

- Using ***CASAustralia08A,*** get the ***Five Number Summary*** for both male and female heights.

Summary statistics for Height:
Group by: Sex

Sex	Min	Q1	Median	Q3	Max
Female	105.5	160	165	170	197
Male	149	169	175	181	198

c. *[Review]* Use the numbers in the ***Five Number Summary*** to calculate the ***Lower Fence*** and the ***Upper Fence*** for the female high school students. The numbers you get should agree with what you see in the box plot.

- Recall that standard deviation is most appropriate for symmetrical distributions, but the height distribution for the females has outliers. We will look at female height in the data without the outliers. Using ***CASAustralia08A,*** filter for Sex = "Female" and Height in the interval: $145 \le Height \le 185$, and get the summary statistics: mean, standard deviation and count (n).

d. Rounded to one tenth of a centimeter, the mean height of the female Australian high school students (without the outliers) is 165 cm and the standard deviation of 7.4. Make a sketch of the Normal model with $\mu = 165$ and $\sigma = 7.4$, showing the Height values at -3, -2, -1, 0, 1, 2, and 3 standard deviations from the mean.

Summary statistics:
Where: Sex="Female" and Height >= 145 and Height <= 185

Column	Mean	Std. dev.	n
Height	164.95922	7.3892249	282

e. Use the Normal model to find the proportion of Australian female high school students whose height is less than 151 centimeters using the mean μ =165 cm. and standard deviation $\sigma = 7.4$ cm . Go through all the steps; show the sketch, do the calculation, consult the ***Normal Probability Chart,*** and interpret your result. (*Hint:* Is this a "Given a value – find a probability" problem, or a "Given a probability, find a value" problem?)

f. For the female Australian students, find and interpret $P(X \ge 170)$ using the Normal model. Draw a picture, do a calculation, and interpret. (*Hint:* Is this a "Given a value – find a probability" problem, or a "Given a probability, find a value" problem?)

g. Comment how your results from questions e and f compare with the proportion of elderly English women whose height is less than 151 centimeters and the proportion whose height is greater than 170 centimeters. (The elderly English women example was discussed in the Notes; the mean for that collection was $\mu = 160$ cm, and the standard deviation was $\sigma = 6$ cm.)

h. Find the height for the tallest 10% of the Australian female high school students using the Normal model. Draw a picture, do a calculation, and interpret. (*Hint:* Is this a "Given a value – find a probability" problem, or a "Given a probability, find a value" problem?)

i. *[Review]* Use ***Five Number Summary*** (not the Normal model) to calculate a good measure of the range of the heights of the "middle" 50% of Australian female high school students.

5. ***Heights of Female Australian High School Students: Exercise 4 using Software.*** Most statistical software is able to bypass the use of the ***Normal Probability Chart***.

- To answer the questions below using software (and not the Normal Distribution Chart) consult the procedures to use for the Normal Distribution in ***Distribution Calculator*** in the software supplement, where the examples are keyed to Exercises 2 and 4 of this section.

   a. For the answer to question 4e above, you should have $z \approx -1.89189$, which was rounded to $z \approx -1.89$ so that the ***Normal Distribution Chart*** could be used. Use software using the values for the mean μ =165 cm. and standard deviation $\sigma = 7.4$ cm. Because of rounding, the answer that software gives may be slightly different from the answer calculated in Exercise 4e.

   b. Use software to find $P(X \geq 170)$ with the Normal model with the mean $\mu = 165$ and standard deviation $\sigma = 7.4$. Compare your answer to what you got for question 4f.

- For a "Given a proportion—find a value" problem, such as question 4h above, consult ***Distribution Calculator*** in the ***software supplement***.

   c. Use software with the Normal model with $\mu = 165$ and $\sigma = 7.4$ to find the height of the female Australian high school students so that 25% of the students are shorter and 75% are taller. Notice that this gives a value for something like Q_1 but based on the Normal Model.

   d. Use software with the Normal model with $\mu = 165$ and $\sigma = 7.4$ to find the height of the female Australian high school students such that 75% of the students are shorter and 25% are taller. Notice that this gives a value for something like Q_3 but based on the Normal Model.

   e. Compare your answers to parts c and d with the Q_1 and Q_3 given in the Summary Statistics in Exercise 4. If the values are close, that is an indication that the Normal Model is a good fit to the height data.

   f. The ***Lower Fence*** for the female students was 145 cm. Use software and the Normal model with $\mu = 165$ and $\sigma = 7.4$ to find the proportion for the Normal model that are "outliers" in a tail.

- Another way to assess how well the Normal Model fits the data is to use a ***Normal Quantile Plot***, also called a ***QQ Plot***. If the fit is good, the data points should be aligned along a straight line; to the extent that the data deviates from the line the model fit is not good. Use software to get the ***Normal Quantile Plot*** for the female students and also for the male students (that is, group by Gender.)

6. ***Heights of Male Australian High School Students***. Exercises 4 and 5 were about the heights of female Australian high school students; this exercise is similar, but it is about the male Australian high school students. The Summary Statistics that were produced for Exercise 4 are relevant as well.

Summary statistics for Height:
Group by: Sex

Sex	Mean	Median	Std. dev.	IQR	n
Female	164.391	165	9.0759367	10	289
Male	174.70336	175	9.4750148	12	268

a. Look at the ***Summary Table*** for the male students and round the mean and standard deviation to the nearest tenth of a centimeter (the nearest millimeter). Make a nice sketch of the Normal model with μ and σ equal to these values showing the Height values at -3, -2, -1, 1, 2, and 3 standard deviations from the mean.

Summary statistics for Height:
Group by: Sex

Sex	Min	Q1	Median	Q3	Max
Female	105.5	160	165	170	197
Male	149	169	175	181	198

b. ***Tall Guys*** Decide how tall a "tall" male student is. (If you think in feet and inches, express your idea first in inches and then use the multiplication factor 2.54 cm/inch to get the figure in centimeters.) Round to the nearest tenth of a centimeter. Using the Normal model, find the proportion of male Australian high school students that are shorter than the height you think is "tall." Follow the steps: make a sketch (or use the one you made), do a calculation, get the proportion from the chart, and interpret. (Or, instead of using the z score and chart, use software, but you still need to draw a picture and interpret.)

c. ***Tall Guys, con't***. Use your answer to part b to calculate and interpret the probability that an Australian male student is taller than the height you think is "tall." Use probability notation in your answer.

- Use software to get a histogram for the heights of just the male students, making a good choice of bin width.

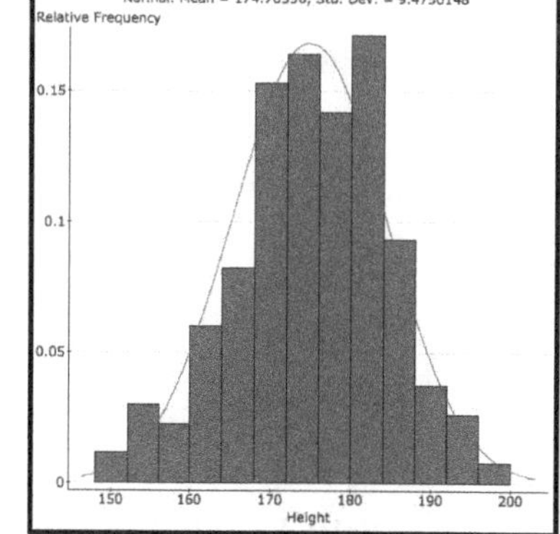

d. ***Tall Guys, con't***. Use the histogram or software to estimate or find the proportion of male Australian high school students that are taller than the height you think is "tall."

e. ***Short Guys*** Decide how tall a "short" male student is. Express this height in centimeters (rounded to the nearest tenth of a centimeter) and using the Normal model for the male Australian high school students, find the probability that a male student is shorter than this height.

f. ***Short Guys, con't***. Use your answer to part b to calculate and interpret the probability that an Australian male student is taller than the height you think is "short." Use probability notation.

g. ***Ordinary Guys*** Find the proportion of male Australian high school students who are neither "tall" nor "short" according to the Normal model. Express your answer using probability notation. Remember that when using a continuous Normal model, there is no difference between "less than" and "less than or equal to."

7. ***Grape Harvests in Europe from the fifteenth through the nineteenth centuries.*** We introduced the Normal distributions as a model to fit an empirical distribution. Here is an example of an empirical distribution for which a Normal distribution fits very well, partly because the data are composed of means. The data show the mean start of harvest after September 1 for northern and central France, Switzerland, and Alsace, and cover the years 1484 to 1879.

The variable *StartHarvest* refers to days after September 1. So, if the grape harvest started on September 3, the value for *StartHarvest* would be 2, and if the harvest started on October 1, the value would be 30, since thirty days hath September.

- Using the file **GrapeHarvestsEurope** and get a histogram for the variable *StartHarvest* using relative frequencies for the vertical scale, and a bin width of five years.
  a. What are the cases for these data? Look at the spreadsheet but mostly just think about the data.
  b. Use the histogram to estimate the proportion of years that the grape harvest took place before September 21 so that *StartHarvest* = 20.
  c. Use the histogram to estimate the proportion of years that the grape harvest started at least forty days from September 1.

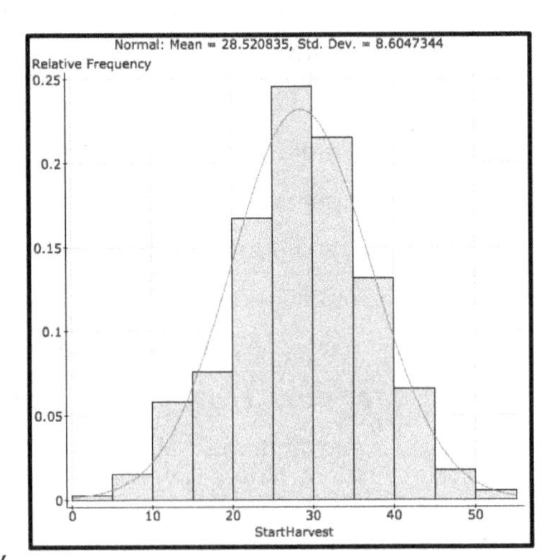

- Have software get summary statistics including the mean and standard deviation for the variable *StartHarvest*.
  d. Round the numbers for mean and standard deviation to the nearest tenth of a day and sketch a Normal distribution model using the mean for μ and the standard deviation for σ. Show the values for -3, -2, -1, 1, 2, and 3 standard deviations on your graph.
  e. Using the Normal model, find the proportion of years that this model predicts that the grape harvest will begin sooner than September 21. (Hint: Which type of probability problem is this?)
  f. Using the Normal model, find the probability that the grape harvest happens at forty days or later after September 1. Show your work.
  g. Using the Normal model, find $P(20 \le X < 40)$ where X stands for the days after September 1 that the grape harvest started (i.e., the number of days in *StartHarvest*). Sketch a picture of the answer.
  h. There is a number of days after September 1 where the grape harvest started in fewer than 5% of the years recorded. Find this number of days and round to two decimal places. (*Hint:* Is this a "Given a value—find a proportion" problem or a "Given a proportion—find a value" problem? Decide and draw a sketch, do a calculation, and interpret.)
  i. Express your answer (and therefore the question) in part h in probability notation using X to denote the days in the variable *StartHarvest*.
- Check your answers to parts e, f, g, and h using software.

Special Exercise A: Normal Foot Lengths

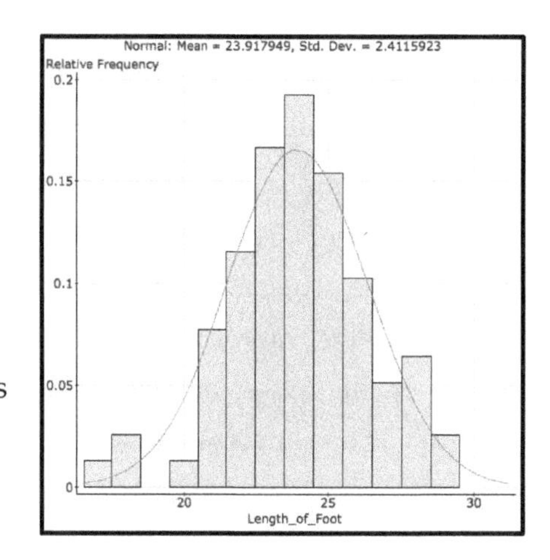

Data were collected on the length of each student's right foot, measured in centimeters. Here is the histogram showing the distribution of right-foot lengths.

- Get this histogram by opening ***CombinedClassDataSpr07.*** And starting the bars at 16.5 cm, and using a bin width of one centimeter.

1. ***Some review using histograms*** On the histogram, the bars start at 16.5 cm. Hence the first bar shows the students whose right feet were in the interval: 16.5 cm ≤ *Length_of_Foot* < 17.5 cm
   a. What is the bin width of each bar in the histogram?
   b. Use the histogram and the relative frequency scale on the left side of the graph to find the percentage of students who have right foot length greater than or equal to 26.5 cm. For your answer, show a sum of proportions. On a copy of the histogram (either hand copied or copied from software) shade in the bars for *Length_of_Foot* ≥ 26.5.
   c. Use your answer to part b to find the percentage of students who have right foot length less than 26.5 cm. (There is a very easy way to do this.)
   d. Use the histogram and the relative frequency scale on the left side of the graph to find the percentage of students who have a right foot length less than 21.5 cm. (You should be adding the heights of four, not five, bars; why?)
   e. We want to know the percentage of students whose right foot is within the interval defined by: 21.5 cm ≤ *Length_of_Foot* < 26.5 cm.
   f. Use your answers to parts c and d (and some subtraction) to get the percentage of students whose right foot length is in the interval: 21.5 cm ≤ *Length_of_Foot* < 26.5 cm.

2. ***Using the Normal Model*** We can answer the same kinds of questions using our Normal model for the distribution.
   a. The histogram above shows the Normal distribution overlaid that has μ =23.9, and standard deviation $\sigma = 2.4$. The choice of μ and σ is from the mean and standard deviation of the data. Work out the values for $\mu - 3\sigma$, $\mu - 2\sigma$, $\mu - \sigma$, $\mu + \sigma$, $\mu + 2\sigma$, and $\mu + 3s$, and make a sketch of a Normal distribution (like the one here) with the values you have calculated, as well as μ. (See ***Special Exercise B*** 1a for guidelines about drawing.)

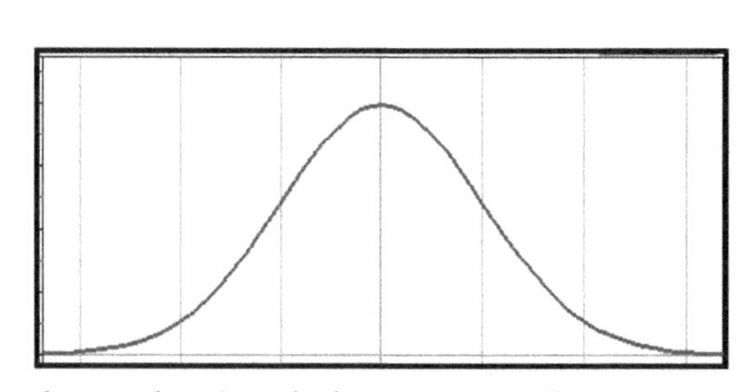

   b. Using the Normal model, find the percentage of students whose right foot length is less than 21.5 cm. Which type of probability question is this? As shown in the ***Notes,*** shade in and label with a question mark on the graph what it is that you are asked to find.
   c. For x = 21.5, calculate the z score: $z = \dfrac{x - mean}{SD}$

d. Use the z score you have calculated and the ***Normal Distribution Chart*** you have been given to find the proportion of students whose right foot is less than 21.5 cm long. Express the answer using the probability notation $P(X < 21.5) =$ _____. (The answer may be checked using software.)

e. Multiply your answer to part d by 100 to get the percentage of students whose right foot is less than 21.5 cm long according to the Normal model and write a short sentence (in the context of the problem) interpreting what you have found.

3. ***Using the Normal Model Again*** Now our question is: "What is the percentage of students whose foot length is greater than or equal to 26.5 cm?"

   a. Is this question a "Given a value—find a proportion" problem or a "Given a proportion—find a value" problem? On the graph you made, label what it is that you are asked to find, using shading.

   b. Using the same mean and standard deviation (it is the same model!) μ =23.9 and standard deviation $\sigma = 2.4$ find the z score for this question.

   c. Your z score should be positive. By looking at the graphic, explain in the context of the problem why it makes sense that the z score is positive.

   d. Look at the Normal table and read off the number for your z score. You should see the number 0.8599, but this is <u>not</u> the answer to $P(X \geq 26.5)$! Explain why is 0.8599 is not the answer, even though the chart shows that number. That is, explain why such an answer makes no sense in the problem and also why the table gives you this answer.

   e. So, to get the answer to the question, what do you have to do with the 0.8599? Do what you have to do and get the answer.

   f. Put your answer in probability notation.

   g. Interpret the answer in the context of the problem.

4. In one "Given a value—find a proportion" problem, you had to subtract what you saw in the ***Normal Distribution Chart*** from 1, and in the other problem you did not. What will you look for in future problems to determine when and when not to subtract from 1? To answer this question, you are to make your own rule to help you remember when and when not to subtract from 1. But it better work! It is actually very logical.

5. ***Using the Normal Model Again*** We want to know the proportion of students whose right foot is in the interval: 21.5 cm $\leq$ *Length_of_Foot* < 26.5 cm and we want to use the Normal distribution model to find the answer.

   a. Is this question a "given a value—find a proportion" problem or a "given a proportion—find a value" problem?

   b. You should be able to use the results of questions 2 and 3 (with some thought about what you have found) to calculate the solution, and you should be able to do this without calculating a z score or looking at the ***Normal Distribution Chart***.

   c. Express your answer using the probability notation, using X to represent *Length_of_Foot*.

6. Questions 2, 3, and 5 repeated what you did in question 1 with the histogram. Compare your results using the Normal model with what you got using the histogram.

Special Exercise B: Normal Foot Lengths

1. ***Being an artist*** In the small table are the answers to question 2a in the first Special Exercise on Foot Lengths. That question asked you to calculate $\mu - 3\sigma$, $\mu - 2\sigma$, $\mu - \sigma$, $\mu + \sigma$, $\mu + 2\sigma$, and also $\mu + 3s$ for the ***Normal distribution*** with μ =23.9 and standard deviation σ = 2.4—the one we have been using to model the foot lengths of the collection of students' feet.

$\mu-\sigma$	21.5	$\mu+\sigma$	26.3
$\mu-2\sigma$	19.1	$\mu+2\sigma$	28.7
$\mu-3\sigma$	16.7	$\mu+3\sigma$	31.1

   a. Use these numbers to make your own beautiful sketch of a Normal distribution. Your sketch
      - should show the location of the mean μ of the model we are using for foot length and also the values for

        $\mu - 3\sigma$, $\mu - 2\sigma$, $\mu - \sigma$, $\mu + \sigma$, $\mu + 2\sigma$, and $\mu + 3\sigma$

        (Remember about the inflection point.)
      - should look like a Normal curve and not like a tent or a mound. Here are two bad examples.
      - need not have a scale on the vertical axis

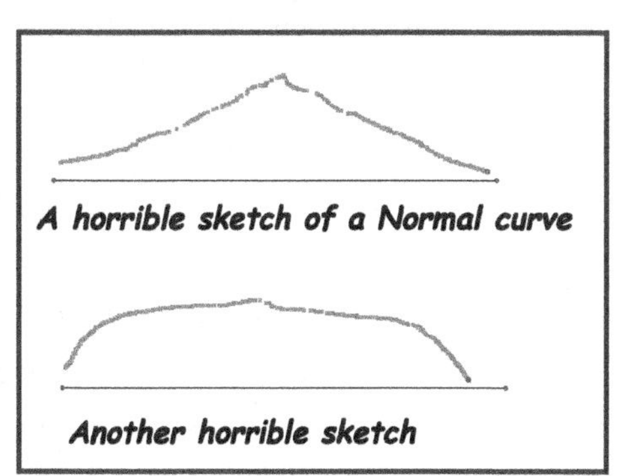

   b. Use the ***Empirical Rule*** to estimate approximately what proportion of students' feet should be within the interval 21.5 cm and 26.3 cm if the foot length distribution is modeled by a Normal distribution with μ = 23.9 cm and standard deviation σ = 2.4 cm.
   c. Use the ***Empirical Rule*** to estimate approximately what proportion of students' feet should be 26.3 cm or longer if we use a Normal distribution with μ = 23.9 cm and standard deviation σ = 2.4 cm.
   d. Is the question asked in part c a "given a value—find a proportion" problem or a "given a proportion—find a value" problem? Give a very short reason for your answer.
   e. Use the ***Normal Distribution Chart*** to determine the proportion of students' whose feet are 26.3 cm or longer, if the foot length distribution is modeled by a Normal distribution with μ =23.9 and standard deviation σ = 2.4. Recall the steps laid out in the ***Notes:*** make a sketch (you may use the one for part a), make a calculation, consult the chart and interpret your result.
2. ***P'tite Feet: Using the Normal Model.*** After taking this course, you plan to open a boutique in Healdsburg (Healdsburg is full of boutiques) for people with small feet. You will cater for the "smallest-footed" 10% of people. So your question is: "What is the foot length for the 10% of people with the smallest feet, according to the ***Normal distribution*** with μ =23.9 and σ = 2.4?"
   a. Is this question a "given a value—find a proportion" problem or a "given a proportion—find a value" problem? Give a reason for your answer.
   b. Make another sketch of a Normal distribution with μ =23.9 and standard deviation σ = 2.4 and indicate on this sketch what you are given and what it is you are being asked to find.
   c. Find the z score that you will use to solve this problem. (Remember that 10% is .1000.)
   d. Use the z score to find the numerical answer to the question. (The answer is 20.8 cm.)
   e. Give an interpretation in the context of the question asked.

3. ***Sasquatch Ambition*** Or, alternatively, instead of a boutique in Healdsburg for small-footed people, you could cater for the 5% of people with the biggest feet. Then your question is: "What is the foot length for the 5% of people with the biggest feet if we use the ***Normal distribution*** with $\mu = 23.9$ and standard deviation $\sigma = 2.4$ as our model?"
   a. Is this question "given a value—find a proportion" or "given a proportion—find a value"? Give a reason for your answer.
   b. Make another Normal distribution sketch using shading and labeling to show what you are given and what you are to find.
   c. Should the z score that you use be a positive or a negative number? Give a reason for your answer based on your sketch.
   d. Get the correct z score and do some algebra to find the numerical answer to the question.
   e. Forgetful Fiona gets 19.9 cm for her answer to part d. Convince FF that this cannot possibly be the answer (refer to your sketch) and determine what mistake she made. (Correct: 27.8 cm.)
   f. Put your answer into the probability notation that we have been using and interpret your answer in the context of the question.
4. ***Just Normal*** Probably the best business plan is to have shoes for 95% of foot lengths. We will again use the ***Normal distribution*** model with $\mu = 23.9$ cm and standard deviation $\sigma = 2.4$ cm.
   a. According to the ***Empirical Rule,*** what is the approximate interval of foot lengths that will accommodate 95% of the people if $\mu = 23.9$ cm and standard deviation $\sigma = 2.4$ cm?
   b. Is the question in part a "given a value—find a proportion" or "given a proportion—find a value"? Give a reason for your answer.
   c. Make another sketch of a Normal distribution and show the mean μ and the points for $\mu - \sigma$, $\mu + \sigma$ $\mu - 2\sigma$,and $\mu + 2\sigma$; show by shading the area of the graph that includes 95% of the data.
   d. The ***Empirical Rule*** gives an approximate answer of two standard deviations on each side of the mean, but there is a more exact answer, which you will now find using "given a proportion—find a value." (You will find that it is almost the same as the approximate answer.) Here are the steps to get this more accurate answer. From your sketch, you should have a shaded in portion in the center of 95%. What percentage does that leave on the left-hand side? Show this percentage and label it the "smallest   % of foot lengths" on your sketch.
   e. Your percentage should be 2.5%. Using the proportion .0250, find the z score from the body of the ***Normal Distribution Chart*** that corresponds to this and write it.
   f. Use your answer to part e and some algebra to get the value of the 2.5% smallest feet.
   g. For the right-hand side (the biggest feet), the fact that the ***Normal Distribution Chart*** gives cumulative proportions (proportions less than a z) means that we have to find the proportion to the left of the end of the 95% middle feet. What is that proportion? Use it to with the body of the chart to find the z score.
   h. Get the upper 2.5% of foot lengths, and thus the upper end of the 95% interval.
   i. You should have found that the answer to part g was positive but the same number as the answer to e. What feature of the shape of ***Normal distributions*** explains why this is so?
   j. Use the results to fill in the blanks: $P(\quad \text{cm.} \leq X \leq \quad \text{cm.}) = 0.95$

Revision for §§1.1 – 1.6: Hobbits and Men

8. *The Fellowship of the Ring* In J. R. R. Tolkien's classic, the heroes of the story are the hobbits. Soon after starting their journey from the Shire the hobbits came to a place named Bree. This town was in an area where both Men and Hobbits lived.

The Big Folk and the Little Folk (as they called one another) were on friendly terms, minding their own affairs in their own ways, but rightly regarding themselves as necessary parts of the Bree-folk. Nowhere else in the world was this peculiar but excellent arrangement to be found.
(J. R. R. Tolkien, *The Fellowship of the Ring*, Ballantine Books Edition, p. 206)

We have managed to get a sample of the heights (measured in cm) of Bree-folk who happened to be in *The Prancing Pony*, the inn in Bree. There are $n = 26$ Bree-folk represented, where $n_{Little} = 12$ and $n_{Big} = 14$. Here are the data. (It was not easy to get the data!)

a. In the data shown on the right, the first entry is 165 cm. Is this number: i) a variable, ii) a value of a variable or iii) a case

b. By hand, make a dotplot of the height data. Start by drawing a line; divide the line into equal intervals, and label the line appropriately. (The resulting dot plot should be neat; do not worry about getting something like 181 exactly correct.)

c. Calculate the mean height for all 26 of the Bree-folk together.

d. Show the symbol and the formula for the mean you have calculated.

e. The correct answer to part a is ii) "value of a variable". Explain why variable is the wrong answer.

f. Explain why case is the wrong answer to part a.

g. Make an ordered stem plot using stems of five, and from that calculate the Five Number Summary for the heights of all the Bree-folk together. For the stems, divide the numbers between the tens and ones digits, so that the stems are 10, 10, 11, 11, . . . ,18, 18.

	Height	Gender
1	165	Female
2	102	Female
3	111	Male
4	164	Female
5	170	Male
6	107	Male
7	109	Male
8	107	Male
9	181	Male
10	104	Male
11	108	Male
12	181	Male
13	174	Male
14	172	Male
15	106	Male
16	115	Male
17	102	Male
18	166	Female
19	104	Female
20	170	Female
21	175	Male
22	106	Female
23	172	Male
24	179	Male
25	172	Male
26	167	Male

- Open the file: ***BreeFolkExpanded*** and check the stem and leaf plot if the software makes stem plots.

h. From your answer to c, make a boxplot. Comment on the usefulness of the boxplot as compared with the dot plot in the context of Bree-folk. (Does the boxplot obscure features of the data?)

- Open the file: ***BreeFolkExpanded*** and get a box plot for the heights of all of the Bree-folk without distinction between the Big Folk and the Little Folk.

i. Calculate means and medians for the height data that make sense to the Bree-folk. (Hint: From the quotation at the head of the exercise, it is clear that the Bree-folk make a clear distinction between Big Folk and Little Folk.)

j. By hand, calculate the Five Number Summary for the heights of the Little Folk.

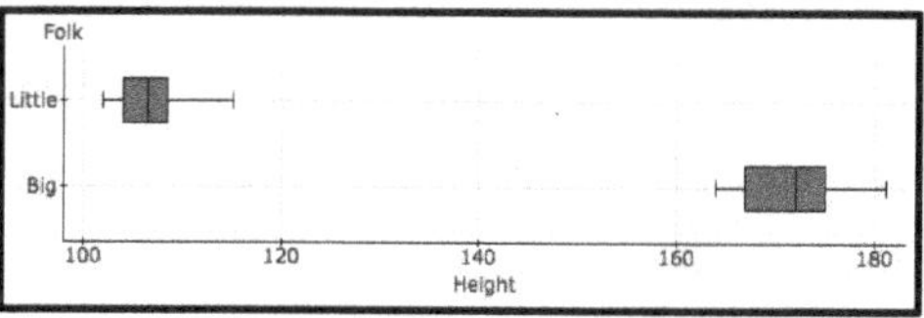

- Check your answers to part I using software and also get the box plots shown here.

k. Here is a histogram showing the heights of the $n_{Little} = 12$ Little folk, the Hobbits. For this histogram, what is the bin width?

Bree-folk | Histogram

Frequency of Height: 1, 2, 3, 4, 5, 6, 7

Height: 100, 110, 120, 130, 140

Folk = "Little"

l. Find the probability that a Hobbit is 105 cm or taller, and express that probability using the correct notation. (Here, "given" notation is optional)

m. You can get the answer for part l by:

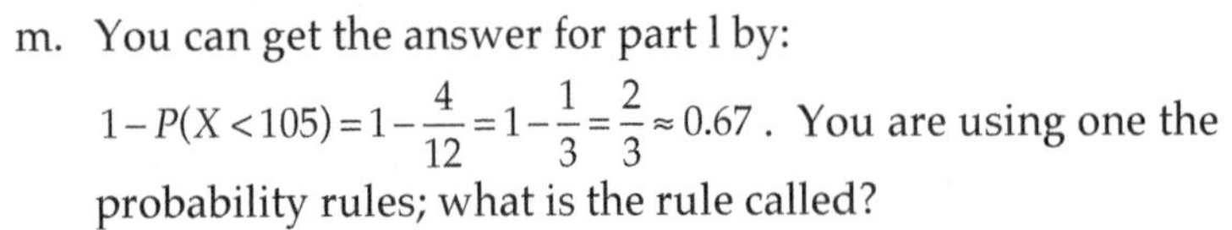

$1 - P(X < 105) = 1 - \frac{4}{12} = 1 - \frac{1}{3} = \frac{2}{3} \approx 0.67$. You are using one the probability rules; what is the rule called?

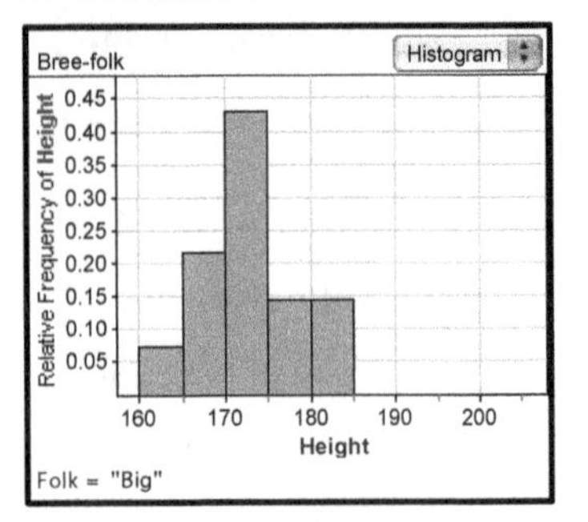

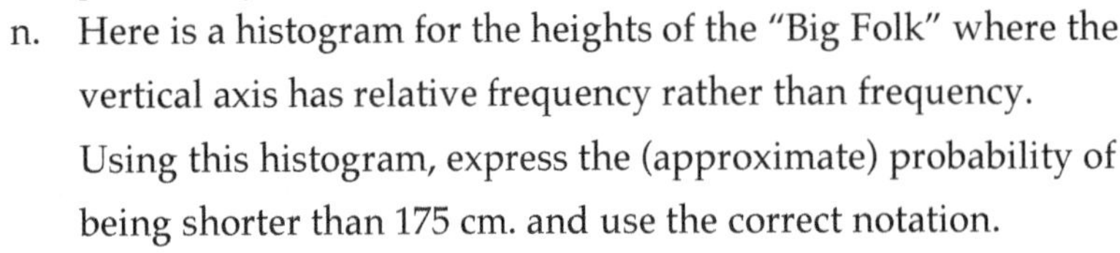

n. Here is a histogram for the heights of the "Big Folk" where the vertical axis has relative frequency rather than frequency. Using this histogram, express the (approximate) probability of being shorter than 175 cm. and use the correct notation.

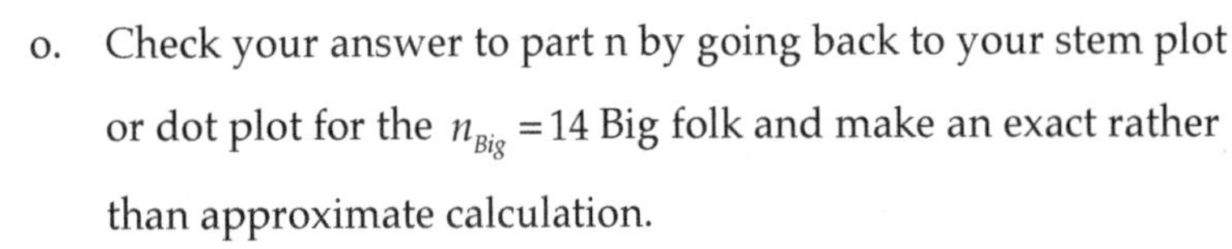

o. Check your answer to part n by going back to your stem plot or dot plot for the $n_{Big} = 14$ Big folk and make an exact rather than approximate calculation.

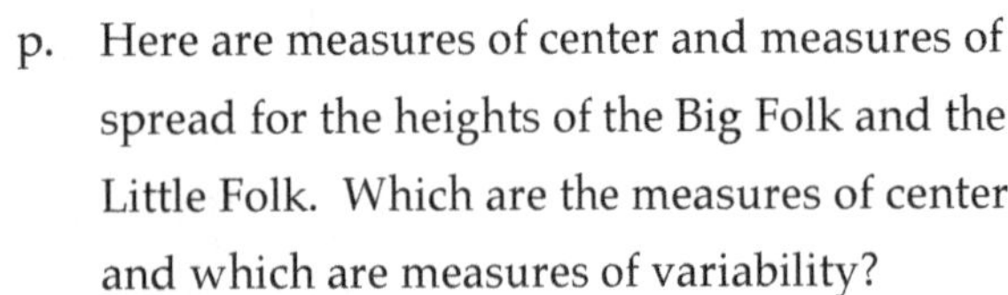

p. Here are measures of center and measures of spread for the heights of the Big Folk and the Little Folk. Which are the measures of center and which are measures of variability?

Summary statistics for Height:
Group by: Folk

Folk	Mean	Median	Std. dev.	IQR	n
Big	172	172	5.5884496	8	14
Little	106.75	106.5	3.7446932	4.5	12

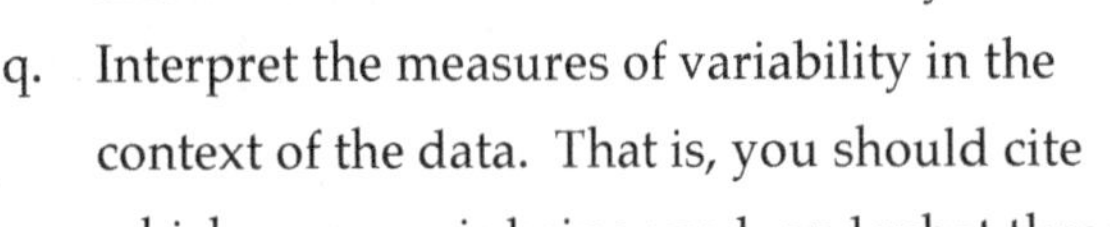

q. Interpret the measures of variability in the context of the data. That is, you should cite which measure is being used, and what these mean in terms of the heights Bree folk.

r. If you calculate $\sum_{i=1}^{n}(x_i - \bar{x})$ will you get: i) the mean, ii) the standard deviation, or iii) zero? Why?

s. The first two heights for the Big folk are 165 and 164, and the last height is 167 of the $n_{Big} = 14$ heights. Using these numbers, set up (but do not complete) the calculation of the standard deviation using the correct formula. Use " . . . " to show the missing parts, but put the numbers in the correct places.

t. Using the correct notation, find the probability that someone in the Prancing Pony is both a male and one of the Big folk, using the table here. [Use F, M, B, L for the events that a female, etc. are in the *Prancing Pony*.]

Bree-folk

		Gender: Female	Gender: Male	Row Summary
Folk	Big	4	10	14
Folk	Little	3	9	12
Column Summary		7	19	26

S1 = count ( )

u. Calculate $P(F|B)$ and give a meaning to the calculation.

v. Find the probability that if a male is found in the Prancing Pony, that male is one of the Big folk. Express the probability using the correct notation.

w. Consider all of the Bree-folk found in the *Prancing Pony*. Calculate probabilities to compare whether females are more likely amongst the Big folk or amongst the Little folk.

x. Determine, with the correct calculation, whether the events F and B are independent or not.

y. Are the events F and B mutually exclusive? Give a reason for your answer.

§2.1 Exercises: Calculation and Interpretation

1. *Interpreting Shape, Center, and Spread: Steam Schooners, California History*

At one time (about a hundred years ago), the primary means of transport to the communities on the coast one hundred miles or so north of San Francisco was by ship. Originally, the ships were schooners (sailing ships), but someone had the idea of installing a steam engine, so between 1875 and 1923 many wooden steam schooners were built. We have data on 216 of the wooden steam schooners that were built. (The data are from: Jack McNairn and Jerry MacMullen, *Ships of the Redwood Coast*, Stanford, CA: Stanford University Press, 1945.) Here is a description of the variables measured.

Steam Schooners

	Name	Tonnage	PlaceBuilt	YearBuilt	TimeBuilt	StateBuilt
1	Argo	210	Ballard, WA	1898	Early: 18...	WA
2	C. G. White	169	San Francisc	1884	Earliest: ...	CA
3	C. H. Wheele	371	Portland, OR	1900	Early: 18...	OR
4	Cleone	197	San Francisc	1887	Earliest: ...	CA
5	Daisy	621	Fair Harbor,	1907	Later: 19...	WA
6	Daisy Mitchell	612	Fairhaven, C	1905	Early: 18...	CA
7	Egaria	2360	Astoria, OR	1920	Latest: a...	OR

Tonnage	*Measures the cargo capacity of a ship; the bigger the tonnage, the more the ship will carry.*
PlaceBuilt	*Place the steam schooner was built*
YearBuilt	*Year the steam schooner was built*
TimeBuilt	*Categorcial variable that groups the year the ships were built into four groups: "Earliest: before 1896", "Early: 1896–1905, "Later: 1906–1915", "Latest: after 1915"*

a. What are the cases for these data?

b. Which variables are quantitative and which are categorical?

In the table here there are measures of the center and the spread of the tonnage for the ships built in the four different time periods.

Summary statistics for Tonnage:
Group by: TimeBuilt

TimeBuilt	Mean	Median	Std. dev.	IQR	n
Earliest: before 1896	284.97674	258	106.58229	108	43
Early: 1896-1905	536.94915	469	209.08655	299	59
Later: 1906-1915	758.07353	749	154.05652	249	68
Latest: after 1915	1389.9773	1253	543.11751	499.5	44

c. Which variable in the *Summary Table* is the explanatory variable and which variable is the response variable? Give a reason for your answer. ("We think that...depends on...")

d. Shape of the tonnage distributions. What do the measures of mean and medians of *Tonnage* tell you about the shapes of the four tonnage distributions? Give reasons for your answer.

e. Comparing average tonnage of the ships among the time periods. Compare the measures of center for Tonnage across the four categories of *TimeBuilt*. What do these measures tell you about the average *Tonnage* for the ships built in the different periods?

f. Comparing variability in tonnage among the time periods. Compare the measures of spread for Tonnage and write what these measures tell you about the *Tonnage* of ships and the time they were built. Words that are good to use to describe difference amounts of spread are: "variability," "more or less different," "diverse." (How would the shipyards before 1896 be different from the shipyards after 1915 in the variety of ships being built?)

2. *Household Size in Past Time and Now* There are just two variables in this analysis.

NumberinHousehold *Number of people in a household* *Place* *The places we are comparing are:*

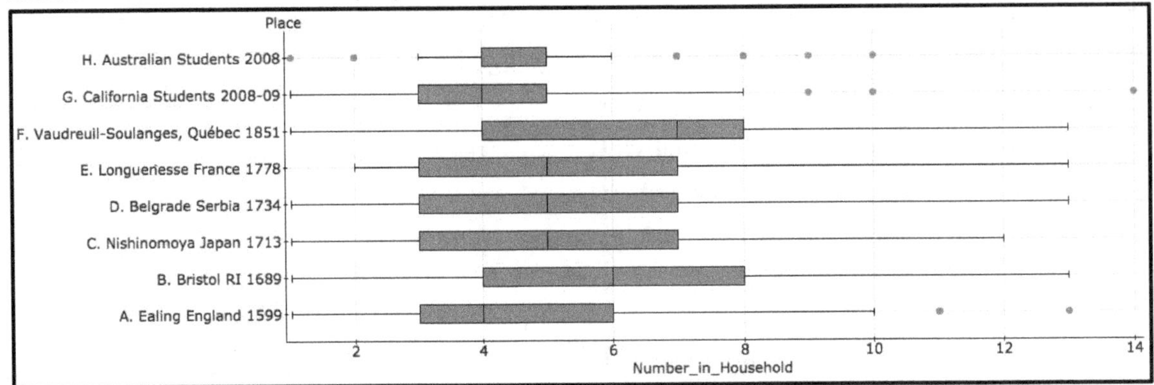

Notice that the times and the cultures for the data sets are quite diverse. The earliest data set is from 1599, and the latest are data on college students in California (2008-2009) and high school students in Australia (2008). (The data from Québec are primarily Gaelic speaking Highland Scots.)

a. You meet someone who has just arrived via a time machine from one of these places. Assuming that you can communicate use the measures of center and spread and the box plots to explain to the time traveler how your experience is different from theirs. You can choose the place your time traveler comes from, but identify it.

Summary statistics for Number_in_Household:
Group by: Place

Place	Mean	Median	Std. dev.	IQR	n
A. Ealing England 1599	4.6823529	4	2.7351842	3	85
B. Bristol RI 1689	5.8194444	6	2.7492708	4	72
C. Nishinomoya Japan 1713	4.9924242	5	2.5095123	4	132
D. Belgrade Serbia 1734	5.4101124	5	2.6222234	4	178
E. Longuenesse France 1778	5.0151515	5	2.5145266	4	66
F. Vaudreuil-Soulanges, Québec 1851	6.628866	7	2.7549939	4	97
G. California Students 2008-09	3.9889706	4	1.5450993	2	272
H. Australian Students 2008	4.4622144	4	1.2513638	1	569

- What do the mean and medians tell you about the sizes of households in the past and in other cultures compared with the experience of Californian college students?
- What do the standard deviations and the IQRs tell you about the sizes of households in the past and in other cultures compared with the experience of Californian college students?

b. Of the two variables, *Place* and *Number in Household*, which should be regarded as the explanatory variable and which should be regarded as the response variable?

c. Here is a boxplot and statistics from the US Census showing the distribution of household sizes for households in California. Technical question: there is a good reason that the bar for the median is not seen. What is the reason?

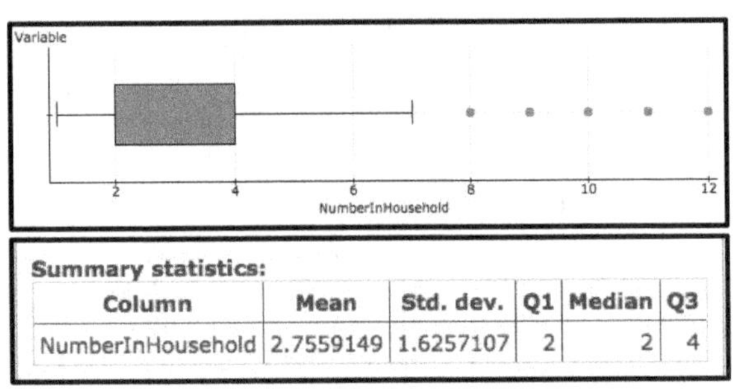

Summary statistics:

Column	Mean	Std. dev.	Q1	Median	Q3
NumberInHousehold	2.7559149	1.6257107	2	2	4

d. Notice that the mean and median household size for the California student sample than for the US Census data. Give a plausible reason why the mean and median household size should be larger for students than for the population in general.

Ref: Laslett, Peter (ed.), *Household and Family in Past Time,* Cambridge, Cambridge University Press, 1972

3. ***Interpreting Measures of Center and Spread: Gearhead Cars*** Road & Track is a magazine that is published for people who like cars. Hence, the cars they write about do not represent all cars that are being made but rather the cars that are interesting to "gearheads"—that is, people who like cars. Here are some of the variables measured:

Continent	*Place manufactured: America, Asia, or Europe*
Horsepower	*A measure of how powerful the engine is*
MPG	*Miles per gallon:. The larger the MPG, the less fuel is used.*
Acc_060	*The number of seconds taken to reach 60 mph from a standing start. The smaller the number, the quicker the car can accelerate.*

a. What are the cases for these data?

b. Which of the variables are quantitative?

c. Which of the variables are categorical?

Road&Track Jan09 Cars

	Make	Model	Continent	Horse...	MPG	Acc_060...
11	Audi	S6	Europe	435	16	5.1
12	Bentley	Continen...	Europe	552	12.5	4.3
13	Bentley	Continen...	Europe	600	11	4
14	BMW	M3 Coup...	Europe	414	13.1	4.3
15	BMW	M3 Sedan	Europe	414	14.6	4.6

In the table are measures of the center and the spread of the three distributions of *MPG* for cars that come from America, from Asia, and from Europe for the cars featured in Road & Track. Answer the following questions based on this summary table.

Summary statistics for MPG:
Group by: Continent

Continent	Mean	Median	Std. dev.	IQR	n
America	17.26	17	2.8160256	2.9	25
Asia	22.02973	20.4	6.4436388	3.6	37
Europe	16.74127	16	5.3699773	6	63

d. What do the means and medians tell you about the *shapes* of the three *MPG* distributions? Give reasons for your answer.

e. Compare the means for the cars for the three continents, and say what the comparison tells you about how the fuel consumption (which is what *MPG* measures) compares for the cars coming from America, Asia and Europe. Using the medians, do you come to the same conclusion?

f. Compare the standard deviations and IQRs for MPG for the cars from the three continents, and say what the comparison tells you about the variability in *MPG* amongst the cars in the three continents. Words that are good to use to describe different amounts of spread are: "variability," "more or less different," "diverse."

Here is another summary table showing the differences in the variable Horsepower by Continent.

Summary statistics for Horsepower:
Group by: Continent

Continent	Mean	Median	Std. dev.	IQR	n
America	412.03846	400.5	155.00722	277	26
Asia	251.02703	263	84.354571	91	37
Europe	390.15873	414	142.76071	242	63

g. Use the medians to write an interpretation of the differences in average horsepower for cars from the different continents.

h. Use the information given to calculate the IQRs for Horsepower for the cars from the different continents and say what the numbers tell you.

i. For the analyses in this exercise, which variable is (or variables are) being considered as the response variable and which variable (or variables) as the explanatory variable? Give a reason for your answer.

Writing Assignment: Exercises 4–12

Directions: The five instructions saying what you are to do:

1. Choose one of the exercises in the group of Exercises 4–12. (However, if you have an idea that is comparable to these exercises but is different then consult your instructor.)
2. Follow the directions to do the statistical analysis using software.
3. Create a Word (or Pages) document to keep the **graphics** and **Summary Tables** that you may need for your report and in which to write the report.
4. Based upon the statistical analysis that you have done, write a two- to three-page report explaining to a reader who has *not* had a course in statistics what the analysis tells you about the statistical questions. Some or all of the **graphics** and **Summary Tables** that you saved will be incorporated in the paper.
5. You will be given additional instructions about the following, which you should note:

- Whether the assignment is to be done individually or as a team, or in either mode
- The format of the report (double spacing, type of referencing, etc.)
- Mode of submission (whether electronically, or paper, or both)
- Due date or dates

Here are some tips on proceeding:

- Read the example essay posted on-line, especially if you have no idea of the kind of thing that is being asked for with this project. That example should show not only the process but will also give you some idea of what the essay should be at the end.
- In doing the analysis, think whether the statistical question involves two categorical variables or one categorical variable, and one quantitative variable and one categorical variable. Consult this section of the ***Notes.***
- Keep in mind which of the variables is being used as the *explanatory* variable and which variable is the *response* variable.
- Even before looking at the analysis output, sketch out roughly what you *think* about the statistical questions. Break down the questions. How do *you* think the *response* variable is related to the *explanatory* variable? You may wish to discuss with others.
- To incorporate output from software into a Word or Pages document, consult the software supplement for the software that is included with these exercises:

4. *Australian High School Students' Lives*

Introduction and Background The Census at School in Australia collects data from students at all levels of schools (but not universities and colleges) to aid in the teaching of data analysis and statistics in Australia. What happens is that teachers or school administrators opt for the program, and then all of the students in the school or class go on line and answer questions. Although it is called a census, the data collection does not cover all students in Australia; whether a school or class participates or not; however, the coverage over Australia is wide.

- Go to http://www.cas.abs.gov.au/tmp_cas/casq_2011_sample.htm (or browse for "Census at School Australia" and go from there) to view the questionnaire to see what questions were asked.
- For this writing assignment exercise, scroll down to questions 26 and 27, which have to do with what high school students do.

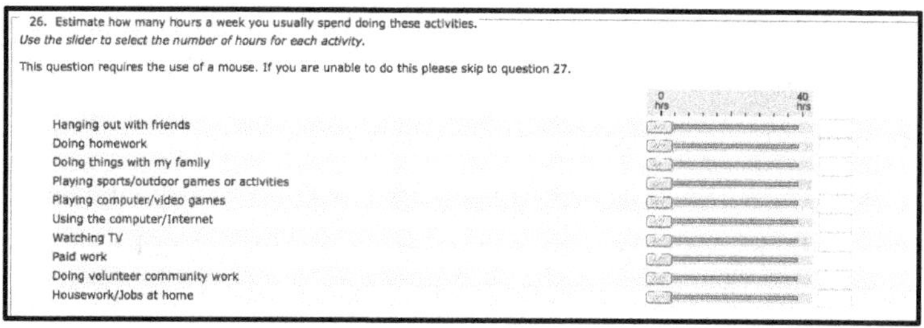

27. How do you usually spend your time on the Internet?
Select a category for each activity.

	Never	*Rarely*	*Sometimes*	*Often*
Social networking sites (e.g. Facebook, Twitter, chatrooms)	O	O	O	O
Researching for school work	O	O	O	O
Emailing friends and family	O	O	O	O
Searching/Browsing for information	O	O	O	O
Uploading/Downloading (e.g. music, photos)	O	O	O	O
Playing games	O	O	O	O
Video Sharing sites (e.g. YouTube)	O	O	O	O
Buying and/or selling things (e.g. eBay)	O	O	O	O

- Based on your own past high school experience, choose at least *two behaviors* from each question (two from 26, two from 27) where you think that the results will vary by either Gender or Year in School (or other explanatory variable), and develop specific statistical questions based on your ideas.
  - For example: if you think that there is a gender difference in the number of hours devoted to playing computer or video games, then the statistical question is: *Is there a difference in the distributions of hours male and female students play computer or video games?*
  - If you think that there may be a difference by Year in School in using the internet for "Researching for School Work" then: *"Is there a difference by school year in the use of the Internet for school work research?"*
- Use software with file ***CASAustralia2011Combined*** to get either summary statistics and graphics, or contingency tables to help answer the statistical questions posed. For the two statistical questions above, the beginning analysis will look like this:

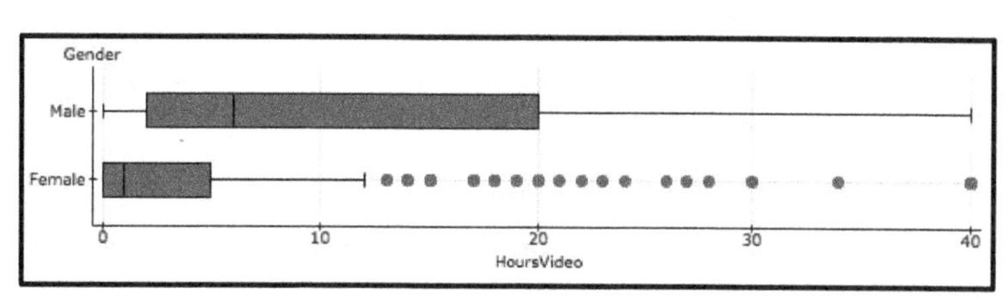

CASAustralia2011Combined

		InternetResearchSchool				Row Summary
		(a) Never	(b) Rarely	(c) Sometimes	(d) Often	
YearSchool	(a) Year 9	18	77	164	82	341
	(b) Year 10	20	58	107	55	240
	(c) Year 11 or 12	19	47	89	46	201
	Column Summary	57	182	360	183	782

S1 = count ( )

Summary statistics for HoursVideo:
Group by: Gender

Gender	Mean	Median	Std. dev.	IQR	n
Female	4.4119241	1	7.6704167	5	369
Male	11.70297	6	12.433834	18	303

5. Australian Students' Concentration Game Response Times

Statistical Questions

- *Is there evidence that students who spend much time playing video/computer games will record shorter response times for a concentration game than students who play video/computer games less often?*
- *Is there a difference between male and female students in the relationship between the response times and the reported hours per week playing video/computer games?*

Background and Data

The general question has to do with the effects of playing computer games; it should be easy to find background material alleging that playing video games has a negative effect (or possibly no effect or even a positive effect) on high school students. Here we can focus on the time to complete a simple game that require some concentration. The data come from Australian Census @ School questionnaire that is done on-line by thousands of primary and high school students. Experience what the students did by going to http://www.cas.abs.gov.au/tmp_cas/casq_2011_sample.htm and scrolling down to Question 25. A tiny version of what you will see initially (before the game starts) is shown here. The reaction time was recorded in seconds (or fractions of a second) and is recoded in our data.

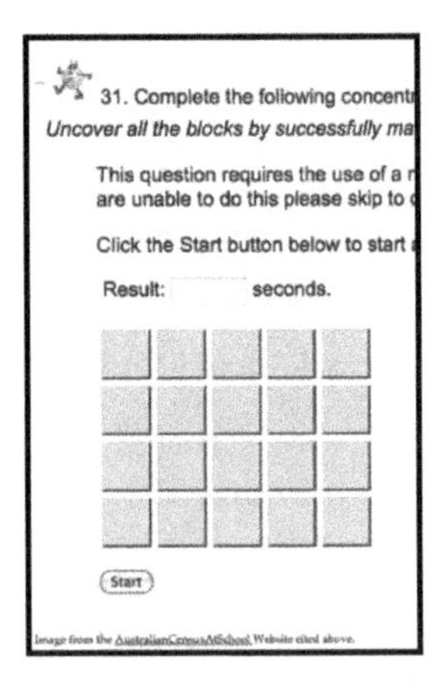

Gender	*Whether the student was male or female*
TimeConcentrationGame.	*Reaction time in seconds to concentration game..*
HoursVideo	*Number of hours per week playing video/computer games.*
FrequentVideoPlayers	*A categorical variable with two categories:* *"(a) More than10 hours per week on video games"* *"(b) Ten or fewer hours per week on video games"*

Data Analysis

- Use software with the file ***CASAustralia2011Combined*** to get a graph showing the distributions of the variable *TimeConcontrationGame* broken down by categories of *FrequentVideoPlayers.*

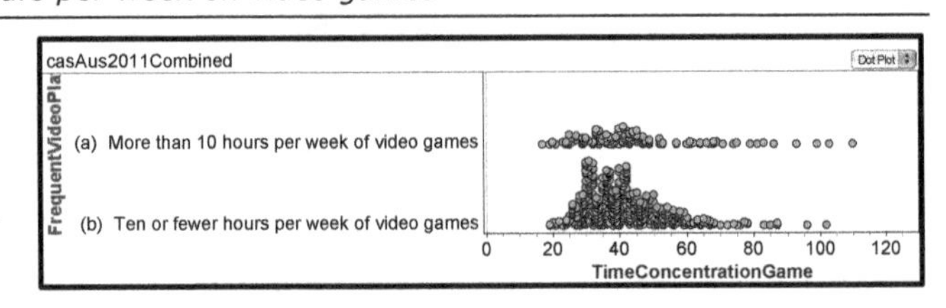

- Get summary statistics -- two measures of center and two measures of spread -- for the variable *TimeConcontrationGame* broken down by categories of *FrequentVideoPlayers.*

Summary statistics for TimeConcentrationGame:
Group by: FrequentVideoPlayers

FrequentVideoPlayers	Mean	Median	Std. dev.	IQR	n
(a) More than 10 hours per week of video games	43.232258	40	16.627242	16	155
(b) Ten or fewer hours per week of video games	41.765873	40	12.717482	16	504

- Get a contingency table that relates the number of males and the number of females who play computer and video games either ten or fewer hours per week or ten or more than ten hours per week.

	Female	Male	Total
(a) More than 10 hours per week of video games	46 (12.47%)	113 (37.29%)	159 (23.66%)
(b) Ten or fewer hours per week of video games	323 (87.53%)	190 (62.71%)	513 (76.34%)
Total	369 (100%)	303 (100%)	672 (100%)

- Filter the basic analysis by *Gender* to determine whether the effect of video game playing (if any) differs by gender.

6. *Internet Used Car Markets for BMW 3 Series in Boston and San Francisco*

Statistical Questions These questions compare the markets for BMW 3 Series cars being sold on www.cars.com in the Greater Boston Area and the San Francisco Bay Area. The questions are whether there are differences in the ages, prices, and miles driven for the cars being sold through the Internet. In both places, the cars chosen were those within a 100 mile radius of a central zipcode.

1. *Do the distributions of the ages and prices differ in center, spread, or shape between Boston and the San Francisco Bay Area or are they similar?*
2. *Are there differences in the way the BMWs depreciate in the two places? (Use the least squares lines for price and age or price and miles to answer this question.)*
3. *Are there differences in the distributions of body styles for the cars being sold in Boston and SF?*

Background and Variables

All of the information is from what is shown on the www.cars.com website. These data only include cars that had been listed in the two weeks prior to July 13, 2009, and within a hundred-mile radius of the zip code for CSM, 94402. Keep in mind that the data only refer to the cars that are being sold; the data do not necessarily represent *all* BMW 3 Series or Mercedes' C Class cars on the road.

Place1	*Whether the car is being sold in the Greater Boston Area or in the San Francisco Bay Area*
Body	*The body style of the car being sold: sedan, coupe, convertible, hatchback, and wagon*
Age	*Estimated age of the car being sold. The only information about age given is the model year. The age is calculated from this information. See the explanation under Question 7.*
Price	*Price of the car listed on the website*
Miles	*The number of miles the car for sale has been driven*

Summer 09 Used Cars — Box Plot

Place1: Boston, SF Bay Area

Age: 0 2 4 6 8 10 12 14 16

Contingency table results:
Rows: Place1
Columns: Body

	Convertible	Coupe	Hatchback	Sedan	Wagon	Total
Boston	10	12	0	101	3	126
SF Bay Area	21	29	1	191	7	249
Total	31	41	1	292	10	375

Summary statistics for Age:
Group by: Place1

Place1	Mean	Median	Std. dev.	IQR	n
Boston	3.095896	2.33333	2.1972454	2	126
SF Bay Area	3.8190705	3.33333	2.9503972	4	263

Summary statistics for Price:
Group by: Place1

Place1	Mean	Median	Std. dev.	IQR	n
Boston	28985.924	29999	8529.2106	7148	119
SF Bay Area	25376.738	25990.5	10177.866	13795	248

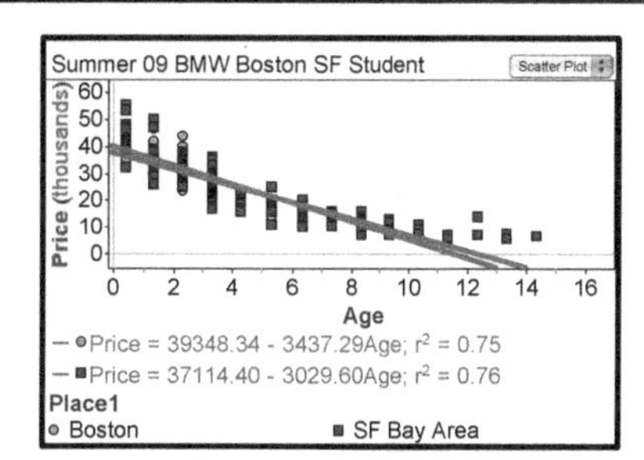

- Use software and the file ***Summer09BMWBostonSF*** to get a graphic showing the distributions of the variable *Age* broken down by categories of *Place1.*
- Get summary statistics -- two measures of center and two measures of spread -- for the variable *Age* broken down by categories of *Place1.*
- Get the scatterplot and linear model for the relationship between *Age* and *Price* by categories of *Place1.* Interpret the slopes and r^2 to say something about the statistical questions.
- Get a contingency table showing the relationship between the body style and whether the car is in Boston or the SF Bay Area, and analyze the numbers using conditional probabilities. Perhaps filter out the wagons and hatchbacks.

7. *Internet Used Car Markets for BMW 3 Series and Mercedes C Class*

Statistical Questions These questions compare the markets for BMW 3 Series cars and Mercedes-Benz C Class being sold on www.cars.com in the San Francisco Bay Area.

1. *Do the distributions of the age, prices and miles driven differ in center, spread, or shape between the BMW and the Mercedes?*
2. *Are there differences in the way the BMWs depreciate compared with the Mercedes'? (Use the least squares lines for price and age or price and miles to answer this question.)*
3. *Are there differences in the distributions of body styles between the makes of car?*

Background and Variables. See the comments on how the data were collected for Exercise 6. Here is a list of relevant variables in the data.

Make1	*Whether the car is a BMW 3 Series or a Mercedes-Benz C Class*
Body	*The body style of the car being sold: sedan, coupe, convertible, hatchback, and wagon*
Age.	*Estimated age of the car being sold. See the explanation of the way Age was calculated in Exercise 6.*
Price	*Price of the car listed on the website*
Miles	*The number of miles the car has been driven*

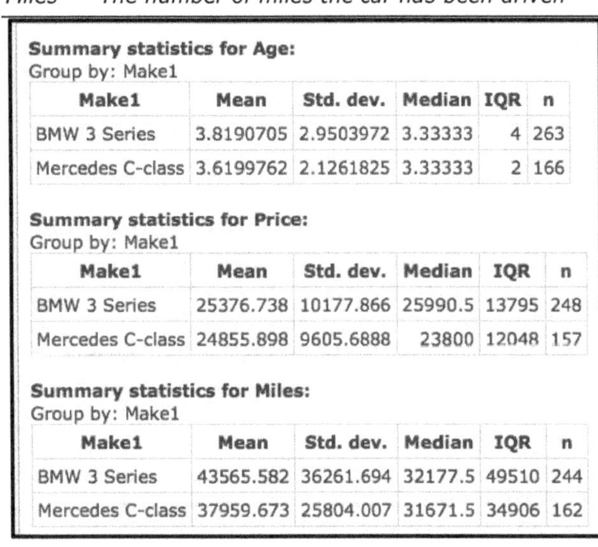

Summary statistics for Age:
Group by: Make1

Make1	Mean	Std. dev.	Median	IQR	n
BMW 3 Series	3.8190705	2.9503972	3.33333	4	263
Mercedes C-class	3.6199762	2.1261825	3.33333	2	166

Summary statistics for Price:
Group by: Make1

Make1	Mean	Std. dev.	Median	IQR	n
BMW 3 Series	25376.738	10177.866	25990.5	13795	248
Mercedes C-class	24855.898	9605.6888	23800	12048	157

Summary statistics for Miles:
Group by: Make1

Make1	Mean	Std. dev.	Median	IQR	n
BMW 3 Series	43565.582	36261.694	32177.5	49510	244
Mercedes C-class	37959.673	25804.007	31671.5	34906	162

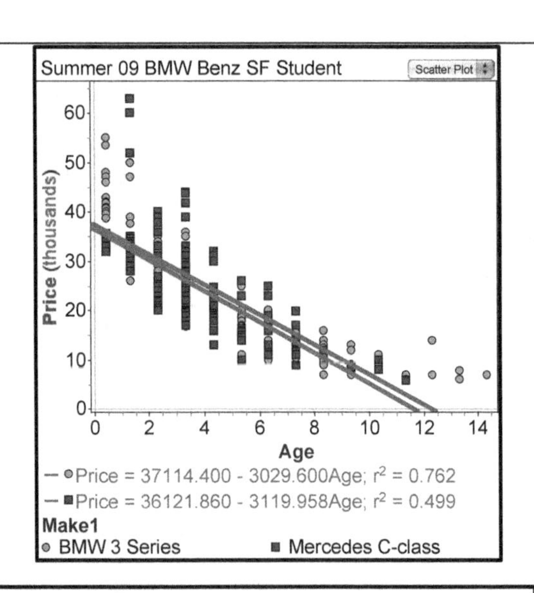

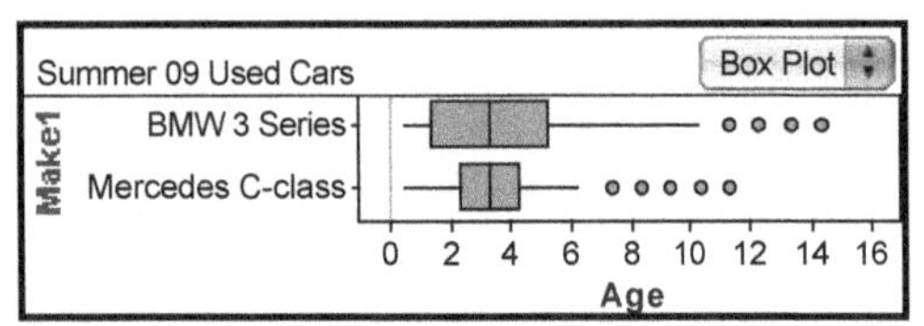

Summer 09 BMW Benz SF Student

		Body					Row Summary
		Convertible	Coupe	Hatchback	Sedan	Wagon	
Make1	BMW 3 Series	21	29	1	191	7	249
	Mercedes C-class	17	32	0	111	3	163
	Column Summary	38	61	1	302	10	412

S1 = count ( )

- Use software and the file ***Summer09BMWBenzSF*** to get graphics showing the distributions of the variables *Age, Price and Miles* broken down by categories of *Make1.*
- Get summary statistics -- two measures of center and two measures of spread -- for the variables *Age, Price* and *Miles* broken down by categories of *Make1.*
- Get the scatterplot and linear model for the relationship between *Age* and *Price* by categories of *Make1.* Interpret the slopes and r^2 to say something about the second statistical question.
- Get a contingency table showing the relationship between the body style and the make of the car, and analyze the numbers using conditional probabilities. It may be helpful to filter out the body styles with a low count.

8. Real Estate Markets in San Mateo County between 2005 and 2008

Background and Variables Starting in 2008 there was a great downturn in the economy compared with the years preceding. This downturn affected real estate market. Since we have data on the housing market fforf the years 2005-2006 and then for the years 2007-2008 we can may be able to assess whether San Mateo County was beginning to feel the effects of the economic downturn or whether at 2007-2008 all was still well. The data are for the houses sold between June 2005 and June 2006 (Y0506) and also for the houses sold between June 2007 and June 2008. The variables that are analyzed here are:

Year	*Whether the house was sold in Y0506 or in Y0708*
ListPrice	*The price the seller originally asked*
SalePrice	*The price at which the house actually sold*
SqFt	*The living area of the house measured in square feet*
SoldOverList	*Two categories: if the SalePrice > ListPrice then "Over List Price" but if the SalePrice ≤ ListPrice then "Not Over List Price"*

Statistical Questions

1. *Are there differences in the centers and spreads of the distributions of the sizes of houses (measured by the variable SqFt) between the houses sold in Y0506 and the houses sold in Y0708?*
2. *Are there differences in the centers and spreads of the distributions of the ListPrice of the houses sold in Y0506 compared with the houses sold in Y0708? If there are differences, what are they?*
3. *Are there differences in the centers and spreads of the distributions of the SalePrice of the houses sold in Y0506 compared with the houses sold in Y0708? If there are differences, what are they?*
4. *Are there differences in the proportions of houses that sold "Over List Price" for the houses sold in Y0506 compared with the houses sold in Y0708?*

Data Analysis

- Use software and the file ***SanMateoREYearComparison*** to get graphs comparing the distributions of the variables *SqFt, ListPrice* and *SalePrice* . There is an example for *SqFt* shown below.
- Get summary statistics for the distributions of the variables *SqFt, ListPrice* and *SalePrice.*
- Get a contingency table showing the numbers of houses that were sold over list price for the year Y0506 compared with the years Y0708 and from that table calculate proportions that give you some insight to question 4.

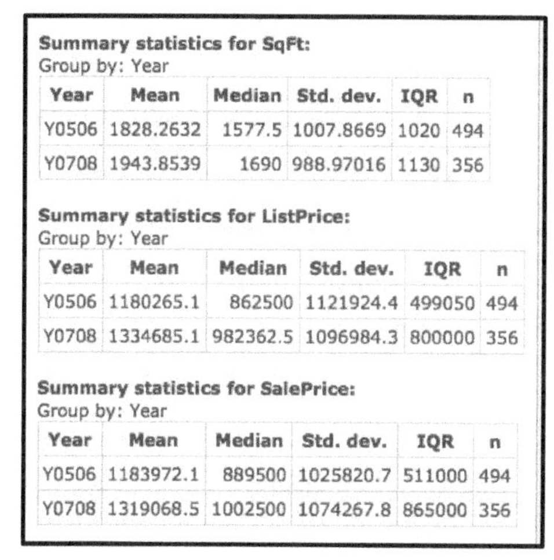

Summary statistics for SqFt:
Group by: Year

Year	Mean	Median	Std. dev.	IQR	n
Y0506	1828.2632	1577.5	1007.8669	1020	494
Y0708	1943.8539	1690	988.97016	1130	356

Summary statistics for ListPrice:
Group by: Year

Year	Mean	Median	Std. dev.	IQR	n
Y0506	1180265.1	862500	1121924.4	499050	494
Y0708	1334685.1	982362.5	1096984.3	800000	356

Summary statistics for SalePrice:
Group by: Year

Year	Mean	Median	Std. dev.	IQR	n
Y0506	1183972.1	889500	1025820.7	511000	494
Y0708	1319068.5	1002500	1074267.8	865000	356

Sample for San Mateo RE Comparison

		SoldOverList		Row Summary
		Not Over List Price	Over List Price	
Year	Y0506	210	284	494
	Y0708	235	121	356
Column Summary		445	405	850

S1 = count ( )

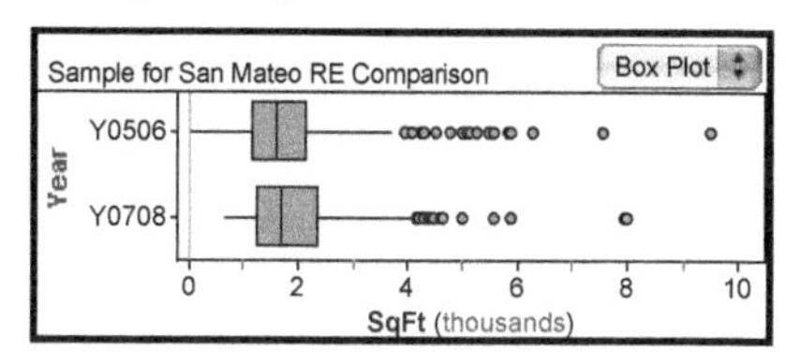

9. Roller Coasters around the World

Background and Variables There is a website (www.rcdb.com) that is a database for roller coasters. The website says that it "is a comprehensive, searchable database with information and statistics on over 2000 roller coasters throughout the world." The data in the file ***RollerCoastersWorldExpanded*** is from this website and includes data for roller coasters in North America, Latin America, and Europe. However, the data in the database is by no means complete; the database compilers are dependent on the information available from amusement park operators and manufacturers of roller coasters, and a look at the data reveal many "missing values." Notice that the measurements are all in metric units, so length and height are measured in meters, and speed is measured in km/hr.

Construction	*Whether the roller coaster is of steel construction or wood construction*
Length	*Length of the roller coaster in meters*
Height	*Height of the roller coaster in meters*
Speed	*Speed in kilometers per hour (km/hr)*
Inversions	*Whether or not the data mentioned the roller coaster having an inversion*
Duration	*How long the ride is in seconds*
Region	*Three categories: "North America," "Latin America," and "Europe"*

Statistical Question In this exercise, there is only one statistical question, but it needs to be broken down by the variables measured.

"What are the differences in the distributions by Region *of length, height, duration, and whether there are inversions?"*

To answer these questions, it will be necessary to break the question into a number of parts.

Data Analysis

- With the file ***RollerCoastersWorldExpanded*** use software to get graphs of the distributions of *Length, Height, Speed* and *Duration* broken down by the variable *Region,* as well as summary statistics for the variables by *Region.* Here are examples, but the analysis will require more than what is shown here
- For the quantitative variables where the relationships between the variables makes sense, get a scatterplot showing a linear model to see how the value of one variable as the explanatory variable is related to a response variable. Here is an example:
- Get a contingency table to analyze the incidence of *Inversions* by *Region,* and then use conditional probabilities to analyze. (For *Inversions* it is important to know that the variable Inversions just measures whether inversions were mentioned. There may be some rollercoasters that have inversions even though the data do not record this!)

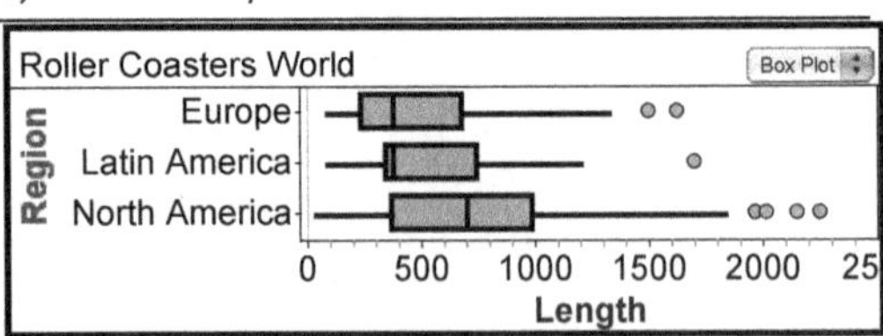

Summary statistics for Height:
Group by: Region

Region	Mean	Median	Std. dev.	IQR	n
Europe	16.028448	13	12.786527	16.85	116
Latin America	21.327692	15	15.663039	23.3	65
North America	29.607641	27.432	21.224443	23.4696	145

Summary statistics for Speed:
Group by: Region

Region	Mean	Median	Std. dev.	IQR	n
Europe	54.955789	47	23.758849	34	95
Latin America	67.171429	75.6	24.803209	40.3	49
North America	81.077143	81	29.871224	34.02	126

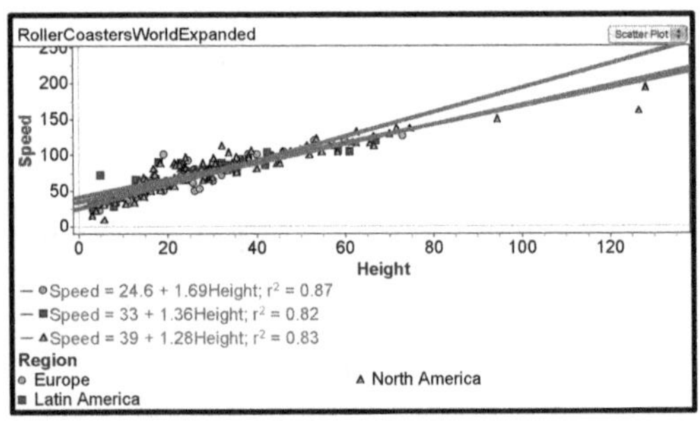

RollerCoastersWorldExpanded

		Inversions No	Inversions Yes	Row Summary
Region	Europe	106	20	126
	Latin America	37	32	69
	North America	165	48	213
Column Summary		308	100	408

S1 = count ( )

10. NHANES Health Data: Obese and Overweight Gender Differences

Background and Variables

The data are part of the National Health and Nutrition Examination Survey, which is a national survey of a random sample of residents of the US. To gain some understanding of the scope of the study, visit http://www.cdc.gov/nchs/nhanes.htm. The main variables to consider for an examination of gender differences in obesity are BMI and waist circumference.

Gender — *Male or Female*

WaistCir — *Waist circumference measured in cm*

BMI — *Body mass index* $BMI = \frac{Weight}{(Height)^2}$, *where weight is measured in kilograms and height is measured in meters. The idea is to measure "fatness" by taking account of height, since taller people will weigh more just because they are taller. It is widely used but not without controversy. The usual definition of obese is BMI > 30, and the usual definition of "overweight" is BMI > 25.*

FitnessSatus — *Three categories: (a) Fit (BMI ≤ 25) (b) Overweight but not Obese (25< BMI ≤ 30) (c) Obese*

Statistical Questions For these data, it is best to analyze a range of restricted range of ages. So first choose a "ten" year age category and the relevant data file (one of the files: ***NHANES2010Age13to20, NHANES2010Age21to30, NHANES2010Age31to40, NHANES2010Age41to50, NHANES2010Age51to60, NHANES2010Age61to70.***). *For the age category you have chosen:*

1. *What differences in center, spread, and shape of the distributions of **BMI** and of waist circumference (WaistCir) are there between males and females?*
2. *What differences are there in the proportions fit, Overweight but not obese, and obese by gender?*
3. *Extra: Work out from your data either an equivalent waist circumference or one of the other measures of "fatness" to the BMI > 30 measure of obesity.*

Data Analysis

- Using software and one of the ten year files, and get graphics and summary statistics for the variables *BMI* and *WaistCir*. Examples are shown here.
- Get a contingency table for the variable *FitnessStatus* broken down by gender, and analyze using conditional probabilities to answer the second statistical question above.

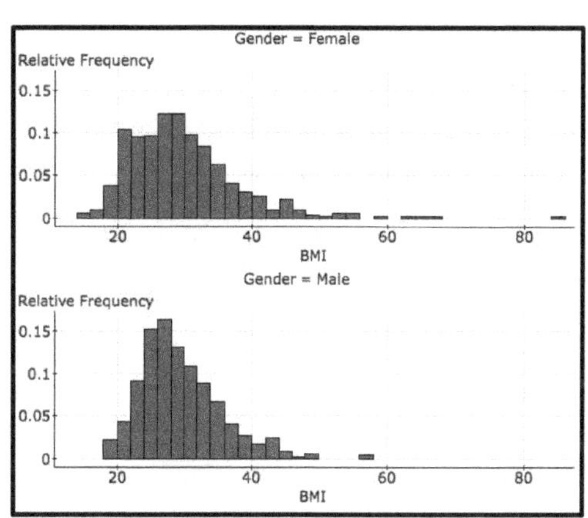

Summary statistics for WaistCir:
Group by: Gender

Gender	Mean	Median	Std. dev.	IQR	n
Female	96.57658	94.1	16.436588	22.1	538
Male	102.37131	99.8	15.416147	18.2	481

NHANES2010Combined

		Gender: Female	Gender: Male	Row Summary
FitnessStatus	(a) Fit	165	112	277
	(b) Overweight but not Obese	167	186	353
	(c) Obese	227	194	421
	Column Summary	559	492	1051

S1 = count ( )

11. Birth Data Questions I: Influence of Smoking

Background and Variables

Almost every birth is recorded on a birth certificate. There are millions of records, and they are available to the public. We have data on $n = 100{,}000$ births randomly selected for the year 2006. For this exercise, a random sample of $n = 9000$ has been drawn from the larger random sample and these data are available in the file ***NoveMilBirthSample*** Here are the variables measured.

Sex	*Sex of the child born*
BirthWeight	*Weight of the child at birth in grams*
Plurality	*Whether the birth was single birth (1), a twin (2), or triplet (3)*
Gestation	*Estimated length of pregnancy in weeks*
Premature	*Whether the birth was premature (gestation < 37 weeks) or full-term*
TotalBirthOrder	*For the mother, whether this birth was number 1, 2, 3, etc.*
FirstBirth	*two categories: (a) First Birth (b) later birth.*
AgeMother	*Age of the mother of the birth.*
EducationMother	*Mother's education: (a) Less than HS, (b) high school, (c) college, (d) post-graduate*
ParentsMarried	*Whether the parents of the child are married*
MotherSmokes	*Whether the mother of the child is a smoker or not. The data on smoking is not collected on all birth certificates in the same manner. For this reason, the count for this variable is lower.*

Statistical Questions on Mothers' Smoking

1. *Is there a difference in average* BirthWeight *for full-term babies depending upon whether the mother smokes or not? Is there more variability in* BirthWeight *for full-term babies depending upon* MotherSmokes?

a. *Is the relationship between* BirthWeight *and* MotherSmokes *the same for first births as for all births? Do the analysis separately for first births and later births. (That is filter by* FirstBirth.*)*

b. *Is the relationship between* BirthWeight *and* MotherSmokes; *the same for boy babies and girl babies? Filter by Sex.*

2. *Are babies born to smoking mothers more likely to be premature? Is it equally true for boy babies and girl babies?*

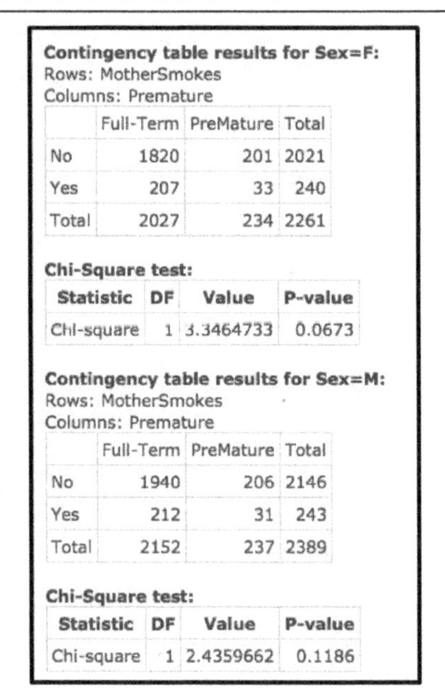

Contingency table results for Sex=F:
Rows: MotherSmokes
Columns: Premature

	Full-Term	PreMature	Total
No	1820	201	2021
Yes	207	33	240
Total	2027	234	2261

Chi-Square test:

Statistic	DF	Value	P-value
Chi-square	1	3.3464733	0.0673

Contingency table results for Sex=M:
Rows: MotherSmokes
Columns: Premature

	Full-Term	PreMature	Total
No	1940	206	2146
Yes	212	31	243
Total	2152	237	2389

Chi-Square test:

Statistic	DF	Value	P-value
Chi-square	1	2.4359662	0.1186

Data Analysis

- Using software and file ***NoveMilBirthSample*** show the distributions of the variable *BirthWeight* broken down by categories of *MotherSmokes.* Also get summary statistics. Make certain that you filter for the full-term births only, using Premature = "Full-Term."

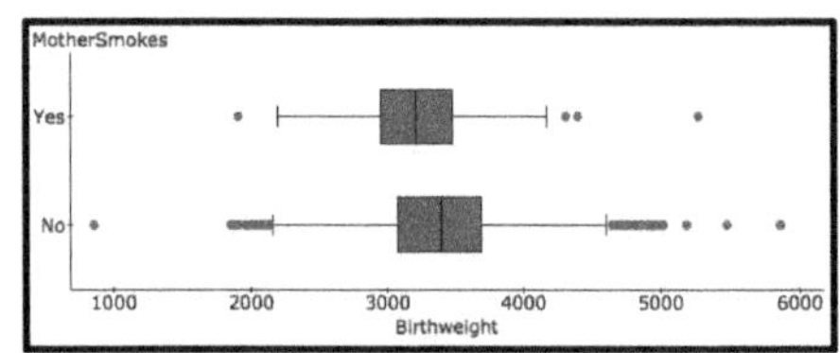

- Get contingency tables showing the numbers of premature births by smoking status of the mother, and calculate conditional probabilities correctly to answer question 2.

Summary statistics for Birthweight:
Where: Premature="Full-Term"
Group by: MotherSmokes

MotherSmokes	Mean	Median	Std. dev.	IQR	n
No	3388.5302	3390	475.2347	618	3757
Yes	3219.2816	3204	415.78573	528	419

12. *Birth Data Questions II: Influence of Education*

Background and Variables The data that you will use is the sample of information on births. The detailed definitions of the variables are shown in question 11 above.

Statistical Questions on Mothers' Education For these data are there many statistical questions that can be asked. Here are the questions to be answered for this exercise. (You may choose others; consult your instructor.)

1. *Is there a difference in average* AgeMother *for these births depending upon her education? Are there differences in the variability of the age of the mother (that is, in* AgeMother*) by the education of the mother?*
   a. *Compare the average ages and the variability in ages of the mother among the different education groups for just the data on first births. (See the data analysis section below.) In other words, answer the questions about averages and variability for just first births.*
   b. *Are the answers about average age of mothers and the differences in variability in age of mother the same for the married mothers as for the unmarried mothers?*
2. *Is there a relationship between mothers' education and whether the mother is married or not?*

Data Analysis

- Using software and file ***NoveMilBirthSample*** show the distributions of the variable *AgeMother* broken down by categories of *EducMother.* Also get summary statistics. Filter for the first births, using *FirstBirth*="First Birth".
- Get a contingency table showing the relationship between education of the mother (*EducationMother*) and whether the parents are married or not (*ParentsMarried*)? Use conditional probabilities, calculating the probabilities in the correct direction, or Chi-Square Test for Independence (See §4.4) to answer question 2.

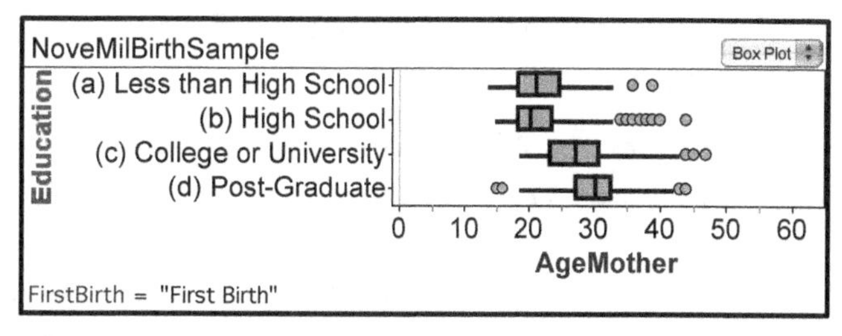

Summary statistics for AgeMother:
Where: FirstBirth="First Birth"
Group by: EducationMother

EducationMother	Mean	Median	Std. dev.	IQR	n
(a) Less than High School	21.937063	21	5.41414	7	143
(b) High School	21.732026	20	4.7358015	6	1530
(c) College or University	27.334697	27	5.3387732	8	1467
(d) Post-Graduate	30.263923	30	4.6634898	6	413

NoveMilBirthSample

		ParentsMarried		Row Summary
		No	Yes	
EducationMother	(a) Less than High School	96	47	143
	(b) High School	1046	484	1530
	(c) College or University	423	1044	1467
	(d) Post-Graduate	46	367	413
	Column Summary	1611	1942	3553

S1 = count ( )
FirstBirth = "First Birth"

§2.2 Exercises on Scatterplots and Correlation

1. *Mexican Roller Coasters* The data for this exercise come from the Roller Coaster Data Base (www.rcdb.com). The data on roller coasters are very incomplete because the website depends upon the owners or the builders of the roller coasters for the information. There are certainly more than thirteen roller coasters in Mexico; there were just thirteen with complete information for our variables. The variables are:

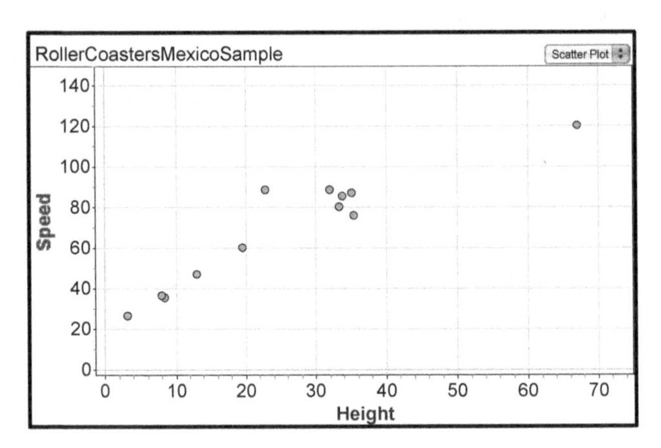

Height (measured in meters)
Length (measured in meters)
Speed (measured in km/hr)

We will examine the relationship between *Height* and *Speed*.

a. What are the cases for these data?

b. Using the scatterplot, make a judgment about the *direction*, the *form*, and the *strength* of the association between the height of roller coasters and their speed.

c. The table shows the calculation of the *correlation coefficient* $r = \frac{1}{n-1}\sum_{i=1}^{n}\left(\frac{x_i - \bar{x}}{s_x}\right)\left(\frac{y_i - \bar{y}}{s_y}\right)$ in the same fashion as in the subsection ***Using the formula***. Label your answers "a" through "i."

	Name	Location	Height x	Speed y	z(x)	z(y)	z(x)*z(y)
1	Batman the Ride	Mexico City	33.3	80.0	0.50	0.47	0.23
2	Boomerang	Mexico City	35.5	75.6	0.62	0.32	a
3	Medusa	Mexico City	32.0	88.5	0.42	0.77	0.33
4	Roller Skater	Mexico City	8.5	34.9	-0.92	b	1.02
5	Superman el Último Escape	Mexico City	67.0	120.0	2.42	1.87	4.54
6	Tsunaumi	Mexico City	8.0	36.0	-0.95	-1.07	1.01
7	Montaña Infinitum	Mexico City	33.8	85.3	c	0.66	0.35
8	Ratón Loco	Mexico City	13.0	46.8	-0.66	-0.69	0.46
9	Catarina	Guadalajara	8.0	36.0	d	e	1.01
10	Titan Cascabel	Guadalajara	22.9	88.5	-0.10	0.77	-0.08
11	Catariños	Monterrey	3.3	26.0	-1.22	-1.42	f
12	Tornado	Guadalupe	19.5	60.0	g	h	i
13	Tsunaumi	Aguascaliente:	35.1	86.9	0.60	0.71	0.43
		Sum	319.9	864.5	0.00	0.00	11.29597
		Mean	24.6	66.5			
		Standard Deviation	17.49	28.57			

d. Write the formula for the correlation coefficient and indicate the part of the formula that corresponds to the number 11.29597—perhaps by enclosing that part of the formula with a kind of circle or enclosure.

e. Use the number 11.29597 to get the value of the correlation coefficient, $r = 0.94133$. Comment on whether the number you got is in accord with your answer to part b.

The vertical line shows the mean of the explanatory variable *Height* and the horizontal line shows the mean of the response variable *Speed*. These lines divide the plot into four quadrants, labeled I, II, III, and IV, as discussed in the subsection ***Zzzzzz***.

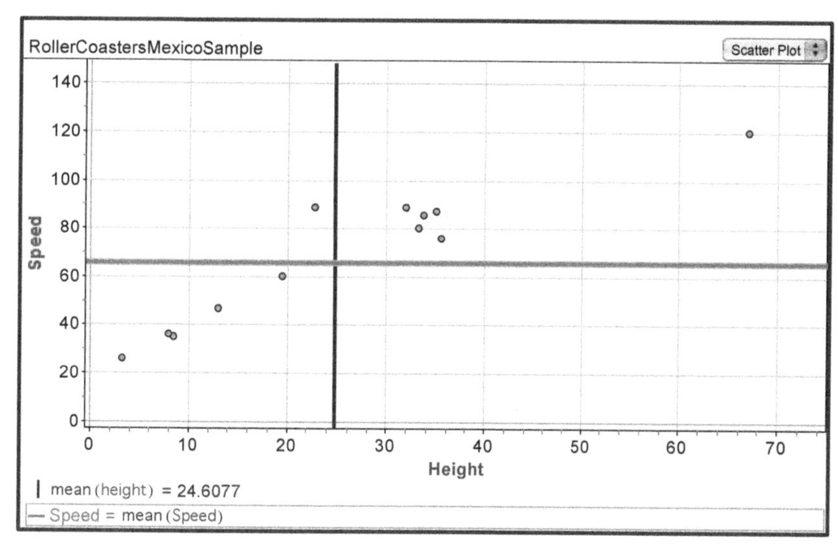

f. Quadrant III has the roller coasters whose *Height* is ________ than the mean *Height* and whose *Speed* is ________ than the mean *Speed*. Fill in the blanks with the words "less" or "more."

g. In quadrant III, the z_x scores will be ___________ and the z_y scores will be ___________ and therefore the contribution to the sum in the formula will be ___________. Fill in the blanks with the words "positive" or "negative."

h. One roller coaster makes a negative contribution to the sum in the formula for r. Find which one it is, either from the plot or from the chart on the previous page to determine the name of that roller coaster. Give a reason for your choice.

i. Translate the value of the correlation coefficient $r = 0.94133$ into a statement about the height and speed of roller coasters without using the word "correlation" or "association" ("The higher a roller coaster is…").

j. The relationship between *Height* and *Speed* for the Mexican roller coasters is strong and positive. Do you think that the relationship between *Length* and *Speed* will also be strong and positive? What about the relationship between the *Height* and *Length* of the roller coasters? Give reasons for your answers. You will check your answers using software: see below.

- With the file ***RollerCoastersMexicoSample*** use software to get a scatterplot with the variable *Length* on one axis and the variable *Height* on the other axis. Also, get the numerical value of the correlation coefficient using software, not by hand calculation. (Decide on which axis the variable *Length* should be: the x-axis or the y-axis.)
- With the file ***RollerCoastersMexicoSample*** use software make the plot but with the axes reversed: that is, if Length was on the y axis and height on the x axis, switch them. Also, get the numerical value of the correlations coefficient with the axes reversed.
- In reversing the roles or the two variables, note whether the correlation coefficient r changes.

k. The correlation coefficient r has the same value for two variables whichever variable is chosen as the explanatory variable x and whichever variable is chosen as the response variable y. State how $r = \frac{1}{n-1}\sum_{i=1}^{n}\left(\frac{x_i - \bar{x}}{s_x}\right)\left(\frac{y_i - \bar{y}}{s_y}\right)$ shows this fact. (*Hint:* If you switch the roles of the two variables would the calculations change?)

l. ***Interpretation.*** Now that you have the correlations between each pair of variables of the three variables, *Height, Length,* and *Speed,* you should be able to write what these correlations mean for visitors to amusement parks or for builders of roller coasters.

[[Questions in square brackets—[...], such as the next two—are usually optional questions; ask your instructor.]]

[m. *Review* There is a variable that distinguishes between the coasters in Mexico City and those in other parts of Mexico. Do an analysis to determine whether there is evidence that Mexico City has higher, longer, and faster roller coasters. Show your work and write up your conclusions.]

[n. *Review* Given what was said about the data in www.rcdb.com, explain why you may want to treat your conclusions stated in part m with some caution.]

2. *Mexican Roller Coasters Not in Mexico City*

Here are the data for the five roller coasters in Exercise 1 that are not in Mexico City. The summary statistics for these five roller coasters are also given. Height and Speed for the $n = 5$ roller coasters for which we have data and that are outside Mexico City. Use this information for these five roller coasters, and the formula for the $r = \frac{1}{n-1}\sum_{i=1}^{n}\left(\frac{x_i - \bar{x}}{s_x}\right)\left(\frac{y_i - \bar{y}}{s_y}\right)$ to calculate the correlations coefficient r for the relationship between height and speed for these five non-Mexico City roller coasters. (Answer: $r = 0.930$)

Name	Location	Height	Speed
Catarina	Guadalajara	8	36
Titan Cascabel	Guadalajara	22.9	88.5
Catariños	Monterrey	3.3	26
Tornado	Guadalupe	19.5	60
Tsunaumi	Aguascalientes	35.1	86.9

Summary statistics:
Where: MexicoCity = "Other Places"

Column	Mean	Std. dev.
Height	17.76	12.59
Speed	59.48	28.58

3. *Road & Track Gearhead Cars: Correlation between Horsepower and MPG*

Here are more data on cars tested by the car enthusiast magazine *Road & Track*. Remember that the cars in the data set do not represent all of the cars for sale but rather cars that would be of interest to "gearheads." The cases are cars; here are the variables measured:

Horsepower	*A measure of the power of a car; the higher the number, the more power*
MPG	*Miles per gallon. The number of miles that a car will travel on one gallon of gasoline.*
Cylinders :	*Number of cylinders: (a) four (b) six (c) eight (d) more than eight cylinders*

a. Before looking at the data, but from what you know about cars, make a guess about the direction of the relationship between Horsepower and MPG—should the direction be positive or negative? Give a reason for your answer.

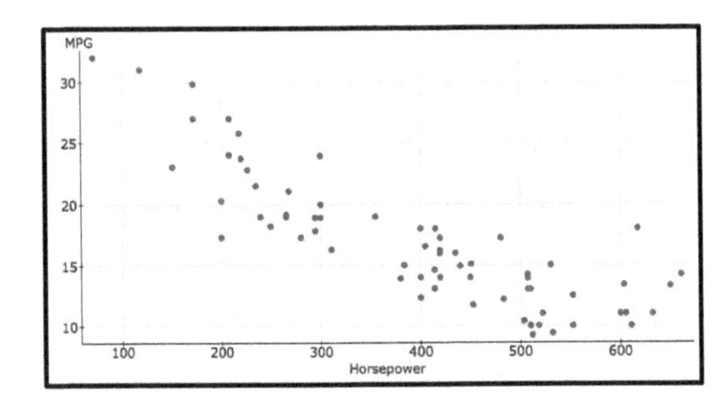

- Using the file ***RoadandTrackEuropeCars*** and software get the scatterplot showing the relationship between the variable *Horsepower* and *MPG*. Try to guess the numerical value of the r but then use software to get the calculated value.)
- Now use software to get the correlation coefficient between *Horsepower* and *MPG* but *within* each category of the variable *Cylinders*. Also get the software to show the plot relating *Horsepower* and *MPG* but by categories of *Cylinders*. Each "dot" should be a color or symbol showing its *Cylinders*.

b. The correlations are not the same for the various numbers of cylinders. Which correlation is strongest and which is weakest? Give a reason for your answer.

c. Describe what you notice about the distribution of the different types of cars (four, six, etc.) in the plot of the relationship between *Horsepower* and *MPG*.

d. Use what you observed and described in your answer to part d to explain how it can be that the overall correlation $r = -.8486$ can show a stronger negative relationship than the correlation for any of the categories that together comprise all the cars. (This question may appear to be difficult. Ask your instructor)

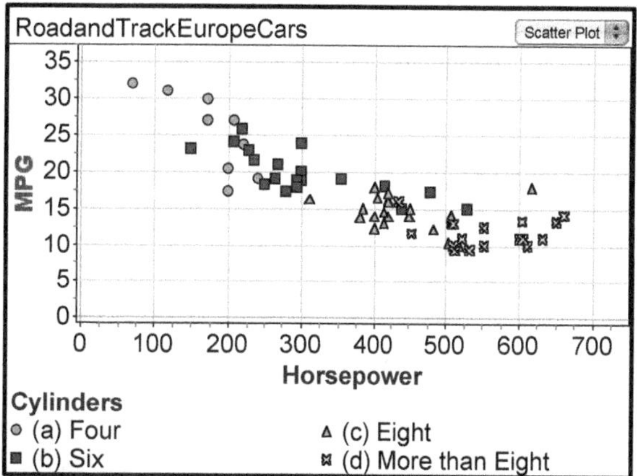

RoadandTrackEuropeCars

Cylinders	(a) Four	-0.7762
	(b) Six	-0.7469
	(c) Eight	-0.1924
	(d) More than Eight	-0.0475
	Column Summary	-0.8486

S1 = correlation (Horsepower, MPG)

4. Mammals: Amount of Sleep, Length of Pregnancy, and Other Tales

Here is a matrix of correlations and four scatterplots from the collection of mammals.

Mammals

	Gestation	BodyWeight	NondreamingSleep	DreamingSleep	TotalSleep	LifeSpan
Gestation	1	0.651102	-0.594703	-0.450899	-0.631326	0.629089
BodyWeight	0.651102	1	-0.375946	-0.109383	-0.307186	0.305518
NondreamingSleep	-0.594703	-0.375946	1	0.514254	0.962715	-0.407499
DreamingSleep	-0.450899	-0.109383	0.514254	1	0.727087	-0.311601
TotalSleep	-0.631326	-0.307186	0.962715	0.727087	1	-0.437657
LifeSpan	0.629089	0.305518	-0.407499	-0.311601	-0.437657	1

S1 = correlation ( )

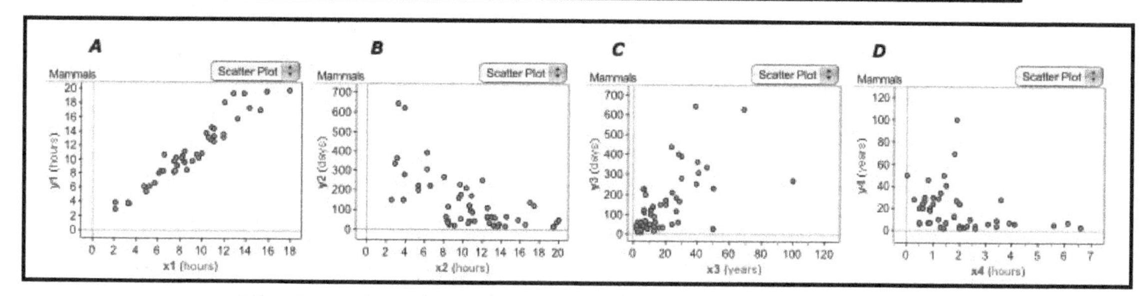

a. ***Detective Work*** The four plots show the relationship between four of the five pairs of variables listed just below. Your job is to determine which pair goes with which plot and which pair of variables has no plot. Use the matrix of correlations, but the units on the plots may also help.

I *TotalSleep, Gestation* *II* *DreamingSleep, LifeSpan* *III* *LifeSpan,Gestation*
IV *NondreamingSleep, TotalSleep* *V* *DreamingSleep, NondreamingSleep*

b. ***Before the Judge***. Having come to a decision about the plots, show the court that you are right. In other words, give reasons for each of your choices.

- Use the file ***Mammals2*** and software to make plots to confirm the answers to part a.

c. For the calculation of the correlation coefficient r for the relationship between *BodyWeight* and *Gestation*, does it matter which variable is considered to be the explanatory variable and which variable is considered to be the response variable? Explain why or why not.

d. The correlation coefficient r for the relationship between *BodyWeight* and *Gestation* is $r = 0.651$. Make a small sketch of what you think the scatterplot will look like for the r that is given.

- Use the file ***Mammals2*** and software to make a scatterplot of the relationship between *BodyWeight* and *Gestation.*

e. Does the actual plot look like your idea? If not, try to explain the magnitude of the correlation coefficient from what you see in the plot and what was said about the correlation coefficient in the *Notes.*

f. For the relationship between *TotalSleep* and *Gestation,* the correlation coefficient is $r = -0.63$. Comment on this interpretation of the correlation coefficient: "Sleeping longer (sometimes as much as 10 to 14 hours a day) *causes* the length of pregnancy to be shorter." (*Hint:* What is the connection between correlation and causation?)

g. Here is a graphic showing the relationship between the quantitative variable *TotalSleep* and the categorical variable *DangerLevel.* Which of the interpretations *I* and *II* is preferable? Or are both or neither good interpretations? Explain why.

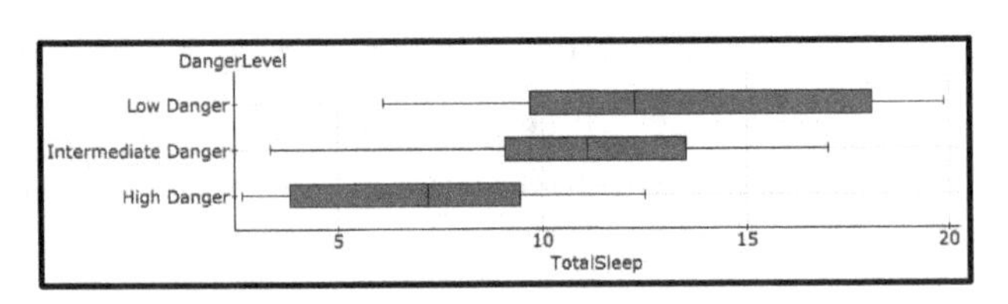

I "There appears to be a correlation between *DangerLevel* and *TotalSleep* for mammals."

II "There appears to be an association between *TotalSleep* and *DangerLevel* for mammals."

5. *Roller Coasters: Height, Length, Speed, Duration, Construction and GForce.*

This exercise uses the data on roller coasters in the file ***RollerCoastersWorldExpanded.*** The measures are in metric units (so meters, km/hr, and not feet and mph), and the variable *Duration* (how long the ride lasts) is measured in seconds. *GForce* measures acceleration compared to gravity, and the variable *Construction* indicates whether the roller coaster is built of wood or steel.

a. Just below are two scatterplots and a number of alternate values for the correlation coefficient for these plots. For each of the alternates, state whether that option is:

- *an impossible value* by the definition of the correlation coefficient
- *a possible value* but *not likely* to be the value for either of the plots shown
- *a possible value* and quite likely to be the value for one of the plots shown

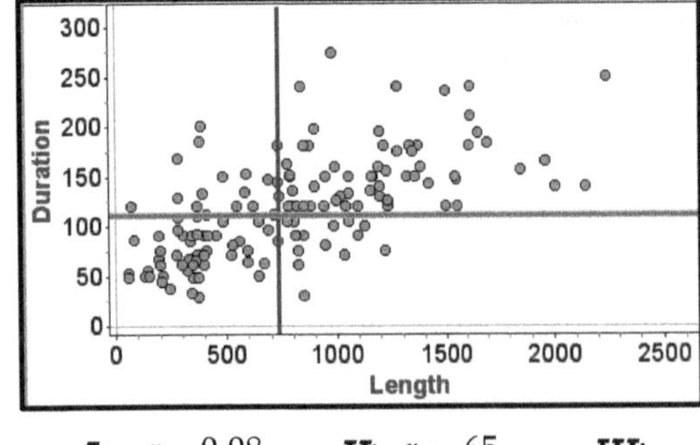

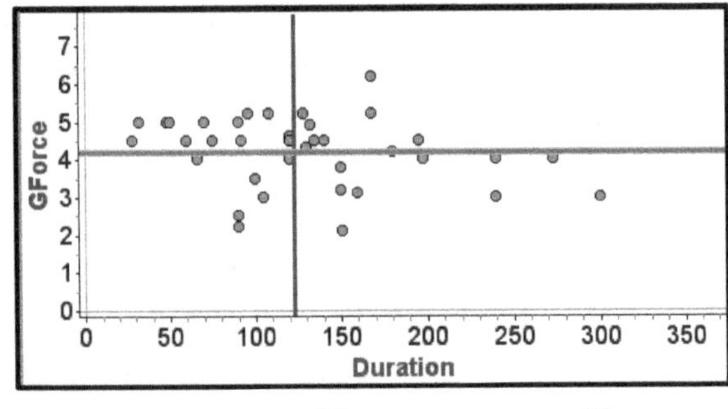

I: $r = 0.08$ ***II***: $r = .65$ ***III***: $r = 1.27$ ***IV***: $r = .98$ ***V***: $r = -.17$

The vertical and horizontal lines indicate the means for the two variables. These may help in identifying the implausible values for r for each plot.

b. Put into words what the scatterplot for the relationship between *Length* and *Duration* means for people comparing roller coaster rides. (*Hint:* A good way to start is: "There is a tendency…").

- With the file ***RollerCoastersWorldExpanded,*** use software to get a correlation matrix with the variables *Length, Duration* and *GForce.* (The correlation matrix should be something similar to what is seen at the beginning of Exercise 4.)

c. The correlation matrix just made should also give an *r* for the relationship between the *Length* of a roller coaster and the variable *GForce.* Which relationship (according to the correlation coefficient) has the stronger linear association, even if by just a little? Give a reason for your answer.

d. Here is the plot for the relationship between *Length* and *Height* of the roller coasters in the sample. The plot also shows the means of the two variables. Describe the relationship that you see in the plot, using the ideas of *direction, form,* and *strength.* Make a guess about the value of the correlation coefficient.

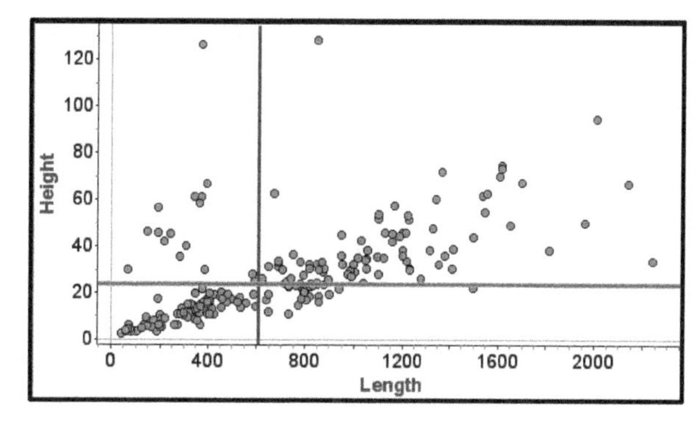

- Using software (and not by hand calculation) get the correlation coefficient *r* for the relationship between the variables *Length* and *Height* to check your guess in part d. (Not to worry: there is no penalty for being wrong; although getting the sign wrong may be worrisome.)
- Make (using software) separate *Length* and *Height* scatterplots for the wooden and steel roller coasters. Or, get just one scatterplot for *Length* and *Height* where the wooden and steel coasters are shown by different symbols. This will show whatever differences there are between roller coasters constructed with wood and steel in the relationship between *Length* and *Height.*

e. Use the plots and the values of *r* you have just made to explain how the relationship between *Length* and *Height* for roller coasters differs between the wood and steel roller coasters.

§2.3 Exercises on Fitting Lines

1. ***Used Cars, Including BMWs*** This exercise uses the same data set that was used for the example in the ***Notes*** except that used BMWs have been added. Here are the same summary statistics that were presented in the ***Notes*** with the addition of the mean, standard deviation, and correlation coefficient for the BMW as well as for the Lexus and Infiniti.

Summary statistics for Price:
Where: !(isnull(Age) or isnull(Price))
Group by: Make1

Make1	Mean	Std. dev.	n
BMW 3 Series	24938.964	9784.7944	756
Infiniti G-series	22621.214	6515.3981	308
Lexus IS	27571.558	6426.065	181

Summary statistics for Age:
Where: !(isnull(Age) or isnull(Price))
Group by: Make1

Make1	Mean	Std. dev.	n
BMW 3 Series	3.9426764	2.9298073	756
Infiniti G-series	3.5884707	1.5739101	308
Lexus IS	2.9033116	1.5230526	181

Correlation between Price and Age for Make1 = BMW 3 Series
-0.85901774
Correlation between Price and Age for Make1 = Infiniti G-series
-0.90257907
Correlation between Price and Age for Make1 = Lexus IS
-0.85135609

a. Using the information in the summary table and the formulas in the box in the ***Notes*** for the slope b and y-intercept of the ***Least Squares Regression Equation,*** do (and show) the calculations and get $\hat{y} = 38000.40 - 3592.04x$ for the equation predicting *Price* from *Age* for the Lexus IS.

b. Careless Carrie has $b = -0.851356\dfrac{1.52305}{6426.06} \approx -0.0002018$ for her calculation. How has CC been careless?

- Check the calculations just made with the slope and the y intercept shown in the plot below for the Lexus (the equation with the triangles in the graphic shown below). Use may also check by getting the least squares regression using software and the file ***Summer09BMWLexusInfiniti***.

c. Interpret the slope of the ***Least Squares Regression Equation*** in the context of the data collected. Include units in the answer.

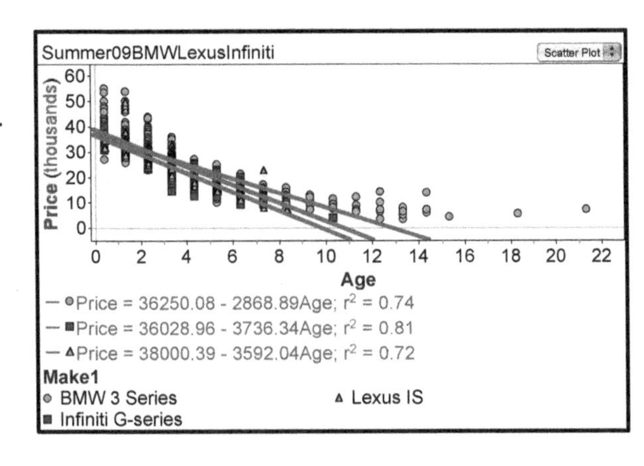

d. The Lexus IS cars are indicated by triangles. There is one Lexus IS that is 7.333 years old and is being sold for \$22,900. Use the ***Least Squares Regression Equation*** to find the predicted price for this Lexus IS. and also find the *residual* for this car.

e. Find the *residual* between the actual and predicted price for the Lexus that was 7.333 years old in age, and was being sold for \$22,900.

f. From the ***Least Squares Regression Equations*** shown, what can you say in context about how the BMW loses value compared with the Lexus or Infiniti?

g. Compare the slopes of the ***Least Squares Regression Equations*** now and see if your answer to question e would change.

2. ***Roller Coasters and the Least Squares Equation.*** In this exercise you will use software to try to "eye-ball" the best-fitting line. You will then compare your attempts to the line found by the equations for the least squares line. We will look at the relationship between the *Duration* of the ride, recorded in seconds, and the *Length* of a roller coaster, measured here in feet (note: not meters).

   a. Do you expect to see a positive or a negative relationship between the *Length* of a roller coaster and the *Duration* of the ride? As well as answering "positive" or "negative," put your idea in the form: "The greater the… the greater (or smaller) will be the…"

   b. Which variable should be the *explanatory* and which variable the *response*? Give a reason.

- Use the file ***RollerCoastersAmericaSample***; If your software has a ***Movable Line Applet*** follow the directions for that applet using the variables *Length* and *Duration*, and filtering for just the roller coasters built of wood, that is, *Construction* ="*Wood*". Answer questions c and d from the results of the ***Movable Line Applet***. If the software does not include the ***Movable Line Applet*** then get the least squares regression equation for the variables *Length* and *Duration*, and filtering for *Construction* ="*Wood*" and answer the questions starting with question e. (See ***Least Squares Regression*** in the software supplement.)

   c. Write the equation of the line that *you* have determined fits the data well using the ***Movable Line Applet***; use the form $\hat{y} = a + bx$.

   d. Copy the value of "*Sum of Squares*" or "*SSE*" from the ***Movable Line Applet***.

   e. Write (for your own use and understanding) a sentence or two that explains what the "*Sum of Squares*" or "*SSE*" is in the context of fitting a linear model to data.

   f. According to the ***Notes,*** the line that makes the ***Sum of Squares*** the *smallest* possible for a collection of data is the best-fitting line and is called the_________________________.

   g. Write the equation of the least squares line and record its "*Sum of Squares*" or "*SSE*". (The Sum of squares should be smaller than the Sum of Squares you managed to get.)

   h. Calculate the equation of the least squares line by using the formulas for the slope and the y-intercept shown in the ***Notes*** with the numbers for the means, standard deviations, and correlation coefficient for the relationship between *Length* and *Duration* is: $r =$ 0.772724.

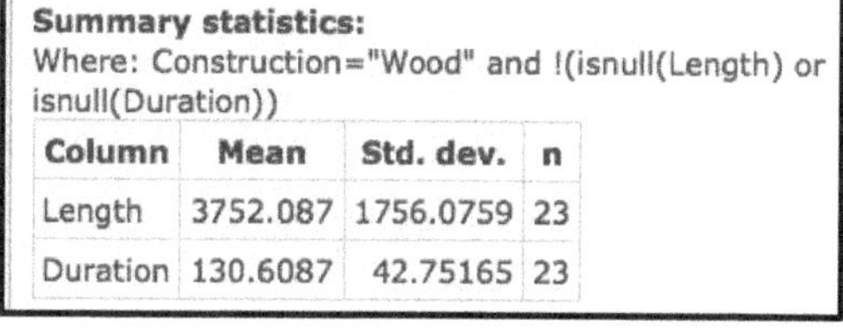

Summary statistics:
Where: Construction="Wood" and !(isnull(Length) or isnull(Duration))

Column	Mean	Std. dev.	n
Length	3752.087	1756.0759	23
Duration	130.6087	42.75165	23

   i. Use the least squares equation to predict the *Duration* in seconds of the coaster called "Grizzly" (in Santa Clara, California) that has a *Length* of 3,250 feet and (actual) *Duration* of 160 seconds.

   j. Find the residual for the prediction you found in part i.

   k. Interpret the slope of the equation of the least squares line in the context of the data.

   l. Interpret the y-intercept if you think that it makes sense, or say why it makes no sense.

3. *A Tale of Three Towns: San Bruno, Redwood Shores, and Hillsborough*

This exercise compares list prices and its relationship with the size of the house for three places in San Mateo County. The variable *List* is the variable *(List_Price/1000);* the first house in the ***Case Table*** for the San Bruno data had a list price of $685,000 and that is recorded as 685. This exercise is a combination of calculation and interpretation, but the end product should be a comparison of the houses for sale in the three towns—a tale of three towns.

Look at the plot showing the relationship between *SqFt* and *List* for the houses from the three towns and also compare the correlation coefficients that are given here. The first number is the correlations coefficient *r* and the second number is the number of cases in each town, *n*. If we were starting from the raw data (shown here for the San Bruno sample) we would have to laboriously calculate the correlation coefficient *r* first. We have let the software take care of this step. However, if you feel a need to calculate, you may apply the formula for the correlation coefficient to these data.

ThreeSanMateoCountyTowns

	City		
	HIL	RS	SB
	0.653291	0.936151	0.874031
	11	11	10

S1 = correlation (SqFt, List)
S2 = count ( )

Three San Mateo County Towns

	SqFt	List
1	1160	685
2	1630	679
3	1410	710
4	720	459
5	1440	770
6	1390	700
7	950	399
8	1390	700
9	1370	695
10	1695	875

City = "SB"

a. What do the three correlation coefficients and the plot tell you about the *linearity* and the *scatter* in the relationship between *SqFt* and *List* for the three towns?

b. Use the formulas for the slope and the y-intercept of the ***Least Squares Regression Equation*** to get the best-fitting equations for the relationship between *SqFt* and *List* for the three towns. All the information that you need is given in the above.

- With the file ***ThreeSanMateoCountyTowns***, use software to check the calculations in part c by getting the ***Least Squares Regression equations*** for predicting *List* from *SqFt* for the three towns. (That is, group by *City*.) Or turn the page for the results.

d. Using the ***Least Squares Regression Equation*** for San Bruno, get the predicted value for the list price of the second house in the San Bruno sub-sample—the one that has 1,630 square feet and was listed at 679 (that is $679,000).

e. Get the residual value for the San Bruno house whose predicted value you got in part d.

f. For the San Bruno sample the least squares regression equation is $\hat{y} = 128.521 + 0.409x$. Identify the slope and give an interpretation in the context of the data. "In the context of the data" means that the interpretation should say something about how much the list price increases for a change in the square feet of space in the house.

Summary statistics for SqFt:
Group by: City

City	Mean	Std. dev.	n
HIL	3516.9091	890.43871	11
RS	2079.0909	414.81212	11
SB	1315.5	296.98906	10

Summary statistics for List:
Group by: City

City	Mean	Std. dev.	n
HIL	2842.7173	691.09998	11
RS	1180.7182	187.71823	11
SB	667.2	139.14006	10

g. For the equation $\hat{y} = 128.521 + 0.409x$ relating the square footage of a houses to their list price, does the y intercept have a meaning? If so, give a good interpretation. If not, say why the y intercept does not have a meaning.

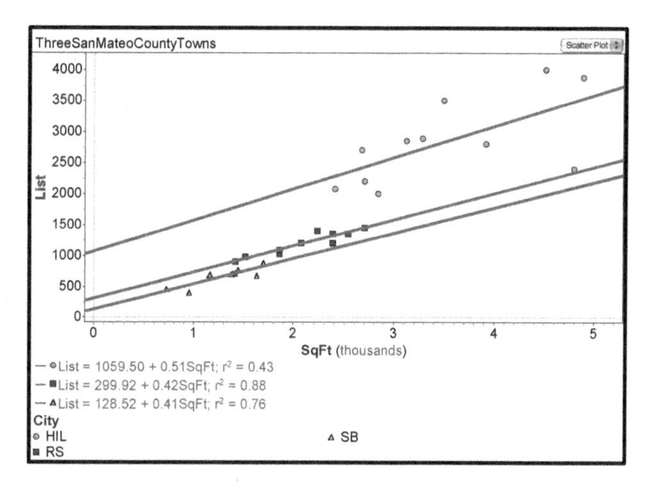

h. Compare the ***Least Squares Regression Lines*** (and their equations) for the three places to discuss the differences and similarities in the relationships between size and price for the houses being sold in the three places: San Bruno, Redwood Shores, and Hillsborough. Write your tale of three towns.

4. *Women's Heptathlon Results for 2008* The heptathlon is a women's event held in the Olympic games since 1984. It has seven events, and each competitor must participate in all of the events.

-	100-meter hurdles; time measured in seconds	*HundredMeterHurdles*
-	High jump; distance measured in meters	*HighJump*
-	Shot put; distance measured in meters	*ShotPut*
-	200-meter run; time measured in seconds	*TwoHundredMeters*
-	Long jump; distance measured in meters	*LongJump*
-	Javelin throw; distance measured in meters	*JavelinThrow*
-	800-meter run; time measured in seconds	*EightHundredMeters*

For each of these events, there is a score and a total score. The highest total wins.

The table shows the means and the standard deviations for the distributions of the results for the *HighJump* and the *LongJump*. The correlation coefficient for the relationship between performance in the *HighJump* and the *LongJump* is [Notice that the summary statistics are only for the $n = 33$ heptathletes who completed both events. Yana Maksimova of Belarus did not complete the long jump, so the mean and sd for the *HighJump* also excludes her.]

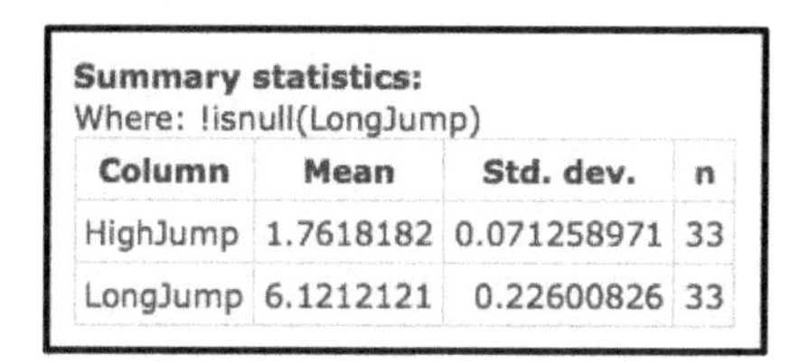
Summary statistics:
Where: !isnull(LongJump)

Column	Mean	Std. dev.	n
HighJump	1.7618182	0.071258971	33
LongJump	6.1212121	0.22600826	33

Correlation between HighJump and LongJump is:
0.70848463

a. Does the direction of the correlation coefficient make sense in the context of these data? Explain briefly.

b. Of the two variables, *HighJump* and *LongJump*, which do you think should be the *explanatory* variable and which should be the *response* variable? Or does the choice not matter, as neither is "prior" to the other? Choose one as the explanatory variable and the other as the response, and from the means, sd, and the r given, calculate the ***Least Squares Regression Equation*** for your choice.

c. Interpret the slope of the ***Least Squares Regression Equation*** you have just calculated in the context of the heptathlon competitors. ("For every . . .)

d. Calculate the ***Least Squares Regression Equation*** for the opposite choice of explanatory and response variables from the same information given above.

- Use software with the file ***Heptathlon2008*** to check the answers to the calculation of the least squares regression equations done in parts c and d. (In the software supplement, see ***Least Squares Regression***.)

   e. Answer this "typical test question" based on what you have done in this exercise: "True or false, and explain your answer: 'For the relationship between any two variables, there is just one *r* but two possible least squares regression equations, although only one of the two may make sense.'"

   f. Marie Collonvillé of France jumped 6.21 m in the *LongJump* and 1.86 m for the *HighJump*. Use the relevant ***Least Squares Regression Equation*** to predict Ms. Collonvillé's *HighJump* performance from her *LongJump* performance.

   g. Find the residual for the prediction you made in part f. Does Ms Collonvillé jump higher than what you would expect from her *LongJump* performance?

   h. Predict Ms. Collonvillé's *LongJump* from her *HighJump,* and find the residual.

   [i. If Exercise 5 is done then write a short paragraph or two explaining any differences or similarities between the analysis you have done for the 2008 heptathlon and the analysis for the 1988 heptathlon.]

5. *Women's Heptathlon Results for 1988* *[Note: This exercise is paired with exercise 4.]*

The heptathlon is a women's event held in the Olympic games since 1984. It has seven events, and each competitor must participate in all of the events. See Exercise 4 for the variables measured. The analysis and questions for this exercise parallel the analysis and questions for Exercise 4. The goal is to compare the two heptathlons (1988 and 2008), twenty years apart.

On the right are the means, the standard deviations, and the correlation coefficient *r* for the relationship between performance in the *HighJump* and performance in the *LongJump*

Correlation between HighJump and LongJump is:
0.78244227

Summary statistics:

Column	Mean	Std. dev.	n
HighJump	1.782	0.077942286	25
LongJump	6.1524	0.47421233	25

   a. Does the direction of the correlation coefficient make sense in the context of these data? Explain briefly.

   b. Of the two variables, *HighJump* and *LongJump,* which do you think should be the *explanatory* variable and which should be the *response* variable? Or does the choice not matter, as neither is "prior" to the other? Choose one as the explanatory variable and the other as the response, and from the means, sd, and the *r* given, calculate the ***Least Squares Regression Equation*** for your choice.

   c. Interpret the slope of the ***Least Squares Regression Equation*** you have just calculated in the context of the heptathlon competitors.

   d. Calculate the ***Least Squares Regression Equation*** for the opposite choice of explanatory and response variables from the same information given above.

- Use software with the file ***Heptathlon1988*** to check the answers to the calculation of the least squares regression equations done in parts c and d. (In the software supplement, see ***Least Squares Regression***.)

e. Answer this "typical test question" based on what you have done in this exercise: "True or false, and explain your answer: 'For the relationship between any two variables, there is just one *r* but two possible least squares regression equations, although only one of the two may make sense.'"

f. Dong Yuping of China jumped 6.40 m in the *LongJump* and 1.86 m for the *HighJump*. Use the relevant ***Least Squares Regression Equation*** to predict Ms. Dong's *HighJump* performance from her *LongJump* performance. (She finished sixteenth overall.)

g. Find the residual for the prediction you made in part f. Does Ms Dong jump higher than what you would expect from her *LongJump* performance?

h. Predict Ms. Dong's *LongJump* from her *HighJump* and find the residual.

[i. If Exercise 4 is done then write a short paragraph or two explaining any differences or similarities between the analysis you have done for the 2008 heptathlon and the analysis for the 1988 heptathlon.]

6. ***Women's Heptathlon Results for 2008*** The heptathlon is a women's event held in the Olympic games since 1984. It has seven events, and each competitor must participate in all of the events, as explained in Exercise 4. For the variables, see Exercise 4.

a. Think about the data first off: do you expect a positive, negative, or no relationship between the variables *HundredMeterHurdles* and *HighJump*? Give a reason for your answer.

b. Which variable should be regarded as the explanatory variable and which variable should be regarded as the response variable here? Or does it not matter? Give a reason for your answer.

Here is a scatterplot showing the relationship between the results for *HighJump* with *HundredMeterHurdles* chosen as the response variable. The least squares regression equation thus predicts the time on the *HundredMeterHurdles* from the distance jumped in the *HighJump*. The means for the two variables (the vertical and horizontal lines.) are also shown on the plot.

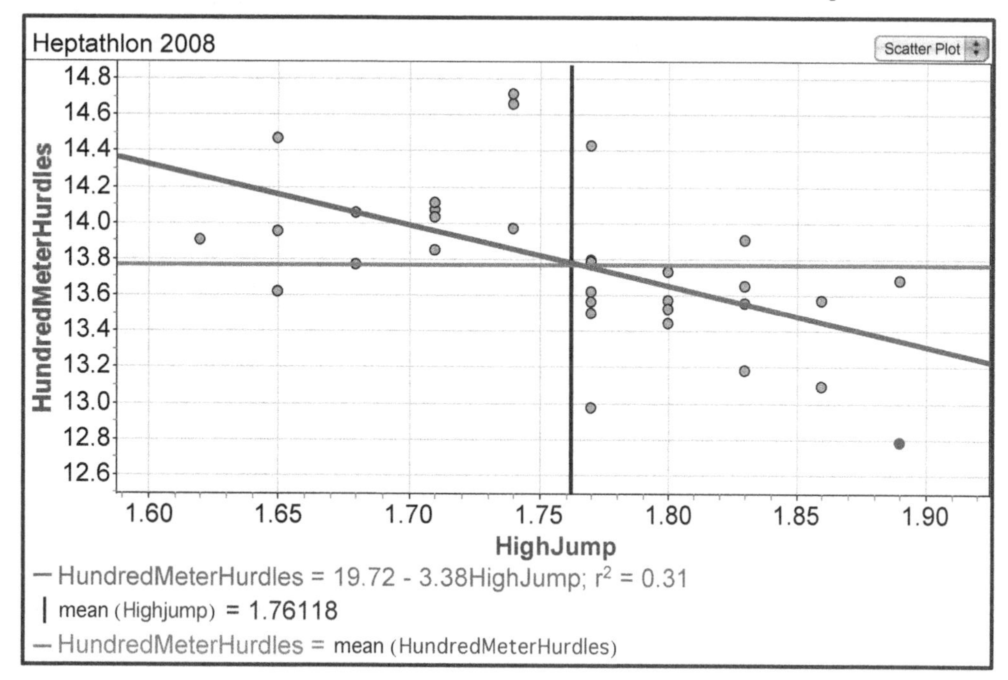

c. Using information in the table of summary statistics and the formulas for the slope b and y-intercept a given in the ***Notes,*** confirm that $\hat{y} = 19.72 - 3.38x$. (You may have a rounding error, however.)

Summary statistics:

Column	Mean	Std. dev.	n
HighJump	1.7611765	0.070270676	34
HundredMeterHurdles	13.773824	0.42895232	34

Heptathlon 2008

	HundredMeterHurdles
HighJump	-0.553178

S1 = correlation ( )

d. Look back at the scatterplot. In which quadrant are the best athletes? Give a short reason for your answer.

e. [*Review of correlation*] In which quadrants are the contributions to the correlation coefficient negative and in which quadrants are the contributions positive?

f. Find an athlete who makes a positive (instead of negative) contribution to the correlation coefficient r.

g. Notice (in the plot on the previous page) that the ***Least Squares Line*** goes through the means of the two variables. What in our formulas guarantees that the best-fitting line goes through the means?

h. Hyleas Fountain had a high jump distance of 1.89 m and a hundred meters hurdles time of 12.78 secs. Find her on the plot. Judging from where she is on the plot, will the residual predicting time in the *HundredMeterHurdles* from her *HighJump* performance be positive or negative for Hyleas Fountain? Is the direction of the residual good or bad for Ms. Fountain?

i. Find a predicted *HundredMetersHurdles* time for Ms. Fountain based upon the numbers given in part h, and from these calculations find the residual. The direction of the residual should agree with the answer to part g.

j. Think about what direction the relationship between *HundredMeterHurdles* and the *JavelinThrow* should be. Should the relationship be positive, negative, or is there essentially no relationship? Give a reason.

Here is the plot showing the relationship between *HundredMeterHurdles* with the *JavelinThrow* chosen as the response variable.

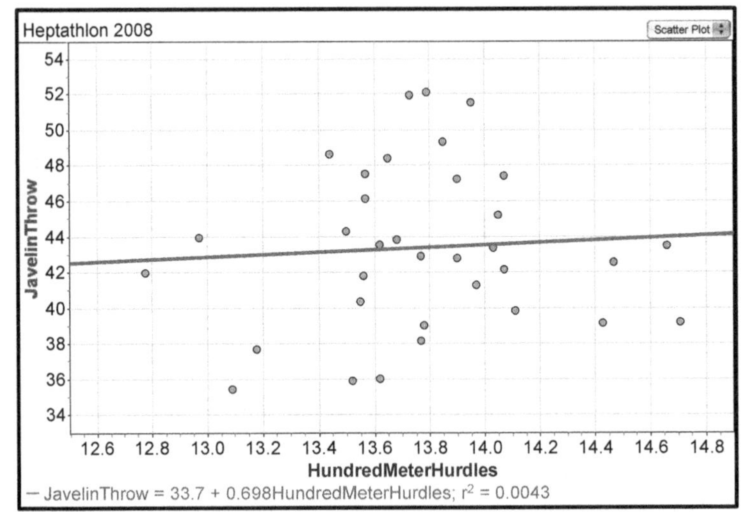

k. Note on the plot that $r^2 = 0.0043$. Take the square root to get r. Compare the correlation coefficient r with your ideas in part j and comment. Were you correct in part j?

7. ***Gearhead Cars: Power and the Quarter-Mile*** For their road tests, *Road & Track®* records the maximum speed for a car accelerating from a standing start to the maximum speed in a distance of one-quarter mile. This is recorded in our data set as *MPHQuarterMIle*. They also record how long it takes the car to reach the one-quarter-mile mark, which is recoded as *SecondsQuarterMile*.

Summary statistics:

Column	Mean	Std. dev.	n
Horsepower	356.47368	147.43204	133
MPHQuarterMile	105.03233	12.348013	133

In this exercise we will look at the relationships between these two measures and the power of the car, measured by *Horsepower*. Here are the means, standard deviations, and the correlation coefficient *r* for the variables *Horsepower* and *MPHQuarterMIle*.

RoadandTrackCars

	MPHQuarterMile
Horsepower	0.940495

S1 = correlation ( )

a. In the way that we usually think about cars, which variable of these two is most naturally the *explanatory* variable and which variable is most naturally thought of as the *response* variable?

b. True or false, and explain: "Since the correlation coefficient *r* is the same whichever variable we choose as the explanatory and whichever variable we choose as the response, the ***Least Squares Regression Equation*** will also be the same."

c. Using the formulas (found in the ***Notes***) for the slope and the *y*-intercept, calculate the ***Least Squares Regression Equation*** according to the choices that you made in part a.

- Check your answers to part c using software and the file ***RoadandTrackCars***. Get the least squares regression equation and also a graph of the relationship between *Horsepower* and *MPHQuarterMIle* with the explanatory and response variables correctly assigned.

d. Identify the slope of the ***Least Squares Regression Equation*** and interpret the slope in the context of the data. (You may find it useful to consider an increase of ten horsepower rather than just one horsepower.)

e. One of the cars in the data is the Dodge Challenger R/T; this car has a *Horsepower* rating of 376 and in the quarter mile had a speed (that is a *MPHQuarterMIle*) of 99 mph. Use the information on *Horsepower* to find the *predicted* value of *MPHQuarterMIle* for the Dodge Challenger R/T.

f. From your calculations in part e, find the residual. Does the Dodge Challenger R/T have a speed that is higher or lower than expected, given its horsepower?

g. Do you expect the relationship between *Horsepower* and *SecondsQuarterMile* to be positive or negative? Give a reason for your answer.

- Use the file ***RoadandTrackCars*** and software to get a graph showing the relationship between *Horsepower* and *SecondsQuarterMile.* Also, check the ***Least Squares Regression Equation*** predicting *SecondsQuarterMile* from *Horsepower.*

h. The Dodge Challenger R/T Use the ***Least Squares Regression Equation*** and information from the ***Case Table*** to find the *predicted value* of the time taken for the Dodge Challenger R/T to cover a quarter-mile.

i. Find the residual for the *SecondsQuarterMile* for the Dodge Challenger R/T. Is the result that you get in accord with the result you got for part f? Explain.

8. ***Gearhead Cars: Power, Displacement, and MPG***

Horsepower measures how powerful a car is, *MPG* (or *m*iles *p*er *g*allon) measures the amount of fuel a car uses (that is, the higher the *MPG*, the more miles the car travels on a gallon of fuel, and so less fuel is being used), and *Displacement* measures how big the engine is. We naturally expect that there is a tendency for cars with larger engines (the technical term is "displacement" measured in cubic centimeyers) to have more power and use more fuel, and therefore have lower *MPG.*

a. With the description given just above (or from what you know about cars), determine which relationships should be positive, which should be negative for each pair below.

- *Displacement* and *Horsepower,*
- *Displacement* and *MPG,*
- *Horsepower* and *MPG*

Summary statistics:
Where: !isnull(Displacement)

Column	Mean	Std. dev.	n
Horsepower	349.8254	142.08725	126
Displacement	3851.1984	1605.128	126

b. Here are the summary statistics for the relationship between *Displacement* and *Horsepower* and the correlation coefficient *r*.

Correlation between Horsepower and Displacement is: 0.87934235

Calculate, using the formulas for slope and *y*-intercept the ***Least Squares Regression Equation*** to predict *Horsepower* from *Displacement.* (First decide which variable is explanatory.)

- Check your answers to part c using software and the file ***RoadandTrackCars***. Get the least squares equation and also a graph of the relationship between *Horsepower* and *Displacement* with the explanatory and response variables correctly assigned.

c. The Ford Mustang Bullitt has 315 horsepower, and a displacement of 4601 cc. Use the values of the correct variable and the ***Least Squares Regression Equation*** to predict the *Horsepower* the Ford should have according to our equation, given its displacement.

d. Find the residual for the Ford Mustang Bullitt. Is the residual positive or negative? Is the Ford above or below the ***Least Squares Line***?

e. Confused Conrad answers question c ("Find the residual…") by writing "$r = 0.879342$." What is CC's confusion?

f. Identify the slope of the ***Least Squares Regression Equation*** that relates *Horsepower* and *Displacement* and interpret it in the context of the data.

g. Identify the *y*-intercept and interpret this number if you think that it makes sense to interpret it; if you think it makes no sense, explain why it makes no sense.

h. Here is the plot and the least squares line to predict *MPG* from *Horsepower*. For the Ford Mustang Bullitt, use the ***Least Squares Regression Equation*** to predict the *MPG* from the *Horsepower,* given that the Ford has 315 horsepower. Its MPG is recorded at 16 mpg.

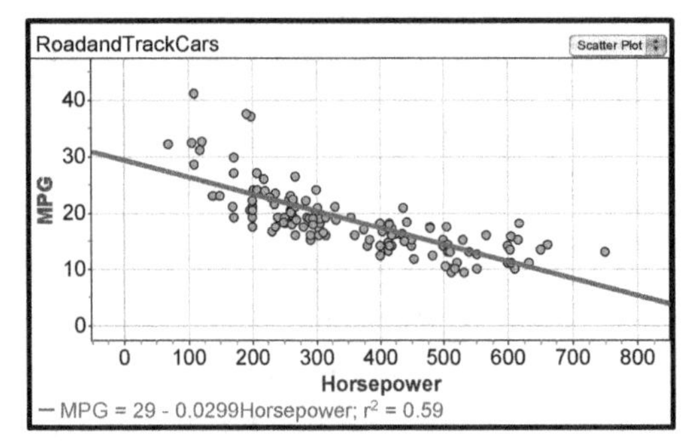

i. Choose any other car you think is interesting and use the two equations you have to predict the *Horsepower* from *Displacement* and *MPG* from the *Horsepower.*

§2.4 Exercises on Evaluating Models

1. ***Used Cars: How well does the linear model fit?*** This exercise uses the same data set that was used for Exercise 1 in ***Exercises §2.3.***

- Use the file ***Summer09BMWLexusInfiniti*** and get the plots of *Price* as a function of the *Age* of the used cars being sold. Also, use the software to show the *residual plots* by the x variable, *Age*, and also confirm that the least squares regression equations are as they are shown here. Depending upon the software used, the plots may differ; for example, you may not be able to get the grid shown, and the least squares equations may be displayed on a different page.

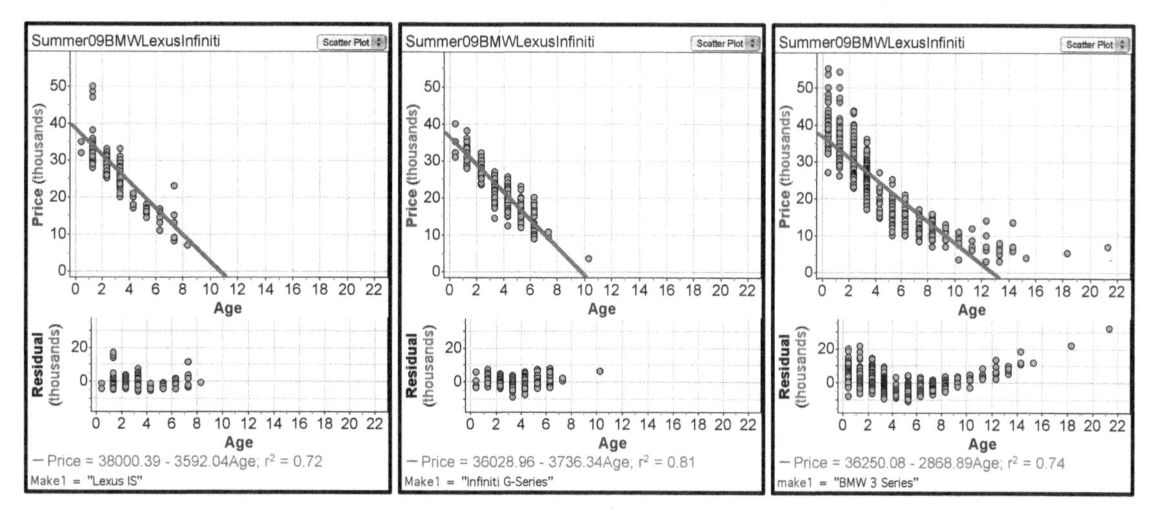

a. The model that we are fitting, $\hat{y} = a + bx$, is a straight-line model – that is, a linear model. But our data may not be linear. You can detect non-linearity by examining the shape of the original plot, or by examining the "residual plots" If the residual plots have fairly well balanced or random scatter on the positive and negative sides of the horizontal line, then that balanced plot is evidence of linearity. On the other hand, if the residual plot shows a non-random pattern, with a large number of the dots on one side of the zero horizontal line, then that is evidence of non-linearity. Do any of the plots for the three cars show evidence of non-linearity? If so, which one? Does the original *Price* and *Age* plot reflect this non-linearity as well? How?

b. There is an important formula for the ***Coefficient of Determination R^2*** but for simple linear models, it is also true that $R^2 = r^2$, that is the square of the correlation coefficient. Confirm that $R^2 = r^2$ for the models of the *Price Age* relationship for the three cars using the correlations shown here.

Correlation between Price and Age for Make1 = BMW 3 Series
-0.85901774
Correlation between Price and Age for Make1 = Infiniti G-series
-0.90257907
Correlation between Price and Age for Make1 = Lexus IS
-0.85135609

c. There is also a standard interpretation of the ***Coefficient of Determination R^2*** (it starts: "The percentage of variability in …") For the BMWs, write the standard interpretation in the context of the data. Context of the data means that something must be said about the prices and the ages of the used cars being sold.

c. For which of the three cars does the linear model fit the best? Give a reason for your answer.

d. Are there any observations that appear to be influential observations? If so, why? If not, why?

2. *Heptathlon for 1988 for the Eight Eastern Bloc Competitors: Calculating R^2 from the Total Sum of Squares and the Sum of Squared Residuals*

This exercise follows the argument of the first part of **§2.4** but for a very small set of data. The data are just the eight competitors in the 1988 heptathlon who were from what was known as the Eastern Bloc. Here (on the right) are the means and standard deviations for the times in the *HundredMeterHurdles* race and the *TwoHundredMeters*. The correlation coefficient between these two variables is given here as well.

Summary statistics:

Column	Mean	Std. dev.	n
HundredMeterHurdles	13.38	0.29736942	8
TwoHundredMeters	23.91125	0.5829834	8

Correlation between HundredMeterHurdles and TwoHundredMeters is: 0.73026622

$R^2 = 1 - \dfrac{\sum_{i=1}^{n}(y_i - \hat{y})^2}{\sum_{i=1}^{n}(y_i - \bar{y})^2}$ is the formula for the coefficient of determination. The formula is made up of two summations. The sum in the numerator is the ***sum of squared residuals*** $\sum_{i=1}^{n}(y_i - \hat{y})^2$ and the one in the denominator is the ***total sum of squares,*** $\sum_{i=1}^{n}(y_i - \bar{y})^2$. We will calculate $R^2 = 1 - \dfrac{\sum_{i=1}^{n}(y_i - \hat{y})^2}{\sum_{i=1}^{n}(y_i - \bar{y})^2}$ for our small data set, after getting $R^2 = r^2$.

a. The easiest way to calculate R^2 is to simply square the correlation coefficient r. Find the R^2 for these data by doing that. At the end of the exercise you will compare the two calculations.

- With the file ***Heptathlon1988EastBloc*** use the ***Movable Line Applet*** to get a scatterplot similar to the one shown above with *TwoHundredMeters* as the response variable (y-axis) and with *HundredMeterHurdles* as the explanatory variable (x-axis).

b. Which sum does the Sum of Squares or SSE = 2.379 represent? Does it represent the ***Sum of Squared Residuals*** $\sum_{i=1}^{n}(y_i - \hat{y})^2$ or the ***Total Sum of Squares*** $\sum_{i=1}^{n}(y_i - \bar{y})^2$? Give a reason for your answer.

Num	Athlete	Team	Hundred MeterHurdles	TwoHundred Meters	y - y-bar	(y - y-bar)2
1	Sabine John	GDR	12.85	23.65	-0.2613	0.0683
2	Anke Behmer	GDR	13.2	23.10		
3	Nataliya Shubenkova	URS	13.61	23.92	0.0087	0.0001
4	Remigija Sablovskaité-N	URS	13.51	23.93	0.0187	0.0003
5	Ines Schulz	GDR	13.75	24.65	0.7387	0.5457
6	Zuzana Lajbnerová	TCH	13.63	24.86		
7	Svetlana Buraga	URS	13.25	23.59	-0.3213	0.1032
8	Svetla Dimitrova	BUL	13.24	23.59	-0.3213	0.1032
						2.3791

c. The calculation for the Sum of Squares = 2.379 is shown at the bottom of the table (on the previous page) for our small data set. This sum of squares is the Total Sum of Squares. In the table, fill in the values for $(y_i - \bar{y})$ and $(y_i - \bar{y})^2$ for Anke Behmer and for Zuzanna Lajbnerová. Find the mean that you need in the Summary Table at the top of the previous page.

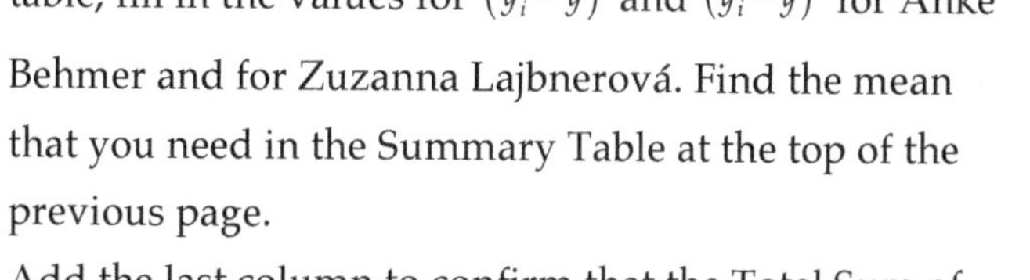

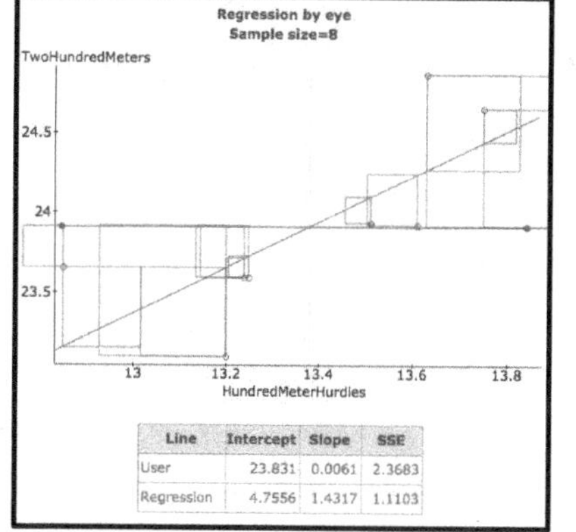

d. Add the last column to confirm that the Total Sum of squares is 2.379.

- Using software, get the least squares line on the plot.

e. The Sum of Squares = 1.110 must represent the ***Sum of Squared Residuals*** $\sum_{i=1}^{n}(y_i - \hat{y})^2$. Print out this graph and indicate the residual for the fastest *TwoHundredMeters* runner (Anke Behmer, the one with the shortest time for that race). Is the residual for this runner positive or negative? Give a reason for your answer based upon where the time is compared to the predicted time.

f. The second table shows the calculation for the Sum of Squares = 1.110, that is, the sum of squared residuals. Fill in the values for $(y_i - \hat{y})$ and $(y_i - \hat{y})^2$ for Anke Behmer and for Zuzanna Lajbnerová. The plot shows that the equation of the least squares regression line is $\hat{y} = 4.76 + 1.432x$; you will need this.

g. Add the last column to confirm that the value for the ***Sum of Squared Residuals*** $\sum_{i=1}^{n}(y_i - \hat{y})^2$ is 1.110.

Num	Athlete	Hundred MeterHurdles	TwoHundred Meters	y-hat	y - y-hat	(y - y-hat)2
1	Sabine John	12.85	23.65	23.1612	0.4888	0.2389
2	Anke Behmer	13.2	23.10			
3	Nataliya Shubenkova	13.61	23.92	24.2495	-0.3295	0.1086
4	Remigija Sablovskaitė	13.51	23.93	24.1063	-0.1763	0.0311
5	Ines Schulz	13.75	24.65	24.4500	0.2000	0.0400
6	Zuzana Lajbnerová	13.63	24.86			
7	Svetlana Buraga	13.25	23.59	23.7340	-0.1440	0.0207
8	Svetla Dimitrova	13.24	23.59	23.7197	-0.1297	0.0168
						1.1110

h. You now have $\sum_{i=1}^{n}(y_i - \bar{y})^2$ and $\sum_{i=1}^{n}(y_i - \hat{y})^2$. Use these to calculate $R^2 = 1 - \frac{\sum_{i=1}^{n}(y_i - \hat{y})^2}{\sum_{i=1}^{n}(y_i - \bar{y})^2}$.

You should get $R^2 = 0.53$. Show your work (there is not much). Does the result agree with part a?

i. Give a good interpretation of the R^2 (that you have calculated) in the context of the data.

j. The Coefficient of Determination $R^2 = 1 - \frac{\sum_{i=1}^{n}(y_i - \hat{y})^2}{\sum_{i=1}^{n}(y_i - \bar{y})^2}$ is known as the "proportion of *explained* variation in the response variable." What would be a good name for just: $\frac{\sum_{i=1}^{n}(y_i - \hat{y})^2}{\sum_{i=1}^{n}(y_i - \bar{y})^2}$, that is the part without subtracting it from one.

3. ***Playing with roller coasters*** This exercise looks at influential observation on the least squares model, as well as interpreting the ***Coefficient of Determination R^2***. We will look at the relationship between the variables *Height* (measured in feet) and *Speed* (measured in miles per hour). The scatterplot with the least squares regression line is shown here with the residual plot.

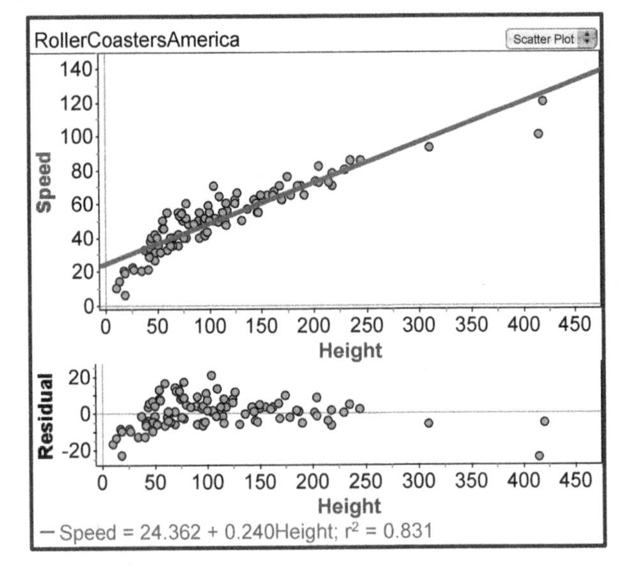

- (***Least Squares Regression*** in a software package can make something similar to these plots, though questions a and b below can be answered with the plots here. Use the file ***RollerCoastersAmericaSample***.)

   a. Judge, from the appearance of the scatterplot and the residual plot that you have, whether there is any evidence of non-linearity (curvilinearity). Give reasons for your answer. If you think there is at least some non-linearity present, indicate the roller coasters that are affected.

   b. Whatever your answer is to part a, explain what the value of $R^2 = r^2 = 0.831$ tells you about the relationship between the variables *Height* and *Speed*.

- Using the ***Movable Line Applet*** with the file ***RollerCoastersAmericaSample***, get the least squares model that results if the two most extreme roller coasters in Height are deleted. Record the equations and the coefficients of determination R^2 that result from deleting the two points. (Consult the relevant software supplement for the exact instruction.)

   c. Compare the equations (both the intercept and the slope) and the R^2 for the least squares regression line with and without the two outliers. Is there some change? Do you think that it is a big change or not a very big change?

- Using the ***Movable Line Applet,*** get the least squares model that results if one adds (or drags) two points to have have extreme *Height* (about 400 ft) but very slow *Speed* (about 20 mph). Record the equations and the coefficients of determination R^2 that result from deleting the two points. (Consult the relevant software supplement for the exact instruction.)

   d. Describe what happens to the slope of the least squares line and the R^2 when with the very high but very slow roller coasters in the data set.

   e. You can now have a lively discussion as to whether we are seeing real influential observations or not. Some of the issues may be whether you would ever actually see very high roller coasters that would have a very slow speed.

- Use the ***Least Squares Regression*** facility with the file ***RollerCoastersAmericaSample***, get the least squares model predicting *Speed* from *Height* for the steel and wooden roller coasters separately (That is, group by the variable *Construction*.)

f. Compare the slopes and also R^2 of the wooden and the steel roller coasters for the model predicting *Speed* from *Height*. Are there differences between the wooden and steel coasters, or are the results basically similar?

g. Do the y-intercepts have a meaning? If so, what meaning? If not, why not?

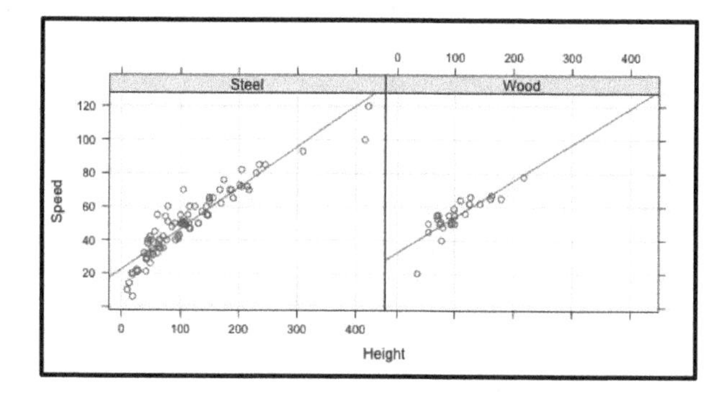

4. ***Australian High School Students' Body Measurements*** This exercise is about the relationship between three body measurements for Australian high school students. Our statistical questions are:

- Can a linear model be used to predict *Armspan* from *Height*?
- If so, is the relationship different for male and female students?
- Can a linear model be used to predict *RightFoot* from *Height*?
- If so, is the relationship different for male and female students?

- ***Arm span and Height*** With the file ***CASAustralia2008B***, use software to get (Consult the software supplement for ***Least Squares Regression***):
  - A scatterplot showing *Armspan* (the response variable) and *Height* (the explanatory variable).
  - A plot of the residuals against the explanatory variable.
  - The ***Least Squares Regression Equation*** predicting *Armspan* from *Height*

a. Identify and interpret the slope for the relationship between arm span and height.

b. Identify and interpret the R^2 (= r^2) for the relationship between arm span and height.

c. The ***R^2 (= r^2)*** is not 1, so the linear model does not fit perfectly. Judging by the plot itself and by the residual plot, do you think that the fit is not perfect:

   (i) because we should be using some kind of (non-linear) curvilinear model, or

   (ii) because there is scatter (variability) in what seems to be a linear relationship?

   Give at least one reason for your answer.

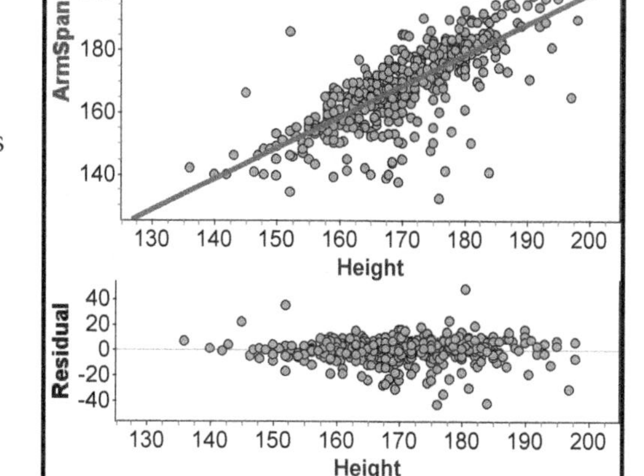

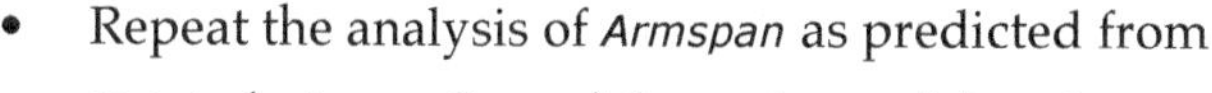

- Repeat the analysis of *Armspan* as predicted from *Height* (using software) for males and females separately. Use ***Least Squares Regression***. [Note: part d is extra credit and is shown at the foot of the exercise.]

e. What differences do you notice in the slopes of the least squares lines for males and females?

f. What differences do you notice in the ***R^2 (= r^2)*** for the least squares lines for males and females? Does the linear model appear to be a good fit? Give a reason for your answer.

- ***Right Foot Length and Height.*** Using software, replace the response variable with *RightFoot* (but keep the explanatory variable as *Height* and explore the prediction of *RightFoot* from *Height* separately for males and females. This should give you the least squares lines for males and females for the relationship between height and the length of right foot. Depending upon the software used, the analysis may resemble the plot on the previous page or resemble just the part shown here for females.

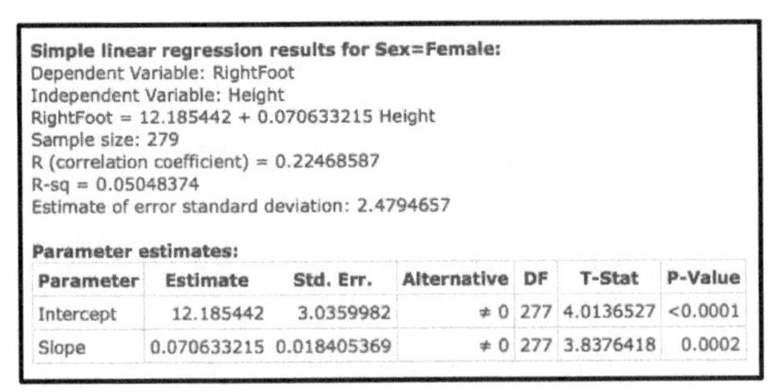

Simple linear regression results for Sex=Female:
Dependent Variable: RightFoot
Independent Variable: Height
RightFoot = 12.185442 + 0.070633215 Height
Sample size: 279
R (correlation coefficient) = 0.22468587
R-sq = 0.05048374
Estimate of error standard deviation: 2.4794657

Parameter estimates:

Parameter	Estimate	Std. Err.	Alternative	DF	T-Stat	P-Value
Intercept	12.185442	3.0359982	≠ 0	277	4.0136527	<0.0001
Slope	0.070633215	0.018405369	≠ 0	277	3.8376418	0.0002

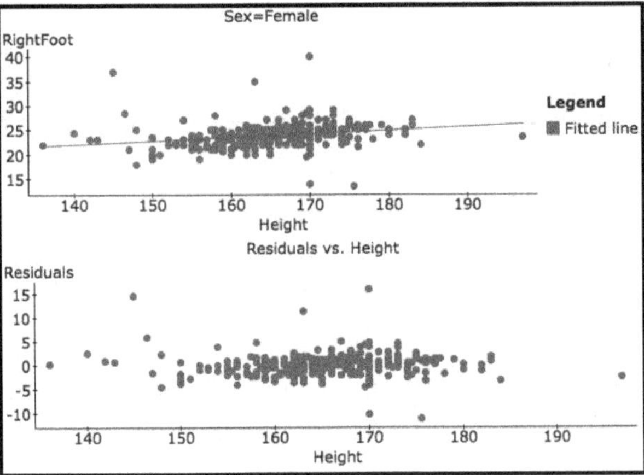

g. Identify and interpret the slopes of the lines for males and for females.

h. Identify and interpret the ***R^2 (= r^2)*** of the lines for males and for females.

i. Is height a good predictor of foot length for either males or females? Give a reason.

[d. Extra Credit: Download an image of Leonardo da Vinci's Vitruvian Man. (A good site to visit may be http://leonardodavinci.stanford.edu/submissions/clabaugh/history/leonardo.html.) Explain in a concise paragraph or two how your linear model predicting arm span from height agrees or disagrees with da Vinci's picture.]

Clean data: Read this, please. The measurements (hence, variables) are *Height, ArmSpan*, and *RightFoot* (the length of the student's right foot), all measured in centimeters. The data have been "cleaned", which means that mistakes and clearly impossible value have been changed to "missing" data; some of the Australian high school students reported being one thousand centimeters tall or said that they had feet one hundred centimeters long. All "measurements" clearly impossible for humans (even in Australia) were cleaned in this way. When working with "raw" data, considerable cleaning may be necessary.

§3.1 Exercises on Trusting Data: Collecting Data

1. Opening of Johnston Road Fraser Suites McDonald's in Hong Kong

The file ***Johnston Road/Fraser Suites McDonalds.pdf*** shows a small crowd at the opening of the McDonald's on Johnston Road at the Fraser Suites Hotel in Hong Kong (the WikipediaCommons reference is: File:HK Wan Chai 莊士敦道 Johnston Road 輝盛閣 Fraser Suites McDonalds grand open 2010-May 泰昌餅家 Tai Cheong.jpg). Here it is reproduced:

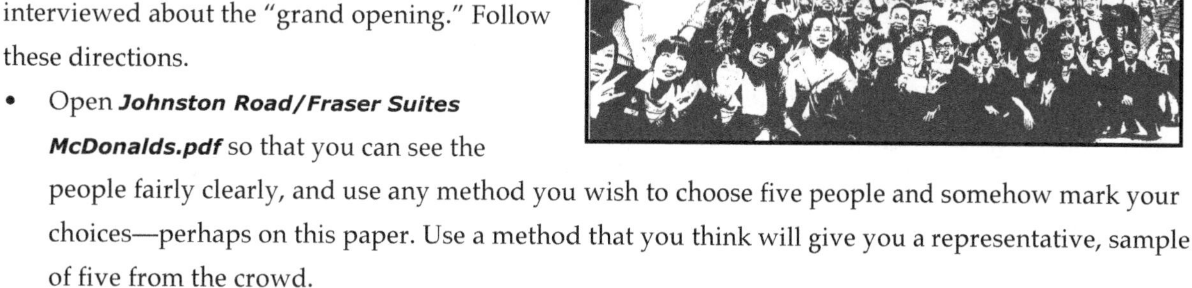

We want to see what happens when we, using personal choice, choose individuals from this crowd. We actually do not know what will happen; it may be that people are able to choose representative samples in some sense. The only way to find out is to collect some data!

Your task is to choose $n = 5$ people from the "crowd" of 52 people. The five people will be interviewed about the "grand opening." Follow these directions.

- Open ***Johnston Road/Fraser Suites McDonalds.pdf*** so that you can see the people fairly clearly, and use any method you wish to choose five people and somehow mark your choices—perhaps on this paper. Use a method that you think will give you a representative, sample of five from the crowd.
- Open the file ***NumberedJohnstonRoadCrowd.pdf*** and record the numbers of the five people that you chose. (The numbers are not shown here on this small black and white picture, but they are on the file that you are to open.)
  a. How did you choose the five people? Did you attempt some kind of randomization? If so, how? Did you attempt some kind of judgment sample? If so, explain what categories you used. Write down what you did.
- Using software that has a ***Random Number Generator*** (it may go by the name of ***Random Integer Generator***) get five numbers chosen randomly between 1 and 52.
  b. Are you able to perceive any differences between your personal choice and the software random sample? For example, did software give you any people standing side by side? Hard-to-see people?
- Record your own personal choice on the file for this purpose on the website or in the classroom.
  c. Record how the attempts of your class to choose a representative sample deviated from random samples chosen by software.
  d. If the choices are truly random, what should the probability of choosing person 14 be?

2. *Movie music on a classical music station.* In the San Francisco Bay Area, KDFC is the classical music station. Classical music stations have a difficult time because there is only a fraction of the population who enjoy classical music, and, at the time, KDFC was a commercial station. As a way to attract more listeners, the station wanted to play more movie music. However, they also did not want to alienate their "traditional" classical music listeners. The program manager of the station, in his blog, invited listeners to comment on whether the station should play more "movie music"—music written for films. Here is how the blog invited comments:

> **Movie Music** in small doses on KDFC was added almost 2 years ago. We've had a wide variety of feedback. In this survey we heard comments like "I would think that movie theme music is not quite for your targeted audiences. I personally detest it" to "I get a real kick out of hearing some of movie soundtrack music you throw in occasionally." Do movie soundtracks have a place on a classical station? Is it the new classical for a new audience, or should it be left out? If it's enjoyable, does it matter where it comes from? *What do you think?*

a. Which of the types of sampling discussed in the ***Notes*** will result from this invitation to comment on movie music? Give a reason for your answer.

b. You had to be a member of "Club KDFC" to actually participate in the survey; however, anyone with an e-mail address can sign up to be a Club KDFC member. (When you are a member of Club KDFC, you get e-mails and can participate in contests, etc.) Which of the following is the best description of the *population* from which the sample of responses is taken?
   - People who go to movies in the Bay Area and listen to KDFC
   - All KDFC listeners
   - Club KDFC members
   - People in the Bay Area who like classical music
   - People who like movie music, whether or not they go to movies

c. Rank the following groups of people from the *least* likely to those *most* likely to respond.
   - Those who hate movie music and think KDFC should not play it at all
   - Those who love movie music and want more of it played
   - Those who think that the present amount of movie music is about right
   - Those who have no strong opinions about movie music on the station

d. Rank the following groups of people from the *least* likely to those *most* likely to respond.
   - Those aged eighty-five years or above
   - Those aged 60 to 84 years old
   - Those aged 45 to 60 years old
   - Those aged 30 to 45 years old
   - Those aged less than 30 years

e. The sample generated is not a random sample. Explain why it is not a random sample based upon what is said about random samples in the ***Notes.***

f. Using the information in part b, devise a method for KDFC to get a random sample of Club KDFC members. (*Note:* KDFC are probably aware of the biased nature of their sampling design and probably do not need a random sample for their purposes, but they can't apply inferential statistics!)

3. *Mobile telephone use among Melbourne drivers*

This exercise will use a study conducted by David McD Taylor and others about the incidence of mobile telephone use among drivers of vehicles in Melbourne, Australia, where there has been a law prohibiting the use of handheld mobile phones since 1999. (The study is in the pdf file: ***Handheld mobile telephone use among Melbourne drivers*** or on line at www.mja.com.au/journal/2007/187/8/handheld-mobile-telephone-use-among-melbourne-drivers.) The exercise will concentrate on the methods used to collect the data.

We performed an observational study of motor vehicle drivers in metropolitan Melbourne, Australia, during October 2006... In brief, we used 12 sites to observe four major metropolitan roads, four central business district (CBD) roads, and four freeway exit ramps. Data were collected on three consecutive Tuesdays at exactly the same sites and by the same techniques as in 2002. Three observation sessions per day at each site (10:00–11:00, 14:00–15:00, 17:00–18:00) provided 36 hours of observation.

a. The first sentence in the description identifies this as an observational study. If you did not have that identification, what is it about the study that indicates that it is *not* an experiment?

b. The researchers chose to observe drivers rather than ask them about whether they had used their mobile phone while driving. List some advantages and disadvantages of this decision. (To answer, just think; there are no technicalities here and nothing in the ***Notes*** that will give you the "answer." Just think about people.)

c. This is one of those situations in which we (the readers or consumers of statistical analysis) must make a decision about how we can generalize the results. What should we regard as the intended population? Choose from the following and give a reason for your choice as to what is the most reasonable choice of the intended population.

    i) Drivers who drive on the roads at the twelve places in Melbourne where the data were collected on Tuesdays at the hours when the data were collected

    ii) Drivers in Melbourne in the CBD (= Central Business District), on major metropolitan roads and at freeway exit ramps on Tuesdays at the hours the data were collected

    iii) Drivers in Melbourne in the CBD, on major metropolitan roads and at freeway exit ramps on weekdays at the hours the data were collected

    iv) Drivers in Melbourne in the CBD, major metropolitan roads, and freeway exit ramps on weekdays

    v) Drivers in Melbourne

    vi) Drivers in urban areas of Australia having a law banning mobile phone usage while driving

    vii) Drivers in urban areas where there is a law banning mobile phone usage while driving

In this study, as with so many other studies with valuable data, the type of data collection does not clearly fall into any of the "textbook" categories, even though the way the authors collected the data makes sense. However, what they did does have connections with some of those categories, both negatively and positively.

d. For the sake of argument, let us say that the intended population is "Drivers in Melbourne." State in the context of the study why the methods used do ***not*** resemble the following types of sampling:

i) Voluntary response sample ii) Simple random sample

e. The researchers' methods resemble a "random cluster sample" even though what they did is not a true random cluster sample. In a true "cluster sample," the clusters would be chosen randomly. It is unlikely that the researchers chose the twelve locations randomly from a larger list; there must have been many potential locations that posed problems for data collection—such as having nowhere for the researchers to safely stand. But in what way does what they actually did resemble a "random cluster sample"?

f. The researchers' methods resemble a "stratified random sample" even though their sample is not a true stratified random sample. In a true stratified random sample, there would be random sampling within the strata. The researchers attempted to observe all cars and did not sample. What were the "strata" that the researchers were using?

g. [*Review*] Here are some of the results of the study for 2006. Because using a mobile phone is not common, the authors calculated the incidence in per thousands, that is "number of mobile phones in use/1000 drivers" rather than in percents. Calculate incidences in "per 1000"s in such a way as to compare the three kinds of locations as to incidence of mobile phone usage. Give an interpretation of your calculations.

Location	Not using mobile phone	Using Mobile Phone	Number of Drivers Screened
Major Metropolitan Road	4721	56	4777
Central Business District	6651	139	6790
Freeway Exit	8504	136	8640
	19876	331	20207

h. [*Review*] For the Melbourne data, here is the comparison of males and females for the "under 30" drivers for just the morning hours data collection.

Gender	Not using mobile phone	Using Mobile Phone	Number of Drivers Screened
Male	819	18	837
Female	430	11	441
	1249	29	1278

Do a calculation to analyze whether there is a difference in the incidence of mobile phone usage between males and females for drivers under thirty for the morning hours at all of the locations.

i. [*Review*] The same methodology was used to collect data on the campus of one community college in Northern California. That is, for about forty-five minutes in a time period that straddled the start and end of classes, drivers were observed at a stop sign coming in and going out of the campus. Here are some results for comparison with the Australian data above for this college campus. These data are also for drivers under 30 years of age. From the table, calculate so as to compare incidence of mobile phone usage by gender. Interpret your calculations in the context of the data.

California College Campus Data			
Gender	Not using mobile phone	Using Mobile Phone	Number of Drivers Screened
Male	1092	39	1131
Female	853	52	905
	1945	91	2036

j. Compare the results for the California college campus with the results for the Melbourne drivers (who are of the same age group). What differences and similarities do you see?

k. The data for the California college campus were collected in 2011. Which would you assess is the most reasonable population from which these data were drawn?
   i) Drivers who drive on the college campus where the data were collected on the days and the hours when the data were collected
   ii) Drivers on college campuses in Northern California during the hours and days that classes are held
   iii) Drivers on college campuses in Northern California
   iv) Drivers in Northern California
   v) Drivers in California

4. *Sampling Students at the University of Michigan*

The University of Michigan is one of the biggest universities in North America, with about forty-two thousand students. On the UM news service on June 21, 2012, the university announced a 2.8% increase in tuition fees and housing costs for the next 2012–2013 academic year. (See: http://www.ns.umich.edu/new/releases/20608-aid-will-cover-total-increase-in-costs-for-students-with-financial-need as accessed on June 22, 2012.) Part of the news report reads as follows:

> ANN ARBOR, Mich.—The University of Michigan Board of Regents today approved a budget that will increase financial aid enough to cover the full rise in the cost of attendance for the typical Michigan resident student with financial need—and reduce the student's loan burden.
>
> This is the fourth consecutive year an increase in financial aid more than offsets the increase in tuition and room-and-board rates for most state resident students with financial need. Additionally, that increase will come in the form of grant aid —which does not need to be repaid—reducing the educational loan portion of those students' financial aid package.

It is evident from the news release that the university authorities know that a fee increase will not be popular with students. If the university authorities had wished to collect data on how the fee increase may affect students and whether students are able to take advantage of financial aid, the university's institutional research office could do it in several ways. For each of plans (A through D) described below, choose the best name for the type of sampling it appears to be from the list below and give a reason for your answer.

i) Voluntary Response Sample
ii) Convenience Sample
iii) Judgment Sample
iv) Simple Random Sample
v) Random Cluster Sample
vi) Stratified Random Sample

a. ***Plan A***: From the list of all students enrolled in the current semester who are not graduating, they draw a random sample. To these students they send a questionnaire (by e-mail or by posting a letter) containing questions about how the fee increase is likely to affect the student.

b. ***Plan B:*** The plan is the same as Plan A, except that a random sample is taken from the current first-year students, another random sample from the current second-year students, and a third from the current third-year students. A fourth sample is drawn from the fourth-year students who are on a five-year degree program.

c. ***Plan C***: From the schedule of classes, a random sample of classes being taught in the current semester is chosen. A short questionnaire on the effects of a fee increase is taken to the classes chosen for the students to fill in.

d. ***Plan D***: The UM News Service have their own pages on several social networking sites. The responses to the news of the fee increase that are recorded on these pages are analyzed for comments about fees, etc.

e. Of the various plans, which is likely to have the lowest non-response rate?

f. Do any of these data collection plans constitute an experiment? If so, which one or ones, and why? If not (for the ones that are not), state why they are not experiments.

5. ***Negative Political Advertisements*** Read about the negative political ad experiment in the ***Notes***. In the experiment, three different negative ads were used: A, B, and C. (You may be able to read the paper at *http://www.jstor.org/stable/4148088*.) The $n = 700$ subjects that received negative mailings were assigned (randomly) to seven groups, each having $n = 100$ cases. Three groups received just one of the negative mailings (A, B, or C) a week before the election. Three groups received two of the three mailings starting two weeks before the election, and one group received all three negative mailings, starting three weeks before the election. Using this information, diagram the experiment as it was actually done, showing the treatment groups and the response variable by making a diagram similar to the ones in the Notes.

6. ***Sampling Houses from San Mateo County***

This exercise will use data on all houses sold in San Mateo County in 2005–2006. We will regard as collection as a population and from this population draw various kinds of samples. We will also be able to compare what we get from a sample to the population. Researchers almost never have this privilege! Usually we know *nothing* about a population and have to make (albeit educated and well founded) guesses from samples. Be thankful!

- Open our population file ***SanMateoRealEstateY0506*** and make a contingency table like the one shown below showing the numbers of "Recent Houses" in the four regions of San Mateo County.

a. Houses that were built in the ten years before 2006 (Age ≤ 10 years) are considered to be "Recently Built Houses." All other houses are older houses. Calculate proportions to determine which regions have the highest and lowest proportions of "Recently Built Houses" and state your conclusions.

San Mateo Real Estate Y0506 0

		RecentHouses		Row Summary
		Older Houses	Recently Built Houses	
Region	Central	1663	95	1758
	Coast	417	76	493
	North	1247	88	1335
	South	1674	226	1900
Column Summary		5001	485	5486

S1 = count ( )

b. The table (here) and the analysis done in part "a" is done on the population as a whole. Select the best name for the data analysis from:

i) Simple Random Sample; ii) Census;i

ii) Stratified Random Sample; iv) Convenience Sample.

Getting a Simple Random Sample. We will get software to draw a sample of $n = 275$.

- Using the file ***SanMateoRealEstateY0506*** use software to get one sample of size $n = 275$.
- From the sample of $n = 275$ get a contingency table showing the relationship between *Region* and *RecentHouses*. (The table should be similar to but have different numbers from the example shown here.)

Sample of SanMateoRealEstateY0506		RecentHouses		Row Summary
		Older Houses	Recently Built Houses	
Region	Central	83	4	87
	Coast	25	7	32
	North	68	4	72
	South	71	13	84
Column Summary		247	28	275
S1 = count ( )				

c. Calculate proportions to determine which regions have the highest and lowest proportions of "Recently Built Houses" for your sample and state your conclusions. Are the proportions similar to or very different from the population proportions?

Getting a Random Cluster Sample The idea of a random cluster sample is to randomly select "groups" (or "clusters") of cases and then take *all* of the cases in the clusters as the sample. Here is a numbered list of the "cities" in the population of houses sold in San Mateo County. The table also shows the region the city belongs to, and the number of houses sold in the population of houses sold in 2005-2006.

ID	City	Region	Number of houses sold
1	Atherton	South	84
2	Belmont	Central	260
3	Brisbane	North	25
4	Burlingame	Central	238
5	Colma	North	6
6	Daly City	North	483
7	El Granada	Coast	65
8	East Palo Alto	South	171
9	Foster City	Central	167
10	Hillsborough	Central	117
11	Half Moon Bay	Coast	118
12	Milbrae	North	159
13	Montara	Coast	29
14	Moss Beach	Coast	13
15	Menlo park	South	382
16	Palo Alto	South	492
17	Pacifica	Coast	268
18	Redwood City	South	590
19	Redwood Shores	South	112
20	San Bruno	North	311
21	San Carlos	Central	301
22	San Mateo	Central	675
23	South San Francisco	North	351
24	Woodside	South	69

- Using software that has a ***Random Number Generator*** (it may go by the name of ***Random Integer Generator***) get three numbers chosen randomly between 1 and 24 and thus choose three of the twenty four cities.

d. A simple random sample (***SRS***) has two requirements: one is numbered list of the elements we are choosing, and the second is a random method of choosing. State how the instructions for the choice of the three cities from the twenty-four meets these requirements.

- With the file ***SanMateoRealEstateY0506*** use software choose just the houses from the three cities chosen by the random process.
- From the sample of the houses in the three places chosen get a contingency table showing the relationship between *Region* and *RecentHouses*. (The table should be similar to but have different numbers from the example shown here, where the cities Montara [ID 13], Hillsborough [ID 10] and Redwood City [ID 18]).

SanMateoRealEstateY0506		RecentHouses		Row Summary
		Older Houses	Recently Built Houses	
Region	Central	109	8	117
	Coast	23	6	29
	South	554	36	590
Column Summary		686	50	736
S1 = count ( )				

e. Calculate the proportions of "Recently Built Houses" by Region for your sample and compare your results to the population proportions and the proportions from the ***SRS***.

f. [Extra credit]. Compare the simple random sample of n = 275 with the sample that was chosen by random cluster sampling. The random cluster sampling method has some advantages and some definite drawbacks. Discuss what these are.

Getting a Stratified Random Sample The idea of a stratified random sample is to randomly sample *within* groups (the "strata"), usually with the idea of improving how well the sample represents the population. For our strata (our groups) we will use the four regions as our groups and sample with sample sizes as shown in the small table here: These sample sizes represent a 5% sample of the population. That is in the population of houses sold in the South Region, there were 1900, and 5% of that is 95.

Region	Central	88
	Coast	25
	North	67
	South	95
Column Summary		275

- Using the file ***SanMateoRealEstateY0506*** use software to get a stratified random sample of size $n = 275$ by taking random samples within the four regions using the sample sizes shown here..
- From the stratified sample of $n = 275$ get a contingency table showing the relationship between *Region* and *RecentHouses*. (The table should be similar to but have different numbers from the example shown here.)

Sample of SanMateoRealEstateY0506

		RecentHouses		Row Summary
		Older Houses	Recently Built Houses	
Region	Central	83	4	87
	Coast	25	7	32
	North	68	4	72
	South	71	13	84
Column Summary		247	28	275

S1 = count ( )

g. Calculate proportions to determine which regions have the highest and lowest proportions of "Recently Built Houses" for your sample and state your conclusions.

h. Which of the types of random sampling (simple, cluster, or stratified) gave, as far as you can assess, the results closest to the population proportions?

7. A weight-loss study in Boston

This exercise is about a study carried out by researchers in Boston of weight-loss programs. The report can be found at http://jama.jamanetwork.com/article.aspx?articleid=200094, or:

Michael L. Dansinger, MD; Joi Augustin Gleason, MS, RD; John L. Griffith, PhD; Harry P. Selker, MD, MSPH; Ernst J. Schaefer, MD, *Comparison of the Atkins, Ornish, Weight Watchers, and Zone Diets for Weight Loss and Heart Disease Risk Reduction: A Randomized Trial,* Journal of the American Medical Association, 293(1), 2005.

Read the following, which is from the abstract of the paper, and answer the questions based upon this abstract. A few of the terms are explained below the abstract.

Objective *To assess adherence rates and the effectiveness of 4 popular diets (Atkins, Zone, Weight Watchers, and Ornish) for weight loss and cardiac risk factor reduction.*

Design, Setting, and Participants *A single-center randomized trial at an academic medical center in Boston, Mass., of overweight or obese (body mass index: mean, 35; range, 27-42) adults aged 22 to 72 years with known hypertension, dyslipidemia, or fasting hyperglycemia. Participants were enrolled starting July 18, 2000, and randomized to 4 popular diet groups until January 24, 2002.*

Intervention *A total of 160 participants were randomly assigned to either Atkins (carbohydrate restriction, n=40), Zone (macronutrient balance, n=40), Weight Watchers (calorie restriction, n=40), or Ornish (fat restriction, n=40) diet groups. After 2 months of maximum effort, participants selected their own levels of dietary adherence.*

Main Outcome Measures *One-year changes in baseline weight and cardiac risk factors, and self-selected dietary adherence rates per self-report.*

- "Single-center" in the context of medical studies simply means that the study was done at one place (in this instance, in a medical center in Boston) rather than at several different locations ("centers").
- "Body Mass Index" is a common measure used to assess whether people are overweight or obese. It takes account of the fact that taller people are naturally heavier. The formula is $BMI = \frac{mass}{(height)^2}$, where mass is measured in kilograms, and height is measured in meters. Adults having BMI in the range 25–29.9 are considered overweight, and those having BMI ≥ 30 are considered obese.
- The researchers were also interested in heart disease; that is why they chose adults having hypertension, dyslipidemia, or fasting hyperglycemia. The questions below do not depend on understanding the exact definitions of these terms.
- "Baseline weight" means weight at the beginning of the study. The researchers are interested in the *changes* in weight and cardiac (heart disease) risk factors from what they were at the beginning of the study.

a. This is an experiment and not an observational study. What is it about the study that makes it an experiment and not an observational study? Answer in the context of the study.

b. What are the experimental units? (One or more answers may be correct.)
   i) Diets (Atkins, Zone, Weight Watchers, Ornish)
   ii) Overweight or obese adults with a history of heart disease risk
   iii) Changes in weight and cardiac risk factors
   iv) Dietary adherence rates

c. What is the *factor*? (One or more answers may be correct.)
   i) Diets (Atkins, Zone, Weight Watchers, Ornish)
   ii) Overweight or obese adults with a history of heart disease risk
   iii) Changes in weight and cardiac risk factors
   iv) Dietary adherence rates

d. For this experiment, what are the treatment groups? Is there a "control" group?

e. Was random assignment to the treatment groups used?

f. What is (or are) the *response variable(s)*? (One or more answers may be correct.)
   i) Diets (Atkins, Zone, Weight Watchers, Ornish)
   ii) Overweight or obese adults with a history of heart disease risk
   iii) Changes in weight and cardiac risk factors
   iv) Dietary adherence rates

The table on the next page shows the baseline characteristics of the study participants. Baseline refers to characteristics at the beginning of the study. The next two questions are about this table. Here is some help in reading the table.

- Where the variable is quantitative, the mean is given with the standard deviation in parentheses.
- Where the variable is categorical, the "count" is given, and the percentage in that diet is in parentheses.
- For "Exercise," the measure is the number of people with more than "mild" exercise.
- "*P* value" will be introduced in §4.2; for now, know that p value ranges from 0 to 1, and that bigger values are an indication of "no difference" between the groups. A p value > 0.20 is big.

Table 1. Baseline Characteristics of Study Participants

Characteristics	Atkins Diet (n = 40)	Zone Diet (n = 40)	Weight Watchers Diet (n = 40)	Ornish Diet (n = 40)	All Diets (N = 160)	P Value
Demographics						
Age, mean (SD), y	47 (12)	51 (9)	49 (10)	49 (12)	49 (11)	.41
Women, No. (%)	21 (53)	20 (50)	23 (58)	17 (43)	81 (51)	.61
White race, No. (%)	32 (80)	26 (65)	30 (75)	32 (80)	120 (75)	.37
Risk factors, No. (%)						
Smoker*	3 (8)	5 (13)	1 (3)	4 (10)	13 (8)	.41
Hyperglycemia†	16 (40)	8 (20)	8 (20)	12 (30)	44 (28)	.14
Exercise‡	8 (20)	14 (35)	12 (30)	5 (13)	39 (24)	.09
Weight factors, mean (SD)						
BMI	35 (3.5)	34 (4.5)	35 (3.8)	35 (3.9)	35 (3.9)	.60
Body weight, kg	100 (14)	99 (18)	97 (14)	103 (15)	100 (15)	.43
Waist size, cm	109 (11)	108 (13)	108 (11)	111 (13)	109 (12)	.63

g. Look at the means and standard deviations for BMI, Body weight, and Waist size. Are these characteristics similar or markedly different in the four treatment groups?

h. In the context of an experiment, is it desirable that the baseline characteristics be almost the same for the treatment groups? Or should they be different? Or does it not matter? Give a reason for your answer.

i. Here is some of the analysis of one of the response variables, the change in weight. (Ignore the daggers for now.) Summarize the figures in the table in the context of the study. What do they tell you about the four diets?

Table 3. Changes in Weight and Cardiac Risk Factors in an Analysis in Which Baseline Values Were Carried Forward in the Case of Missing Data*

	Diet Group, Mean Change (SD)				
Variable	Atkins (n = 40)	Zone (n = 40)	Weight Watchers (n = 40)	Ornish (n = 40)	P Value for Trend Across Diets
Weight, kg					
2 mo	−3.6 (3.3)†	−3.8 (3.6)†	−3.5 (3.8)†	−3.6 (3.4)†	.89
6 mo	−3.2 (4.9)†	−3.4 (5.7)†	−3.5 (5.6)†	−3.6 (6.7)†	.76
12 mo	−2.1 (4.8)†	−3.2 (6.0)†	−3.0 (4.9)†	−3.3 (7.3)†	.40

j. As explained above, a "big" *p* value indicates no difference between the groups. What do these *p* values suggest to you about the four diets and losing weight?

8. *A weight-loss study in Israel.* Read the description of the weight-loss experiment whose report was in *The New England Journal of Medicine* (**vol 359:** pp 229 – 241) entitled "Weight Loss with a Low-Carbohydrate, Mediterranean, or Low-Fat Diet" by I. Shai, D. Schwarzfuchs *et al*. The report was based upon an experimental comparison of the three types of diets. There were a total of 322 subjects in all, and these were randomly allocated so that 104 followed the low-fat diet, 109 followed the Mediterranean diet, and 109 followed the low-carbohydrate diet. Many response variables were measured, but one of these was change in weight (or weight change) from the baseline weight (the weight at the beginning of the experiment).

a. For the weight-loss experiment by Shai, Schwarzfuchs, *et al,* what are the treatments?

b. What is a good name for the *factor* for this experiment?

c. Diagram the experiment using a diagram as shown in the ***Notes***.

The next bulleted exercise gives you some experience in seeing what happens when subjects (experimental units) are randomly allocated to treatment groups. It is possible that the software that you are using will be able to scramble the cases to the treatment groups. If so, use that ***Scramble Applet*** and skip the allocation by hand.

- Open either the small file ***TwelveObeseFemales.***or the file ***TwelveObeseMales.*** You will randomly allocate these twelve to the three diets. (They are actual people from the NHANES data set, although the names in the file have been invented.)
- It is possible that your software will be able to randomly scramble the treatments. If not, get a ten-sided die. For each person in the data set, you will roll the die once:
  - If the die comes up 0, roll again;
  - If the die comes up 1, 2, or 3, that person goes to the low-fat treatment;
  - If the die comes up 4, 5, or 6, that person goes to the Mediterranean treatment;
  - If the die comes up 7, 8, or 9, that person goes to the low-carbohydrate treatment.
- Do this for every person in the data set so as to assign each person to a group, but... (Read on.)
- Once you have four people in a diet group, do not add more. For example, if you get to the sixth person and you have four people in the LF group then keep rolling to allocate the remaining people in other groups.
- In the file assign LF, Med, or LC for each person for the variable *Treatment* according to your randomization.
- Use software to get the means and standard deviations of *Weight* for the three treatment groups. (Your summary statistics output should resemble this, but of course will have different numbers.)

Summary statistics for Weight:
Group by: Treatment

Treatment	Mean	Std. dev.	n
LC	101.25	9.3214806	4
LF	103.45	13.085743	4
Med	124.925	26.211241	4

d. In the Weight-Loss Study, the baseline mean weights, and standard deviations for the three groups were:

Low-Fat:	$\bar{x}$ *: 91.3 kg*	*s: 12.3 kg.*
Mediterranean:	$\bar{x}$ *: 91.1 kg*	*s: 13.6 kg.*
Low-Carbohydrate:	$\bar{x}$ *: 91.8 kg*	*s: 14.3 kg.*

Explain why for an experiment it is desirable that the baseline weights be almost equal and not desirable if the baseline weights are different.

e. We can allocate our twelve people, but we cannot make them diet. Here is a graphic from the results of the Shai, Schwarzfuchs, *et al* study. What can you say from this graphic about the three diets?

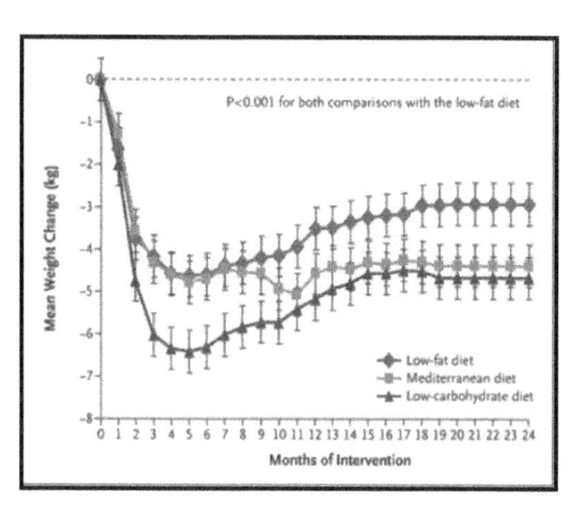

f. The study lasted twenty-four months (two years); what does the curved nature of the graphs for the three diets tell you about the dieting process?

9. *Australian Secondary Schools.* As part of the teaching curriculum, the Australian Bureau of Statistics (ABS) runs a data collection program in many schools called Census @ School (see http://www.abs.gov.au/censusatschool). Here, briefly, are the procedures for data collection:

- Teachers who wish to participate sign up for an account.
- Teachers are given Student Account Numbers (SANs).
- Each student uses his or her SAN to fill in the questionnaire.

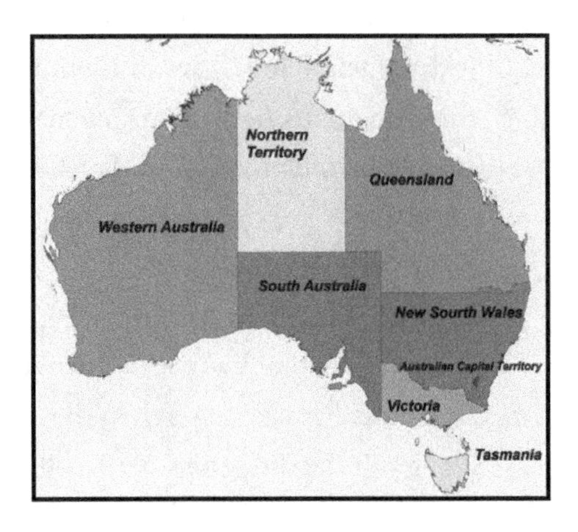

[Image:commons.wikimedia.org/wiki/File:W 1

The sampling plan as described above does not fit clearly in any one of our neat "textbook" categories, but it does resemble some of the categories. Consider how by answering the questions below.

a. Can this plan be a *simple random sample*? Look at the requirements for a SRS and answer.

b. Does this plan have features that make it like a *random cluster sample*? In what way does the plan resemble a random cluster sample and in what way is the plan *not* a random cluster sample?

c. Does this plan have features that make it like a *voluntary response sample*? In what way does the plan resemble a voluntary response sample and in what way may the plan *not* be a voluntary response sample? (It may be that we need more information. For example: it is likely that completing the questionnaire online is something like a required homework assignment. How does this affect the answer?)

d. Does this plan have features that make it like a *stratified random sample*? In what way does the plan resemble a stratified random sample and in what way is the plan *not* a stratified random sample?

When we have data that are not randomly drawn then we may want to assess if the data represents the population. Here is a bar graph provided by the ABS that compares the proportion of questionnaires by state and territory to the national population distribution. This kind of comparison does not guarantee that the data are representative of the population, but they give us a start. ***PTO***

e. Are there parts of Australia that are seriously over- or underrepresented? Give a reason based upon the comparisons given in the bar graph shown.

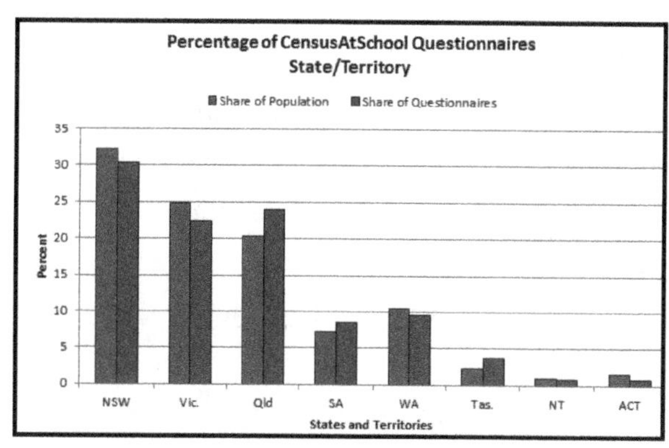

f. [*Review*] Here are data from the Australian C@S for high school students. Do some calculations to compare the proportion of students who "Never or Rarely" use the internet for social networking by region of Australia

casAus2011Combined

		Internet_for_Social_Networking			Row Summary
		(a) Never or Rarely	(b) Sometimes	(c) Often	
State2	NSW and ACT	42	38	143	223
	QLD	47	31	122	200
	Vic and Tas	29	29	97	155
	WA, SA and NT	46	30	128	204
	Column Summary	164	128	490	782

S1 = count( )

g. [*Review*] Here are some data for right foot lengths (in "cm") for male high school students in "Grade 10" or above. For the histogram shown here, what is the *bin width*?

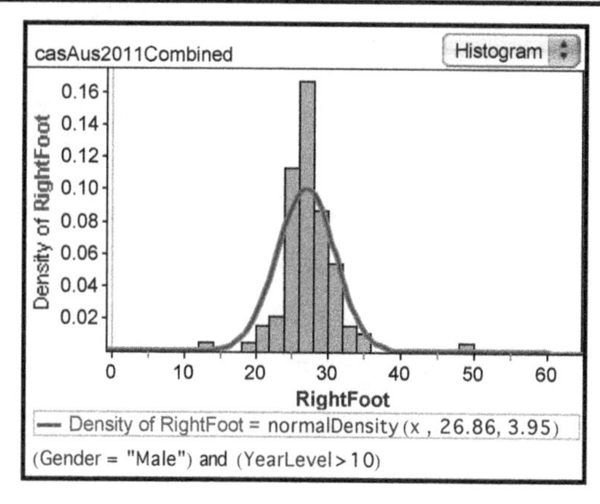

h. [*Review*] Use the histogram to estimate $P(RightFoot \geq 30)$. and then use the Normal distribution with $\mu = 26.86$ and $\sigma = 3.95$ to get $P(RightFoot \geq 30)$. (You may check the Normal Distribution answers using software that has a ***DistributionCalculator***.)

Special Exercise A on Sampling: How Big Were Ontario Farms in 1878?

Introduction: Here is a map from an atlas of one township in the province of Ontario in Canada. The map is from 1878, and it shows each farm, the name of the owner of the farm, and the size of the farm in acres. The map comes from the collection of maps on a site hosted by McGill University in Montreal (http://digital.library.mcgill.ca/countyatlas/). Our statistical question:

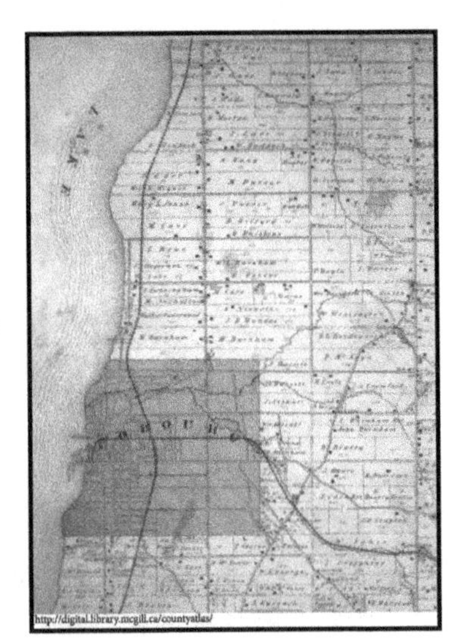

What was the mean size of farms in a Township in Ontario in 1878?

We could go through the map and record the size of each farm then get the mean of these numbers; in other words, we could look at *all* the data in our population.

1. a. What are the cases for these data?
   b. What is a name given to analyzing *all* of the data in a population rather than analyzing a sample of the population?

   Collecting the data from all the farms in the township can probably be done, but doing that is tedious. Instead, we sample.

   – For practice with sampling, we have made a "pretend" version of the township (our population) that has exactly one hundred farms. Look at: ***One Hundred Farms in a "Township in a County in Ontario"*** now.

   – On the map ***One Hundred Farms in a "Township in a County in Ontario,"*** each farm is outlined with solid lines, and within each farm, the dashed lined squares are each ten acres, and each farm is numbered. So, for example, Farm number 1 (in Concession I) has 14 full squares, and so its size is $14 \times 10 = 140$ acres. Farm number 23 (also in Concession I) has two triangles and seven full squares, so $2 \times 5 + 7 \times 10 = 10 + 70 = 80$ acres.

I. Getting a Non-Random "Judgment" Sample

- After studying the map somewhat, use your *judgment* (not the ***dice!***) to pick $n = 5$ farms from the entire township that you think represent the sizes of farms in the township. Record what you have chosen.
- Calculate the mean area of the rectangles (farms) in your sample and record it.

2. Let us think about what we are doing:
   a. For the sample that you have just taken, what is the sample size?
   b. Describe the population from which you chose the sample. Note: this is *not* asking for the size of the population but rather for a description of what the population is.
   c. Describe briefly how you got your sample and what you did (if anything) to try to make it representative. There is no "right" answer to this question; just describe what you did.

II. Getting A Simple Random Sample (SRS)

Read the box in the ***Notes*** about how to get a ***simple random sample*** and then answer question 3.

3. The first requirement for a ***simple random sample*** is to have a numbered list of the population. How does our map of the ***One Hundred Farms in a "Township in a County in Ontario"*** meet this requirement?

   The second requirement for a ***simple random sample*** is to have a chance or random process to choose the units or elements of the sample. For this exercise, our random process uses *ten-sided* dice. To choose five farms randomly, you will roll the ten-sided dice to get numbers that have two digits, corresponding to the numbers of the farms on the map of the farms. Read the bulleted instructions.

- Roll the ten-sided die twice and note the outcome: this will give you the number of the first element in your sample. If 0 and then 9 appears, chose farm 9; if 34 (3 then 4) comes up, choose farm 34, etc. If you get 00, choose farm 100. Roll the dice again and again so as to get $n = 5$ farms randomly and record the farm number and the area of the farm. ***Note:*** If you get the same farm twice, record that farm twice.
- Calculate the mean area of the farms in your sample, and record it in the box with the thick border on your worksheet, if you have a worksheet for the ***First Simple Random Sample (First SRS).***
- Repeat the entire process and record your results in the worksheet for a ***Second Simple Random Sample.*** You may get some of the same farms as you got for the first SRS or you may not; that is randomness!

II. Getting a Random Cluster Sample

The ***Notes*** tell you that in a random cluster sample, we randomly select "groups" (that is, "clusters") and then take *all* the cases in those clusters. We are choosing a very small sample, and so we will randomly choose just *one* cluster from the list on the back of the map and take *all* five farms in the cluster we choose.

- Roll the die twice and choose one farm. The cluster that you choose will be the one that has the farm you choose. Consult the list of clusters on the back of the map to see the cluster you chose. (For example, if you randomly choose farm 37, then your cluster is "H.")
- Record the numbers of the farms in your cluster and calculate the mean farm size for your sample.

III.Getting a Stratified Random Sample

For a ***stratified random sample,*** we first choose groups (or strata) and then randomly sample within the strata.

- *Read carefully:* For our exercise on ***stratified random sampling*** for the Ontario farms we will create just *two* ***strata*** or groups. ***Stratum X***: the sixty farms in Concessions I and II—that is, those farms in the *southern* part of the township, called the ***Southern Farms. Stratum Y***: the forty farms in Concessions III, IV, and V called the ***Northern Farms.*** (Have you noticed that the farms on the northern side of the township are generally larger?)
- To get the ***stratified random sample,*** roll the dice again, but this time do the rolling in two stages. First, roll the dice and *ignore* any farms whose numbers are 61 or greater; only choose farms from the numbers 01 to 60 until you get *three* farms among the ***Southern Farms.***

- Then, after you have the three for the **Southern Farms,** get the *two* for the **Northern Farms** by rolling the dice, but now only paying attention to the numbers 61 through 00 (= farm 100).
- Record the numbers of the farms in your cluster and calculate the mean farm size for your sample.

Thinking about our samples The next two questions are best answered after a discussion.

4. a. If it is 1878 in Ontario (no phones, no cars, only horses), what are the advantages of selecting a *random cluster sample* to collect the data on farm sizes if you have to go from farm to farm (in winter)?

   b. Explain why it is (in the context of Ontario farms) that a *random cluster sample* could result in a very unrepresentative sample.

5. Think of the numbers in ***Stratum X*** and ***Stratum Y*** in the population (60 and 40 respectively) and the numbers we have in the sample of $n = 5$ for the two strata. Explain why it makes sense that we have three from the **Southern Farms** in the sample and two from the **Northern Farms** in the sample.

(• You may be asked to record your means for the various kinds of sampling so that there is collection of the means from many different samples. The mean farm size for the population is 138.65 acres. These questions are to be asked for the collection of the samples collected.)

One Hundred Farms in a "Township in Ontario"

North ---->

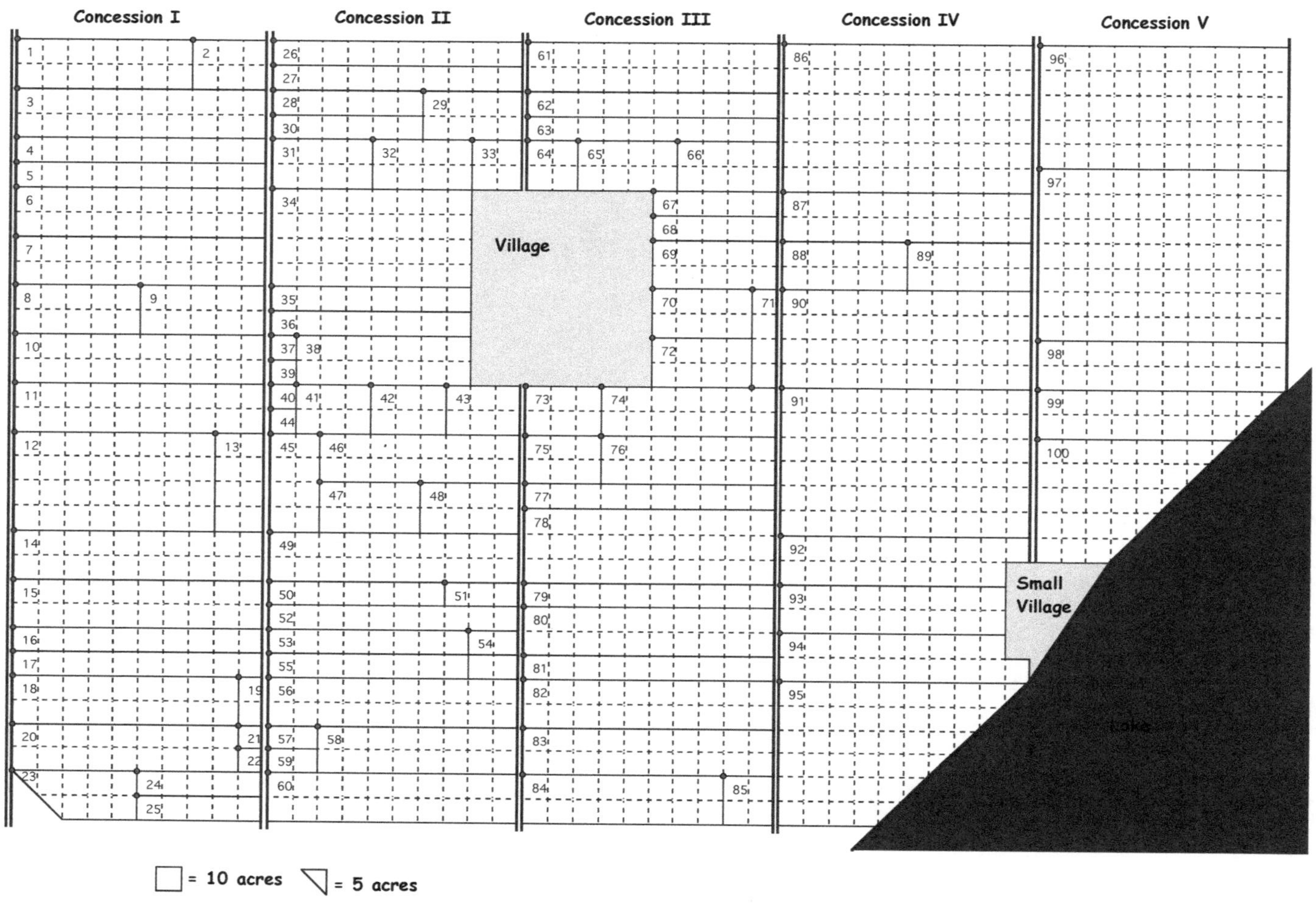

Concession I

1 2 3 4 5 6 7 8 9 10 11 12 13 14 15 16 17 18 19 20 21 22 23 24 25

Concession II

26 27 28 29 30 31 32 33 34 35 36 37 38 39 40 41 42 43 44 45 46 47 48 49 50 51 52 53 54 55 56 57 58 59 60

Concession III

61 62 63 64 65 66 67 68 69 70 71 72 73 74 75 76 77 78 79 80 81 82 83 84 85

Village

Concession IV

86 87 88 89 90 91 92 93 94 95

Concession V

96 97 98 99 100

Small Village

Lake

□ = 10 acres ◸ = 5 acres

Clusters of farms for the Ontario Farms Exercise

Cluster	Farms
A	1 – 5
B	6 – 10
C	11 – 15
D	16 – 20
E	21 – 25
F	26 – 30
G	30 – 35
H	36 – 40
I	41 – 45
J	46 – 50
K	51 – 55
L	56 – 60
M	61 – 65
N	66 – 70
O	71 – 75
P	76 – 80
Q	81 – 85
R	86 – 90
S	91 – 95
T	96 – 100

§3.2 Exercises on Sampling Distributions

1. ***What happens with a bigger sample size?*** This exercise repeats the "interactive" part of the ***Notes*** but also reviews the conclusions about the sampling distribution of the sample mean $\bar{x}$. However, in this exercise, we use a sample size of n = 120 instead of a sample size of n = 40, but with the same population of houses sold in the South Region of San Mateo County. Hence the population mean for *Sq_Ft* is μ = 1949.56 ft^2, and the population standard deviation is σ=1150.01 ft^2.

   a. Read the Summary Boxes ***"Facts about Sampling Distributions..."*** and ***"Central Limit Theorem"*** and consider the sampling distribution of sample means $\bar{x}$ for the variable *Sq_Ft* for samples of size n = 120 from the San Mateo County South real estate population. From the information in the boxes, state what value the mean $\mu_{\bar{x}}$ should have, what value the standard deviation $\sigma_{\bar{x}}$ should have, and what shape the sampling distribution should have. Use the correct notation for your answers. Notice that your overall answer must have three parts.

   b. Confused Conrad gives $\sigma_{\bar{x}} = 1150.01$ as part of his answer to the question in part a. What is Confused Conrad's mistake?

   c. Compare the value of the standard deviation $\sigma_{\bar{x}}$ that you have calculated (from the formula given in the boxes in the ***Notes***) to the value we calculated for samples of size n = 40, which was $\sigma_{\bar{x}} = \dfrac{\sigma}{\sqrt{n}} = \dfrac{1150.01}{\sqrt{40}} \approx 181.83$. Is your value for samples of size n = 120 smaller or bigger? Explain from the formula why your answer differs from 181.83.

- With the file ***SanMateoRealEstateSouth*** and using the facilities of software for simulating sampling distributions of sample means (See ***Simulating Sampling Distributions of Sample Means*** in the software supplement). What comes should be similar to but not exactly as shown here.

   d. Look at the dot plot that you have created. What does each dot in that plot represent?

   e. Confused Conrad answers question d with the statement: "Each dot represents the *Sq_Ft* for a single house." CC is confused and his answer is wrong. Why?

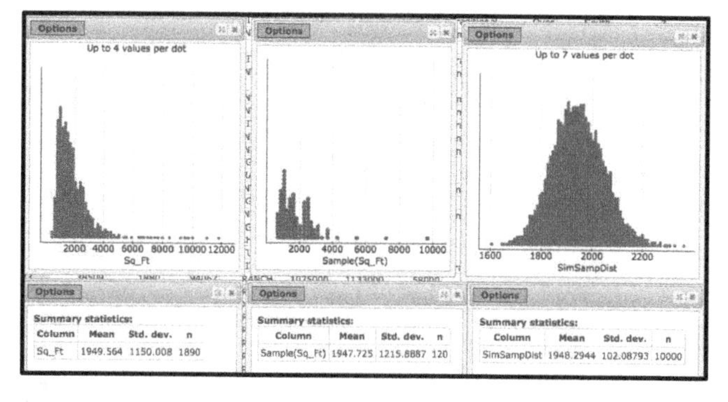

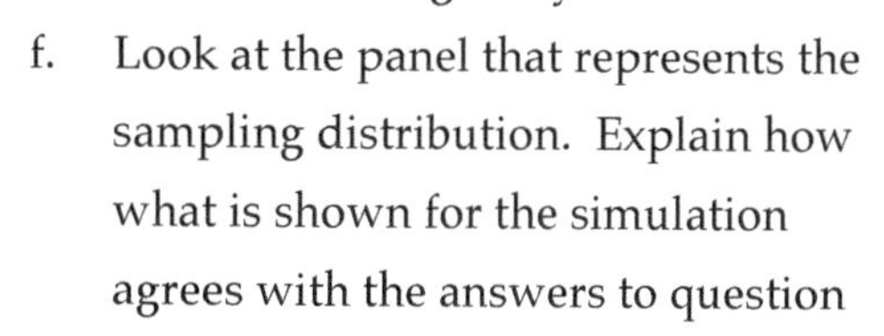

   f. Look at the panel that represents the sampling distribution. Explain how what is shown for the simulation agrees with the answers to question a, even though the numbers are slightly different from the "theoretical" answers.

   g. Judging from comparing what you have done here and the information in the boxes ***("Facts about Sampling Distributions...")***, what changes in the sampling distribution when you have a bigger sample size and how does it change? Does the shape change? Does the center of the sampling distribution change? Does the spread or the variability change? Be as specific as you can.

2. ***What happens with a bigger sample size?*** This exercise is linked with Exercise 1. It also uses the sampling distribution of the xbars of the variable *Sq_Ft* for samples of n = 120 drawn from the population of houses sold in the South Region of San Mateo County. Since we have the same population, the population mean for the variable *Sq_Ft* is μ = 1949.56 ft², and the population standard deviation is $\sigma = 1150.01$ ft².

   a. In the ***Notes*** there was a calculation of the interval of Reasonably Likely values for *xbars* of the variable *Sq_Ft* for samples of n = 40. Rounded to the nearest square foot, we found the interval to be $1593 < \bar{x} < 2306$. That is $P(1593 < \bar{x} < 2306) \approx 0.95$. Repeat this calculation for *xbars* of the variable *Sq_Ft* for samples of n = 120. You will get a different interval.

   b. Is the interval that you calculated in answering part a narrower or wider than the interval found for samples of n = 40?

   c. What part of the formula $\mu - 1.96\frac{\sigma}{\sqrt{n}} < \bar{x} < \mu + 1.96\frac{\sigma}{\sqrt{n}}$ made your interval (which should have been $1744 < \bar{x} < 2155$) narrower than the interval calculated for samples of n = 40? Explain so that you are satisfied that you understand it how the formula made your interval narrower.

   d. In the ***Notes*** we worked with an *xbar* equal to 2,185 ft². For samples of n = 120 (rather than for samples of n = 40), will an $\bar{x}$ = 2185 ft² be a reasonably likely or a rare event (the event is "getting an xbar" in a single random sample of size n = 120)? Give a reason for your answer.

   e. We also calculated the probability of getting a sample mean xbar greater to 2,185 ft² by using the fact that the sampling distribution of the xbars is very nearly a Normal distribution. To calculate $P(\bar{x} > 2185)$ for a sample of n = 120, will we proceed as a "given a value—find a proportion" problem or "given a proportion—find a value" problem?

   f. Make a sketch of the problem and calculate $P(\bar{x} > 2185)$ for a sample of n = 120, using as a pattern the calculations done in the ***Notes.***

   g. For n = 40, we found $z = \dfrac{2185 - 1949.56}{181.83} \approx 1.29$ and $P(\bar{x} > 2185) = P(z > 1.29) = 1 - 0.9015 = 0.0985$.

      For your sample of n = 120, did your probability come out larger or smaller?

   h. Your probability of getting a sample mean greater than 2,185 sq. ft. should be smaller with a sample of n 120 compared with a sample of n = 40. Does this mean that samples of n = 120 are more (or less) accurate than samples of n = 40? Give a reason for your answer that convinces you but based upon your results.

   i. For samples of n = 120, find a sample mean $\bar{x}$ of square feet for the smallest 5% of sample means we expect to see with random sampling. Draw a sketch and find a value for the sample mean $\bar{x}$. (Is this a "given a value—find a proportion" problem or "given a proportion—find a value" problem?)

   j. For samples of size n = 40, we found that the value for the 5% of smallest means we would expect to see was 1,651 square feet. Compare this with your result from part i for samples of size n = 120. Is your result closer or farther away from the population mean μ = 1,949 square feet?

k. You should have found that your answer for the "5% smallest means" is nearer to the population mean μ for samples of size n = 120 compared with samples of size n = 40. Does your finding mean that samples of n = 120 are more accurate than samples of n = 40? Give a reason for your answer based upon your results that convinces you.

3. ***What happens with a Different Population*** For this exercise, we will consider the same variable, *Sq_Ft*, but with a different population. The population will be the houses that were sold in the ***Central Region*** of San Mateo County. Shown here are a graph of the distribution of the variable *Sq_Ft* as well as summary statistics for the distribution.

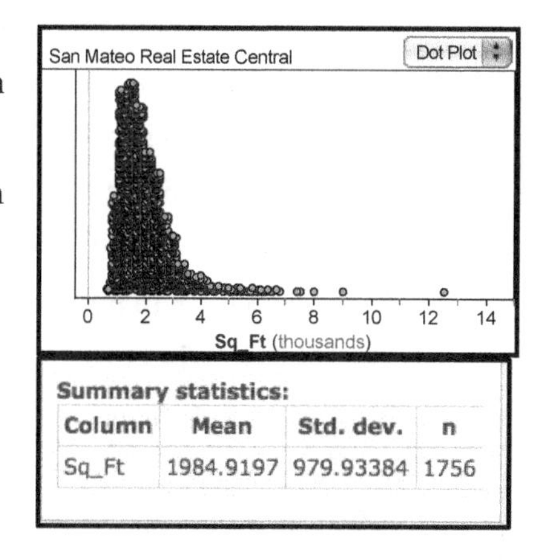

Summary statistics:

Column	Mean	Std. dev.	n
Sq_Ft	1984.9197	979.93384	1756

a. Use the correct symbols to show the mean and the standard deviation of the population.

b. We will draw a sample of size n = 120. For samples of this size, show, with the correct notation, the mean and the standard deviation of the sampling distribution of sample means $\bar{x}$ for the variable *Sq_Ft*.

- With the file ***SanMateoRealEstateCentral*** and software get a random sample of n = 120 that includes the variable *Sq_Ft*.
- Get a graph of the distribution of the variable *Sq_Ft* as well as summary statistics for the for the sample that was drawn.

c. The population distribution of the variable *Sq_Ft* is clearly right-skewed. Do you expect that the random sample of size n = 120 that you draw will also be right-skewed or not? Give a reason for your answer, and check the graph of the distribution that has been produced.

d. The population distribution of the variable *Sq_Ft* is clearly right-skewed. Does this fact mean that the sampling distribution of sample means $\bar{x}$ for the variable *Sq_Ft* will also be right-skewed? Give a reason based on experience with the same variable for the South Region.

e. For part b, you should have that $\mu_{\bar{x}} = 1984.92$ ft^2, and that $\sigma_{\bar{x}} = \frac{\sigma}{\sqrt{n}} = \frac{979.934}{\sqrt{120}} \approx 89.40$. Using this information, find the interval of *reasonably likely xbars* for this sampling distribution of sample means $\bar{x}$ for the variable *Sq_Ft*.

f. Is the sample mean $\bar{x}$ for the random sample of n = 120 that you have drawn reasonably likely or is it rare? Give a reason for your answer.

g. If your sample mean $\bar{x}$ is above the population mean then calculate the probability that you will get a mean greater than the one that you have, using the fact that the sampling distribution is well approximated by a Normal distribution. If your sample mean is less than the population mean then find the probability that you would get a sample mean less than what you have.

h. The population mean for the Central Region for the variable *Sq_Ft* is bigger than the population mean for the *Sq_Ft* for the South Region. Is it possible that if we draw random samples of n = 120 from each of the populations, we would get a sample mean $\bar{x}$ for the South Region that was bigger than the sample mean $\bar{x}$ for the Central Region? Explain why or why not. (Hint: Think *reasonably likely* or *rare*.)

§3.3 Exercises for Binomial Sampling Distributions

1. ***Studying Obesity: Samples of Five.*** To study obesity in a large population of people with a sample of only $n = 5$ does not make sense. However, such a small sample size *is* useful for learning about binomial distributions and how they work. We will graduate to bigger sample sizes.
From studies by the Centers for Disease Control (CDC), the evidence is that approximately 20% of the adult population is obese. We will give this *population proportion* the symbol p, so $p = 0.20$.
   a. If we draw a random sample of $n = 5$, is it possible that we will get five obese people in our sample? Do you think it is likely? Why?
   b. If we have n = 5 in our sample, we could get zero, one, two, three, four, or five obese people in our sample. What number do you think is the most likely? Why?

An imaginary population.

Obese People ID Numbers
000,000
000,001
000,002
000,003
000,004
000,005
000,006
000,007
000,008
•
•
•
199995
199996
199997
199998
199999

We will imagine that we have a population of one million people. Now, imagine that we make the obese people—all 20 percent of the one million, or 200,000 of them—stand on our left so that we can enumerate them in order to take a SRS from the population. The very first obese person gets the number 000,000, the second one gets the number 000,001, and so on, until the last one, whose number is 199,999. On our right we put all the 800,000 non-obese people, and we number them, starting with number 200,000, 200,001, and so on, until the last one, whose number is 999,999.

Now, using just one ten-sided die, we will choose five people and record whether each of the five is obese or not obese. Follow the directions.

Non-Obese People ID Numbers
200,000
200,001
200,002
200,003
200,004
200,005
•
•
•
548,752
548,753
548,754
548,755
548,756
•
•
•
999995
999996
999997
999998
999999

- Roll a ten-sided die; if the die comes up "0" or "1," that will indicate we have chosen one of the obese people (because their "serial numbers" start with 0 or 1). Record an "O" for the first person in this sample. If the die comes up with a number 2 or more (serial numbers 200,000 plus), record an "N" indicating a "not obese" person for the first person.
- Repeat this procedure four more times; that is, roll the die and note each time whether the outcome is "O" or "N" for the second person, the third person, the fourth, and the fifth. This process will produce one sample of $n = 5$. Count the number of "O"s (Obese people in the five) and record the number.
- Repeat the process three more times so that in all there are four samples, each of $n = 5$. Each time, record the number of obese people in each sample.
- When finished, record the numbers of obese people in each of the four different samples either in a survey on line, or if in class, on a common data collection.

   c. The results from this exercise will be combined. In the collection of all the samples, what proportion of the samples of $n = 5$ had just one obese person? What proportion of samples had no obese people? What proportion of samples had five?

Closer to reality: The Binomial Distribution Model. In the exercise above, we had a fixed number of interest; that is, a sample of $n = 5$. We had one of two outcomes ("Obese" and "Not Obese") that we counted as "success" and we set the probability of this success at $p = 0.20$. In the procedure we employed a random process (using the die) to determine the number of "successes" (Obese people) in each sample of $n = 5$.

d. Use the information in the sentences above to show that the conditions ***B, I, N, E*** for the Binomial model are met.

- Use software to get the probabilities using a ***Binomial Distribution Model*** that in a sample of $n = 5$ we will get 0, 1, 2, 3, 4, or 5 obese people if the probability that a person is obese is $p = 0.20$. Record the probabilities in a table something like the one shown here. In the software supplement, see the entry ***Binomial Model.***

k	$P(X = k)$
0	
1	
2	
3	
4	
5	

e. Compare the results to the idea recorded in part a. Does the probability for five obese people in a sample of five agree with that idea. State why (or why not).

f. Write how it makes sense to you that the highest probability is that out of the five people, there is just one that is obese.

g. One of the meanings of the mean of the Binomial Distribution is "Expected Value", or "roughly" the value out of $n = 5$ that we "expect to see." Use the formula for the mean and standard deviation for the Binomial Distribution to show that $\mu = 1$ and $\sigma = 0.8944$.

h. Using the formulas $P(X = k) = \binom{n}{k} p^k (1-p)^{n-k}$ and $\binom{n}{k} = \frac{n!}{k!(n-k)!}$ with $n = 5$ and $p = 0.20$, confirm that $P(X = 2) = 0.2048$.

i. From the results from the software in the table, find (easily) $P(x \geq 2)$.

j. Express in the context of samples of size five the meaning of $P(x \geq 2)$.

k. Another possible meaning to the probability in parts i and j is that it is the probability of concluding that "40% or more people are obese" if we just had our sample of five. Explain where this "40% or more" comes from.

The Binomial Distribution Model with a bigger sample. We have the same scenario about obese people, but now consider a sample of $n = 50$. From software, here is the distribution.

l. What doe you notice about the shape of this distribution compared to the one for $n = 5$?

m. Confirm that the mean $\mu = 10$ and the standard deviation $\sigma = 2.8284$ shown in the figure to the right are correct.

k	P(X = k)
0	0
1	0.0002
2	0.0011
3	0.0044
4	0.0128
5	0.0295
6	0.0554
7	0.087
8	0.1169
9	0.1364
10	0.1398
11	0.1271
12	0.1033
13	0.0755
14	0.0499
15	0.0299
16	0.0164
17	0.0082
18	0.0037
19	0.0016
20	0.0006
21	0.0002
22	0.0001

20 40

$\mu = 10 \quad \sigma = 2.8284$

- n. Use software to find $P(X \geq 20)$ and give a meaning to this probability.

o. Another possible meaning of the probability in part n is that it is the probability of concluding that "40% or more people are obese" if we just had our $n = 50$. Compare this probability with the probability calculated for "40% or more" people are obese for a sample of five. Explain the advantage of having a bigger sample size if actually $p = 0.20$?

2. *Understanding the calculations: left-handed students in an ancient history class.* In unit 4 of the ***Notes***, we will look at the proportion of left-handed people in the general population and see that the proportion is about 12%, although there is much variation among cultures. This exercise focuses on the conditions and the calculations for the binomial distribution. Refer to the sections in the ***Notes*** where the conditions and the calculations are explained.

An imaginary college and an imaginary course. In your college (say), all students are required to take a one-term course on Ancient Etruscan history (or AEH). Sections of the course are offered every hour from 7.00 am to 7.00 pm and are capped at a maximum of twenty-four students. All sections are full, since it is a college requirement. Our goal is to calculate the probabilities of 0, 1, 2, 3, 4, . . . , 24 left-handed students in a single section of twenty-four students, using the ***Binomial Distribution Model***. We make the assumption that the proportion of left-handed is $p = 0.12$.

a. For the scenario about finding the probabilities of 0, 1, 2, 3, …, 23, 24 left-handers in a class section of AEH, identify the values for the symbols n and p.

b. Read over the conditions ***B, I, N, E*** as well as the scenario presented here, and explain how each of the conditions for the binomial distribution are met if you think that they are met. If you think one or more of the conditions are problematical, clearly explain why you believe it is so.

c. Random sampling or random allocation to the classes is not mentioned above. In the example in the ***Notes,*** random sampling is important because it guarantees the independence condition. However, independence of cases (or "trials") is also assured if students sign up for classes essentially randomly *with respect to their handedness*. From your experience as a student, do you think this is true?

d. The meeting of "Left-Handers Action Front" (LHAF) has its weekly meetings at 10.00 am. Does this affect the conditions? If so, explain how; if not, explain why not.

- Use software to get the ***Binomial Distribution*** for this scenario, or use the graphic and the table pribted here to answer the questions below. (Some software uses "spikes" rather than bars of a histogram to show the probabilities of $X = k$.)

k	P(X = k)
0	0.0465
1	0.1522
2	0.2387
3	0.2387
4	0.1709
5	0.0932
6	0.0403
7	0.0141
8	0.0041
9	0.001
10	0.0002
11	0
12	0
13	0

μ = 2.88 σ = 1.592

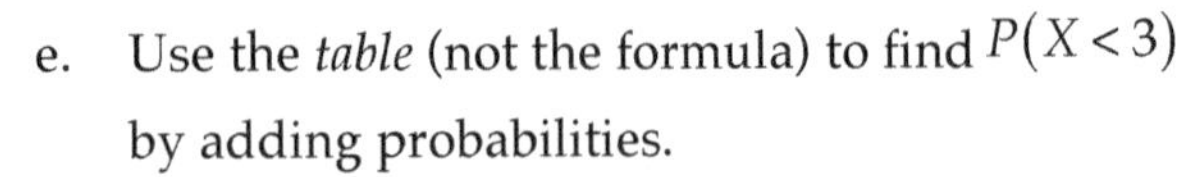

e. Use the *table* (not the formula) to find $P(X<3)$ by adding probabilities.

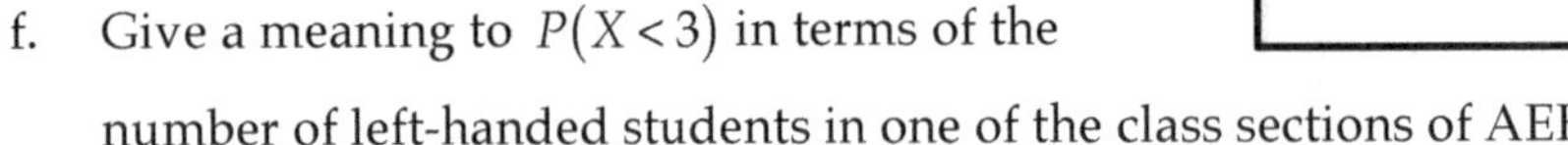

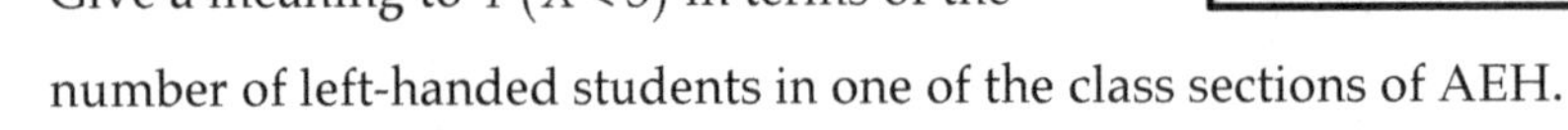

f. Give a meaning to $P(X<3)$ in terms of the number of left-handed students in one of the class sections of AEH.

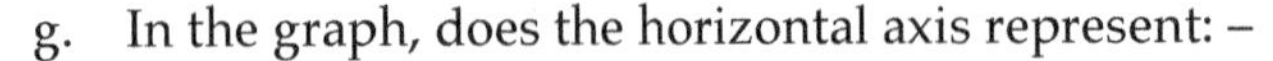

g. In the graph, does the horizontal axis represent: –
   - i) the number of left-handed students, or
   - ii) the probability that a class will have a specific number of left-handed students?

h. In the graph, does the vertical axis (that is, the heights of the bars) represent:
   - i) the number of left-handed students, or
   - ii) the probability that a class will have a specific number of left-handed students?

i. Make a sketch of the graph, and indicate $P(X<3)$ by shading the graph. (the answers to part e and to this question can be checked using software.)

j. From the *table*, calculate the probability that a section will have six or more left-handed students, (by adding probabilities) and use the correct notation to display the result. (You can check the result of your calculations using software.)

k. Make a sketch of the graph and show by shading the answer to part h. (If you checked your answer using software, the shading should also show.)

l. Asked to calculate $P(X=2)$, Confused Conrad calculates $\binom{24}{2}=276$. What is CC's confusion and how should he immediately know that 276 is not the answer to $P(X=2)$?

m. Using the information in part l, correctly use the formula to get $P(X=2)=0.2387$.

n. Do some simple arithmetic to confirm that $\binom{24}{2}=276$. (See the example in the ***Notes***.)

A binomial model for left-handers in seating arrangements.

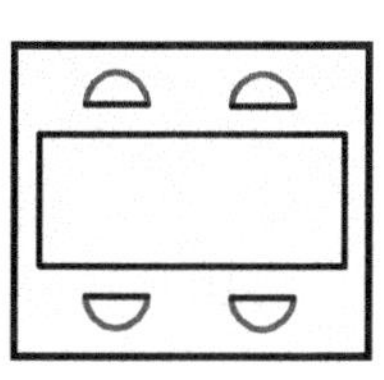

o. Because of the generosity of the benefactor, all of the classrooms have tables with four chairs as pictured here. We are now interested in calculating the probabilities of zero, one, two, three, or four left-handers at a single table. For what we are doing now, what are the values of the symbols n and p?

p. Explain how each of the conditions ***B, I, N, E*** are met if you determine they are met. If you have doubts about whether one of the conditions is met, explain which one and why.

q. What would happen if left-handers had a tendency to seek out their fellow "lefties" and tried to join a table where they saw another left-hander was sitting? Which condition or conditions would be affected?

r. By hand, calculate $P(X=k)$ for $k=0$, $k=1$, $k=2$, $k=3$ and $k=4$ using the formula for binomial probabilities. Show your work. Here are the answers you should get.

k	P(X = k)
0	0.5997
1	0.3271
2	0.0669
3	0.0061
4	0.0002

s. You should have found in the calculation of $P(X=2)$ that

$\binom{4}{2}=\frac{4!}{2!2!}=\frac{4\cdot3\cdot2!}{2\cdot1\cdot2!}=6$ so there are six different placements of the two left-handers at the four chairs at the table. Label the four chairs A, B, C, D and show the six different placements of the two left-handers and the two right-handers. Here is a start showing that one seating arrangement is for chairs A and B to have left-handers, and chairs C and D to have right-handers. (Notice that the identities of the two "lefties" and the two "righties" do not matter in this exercise; only fact of their handedness is important.

	A	B	C	D
1	*L*	*L*	*R*	*R*
2				
3				
4				
5				
6				

3. ***Conditions: Is the Binomial Model applicable?*** For each of the following scenarios, determine whether each of the conditions ***B, I, N, E*** are met. If they are, state why; if not, state why. You may well have doubts about some of the conditions and yet think others are fully met. You may also suggest ways to change the scenario so that the conditions are met. For each scenario that meets the conditions, identify the n and the p.
   a. In classes of thirty-six students, the students are asked whether they have a tattoo or not. We want to calculate the probability that we will see 0, 1, 2, . . ., 36 having a tattoo if we think that 20% of students have tattoos.
   b. The same classes of thirty-six students are also asked what kind of mobile telephone they have. (The choices are iPhone, Blackberry, Android, Other Smartphone, or no Smartphone; the proportions are about 0.50. 0.15, 0.20 and 0.05 and 0.10.) We want the probability of each type of device in a class of thirty-six.
   c. There is a law in California that prohibits talking on a mobile telephone. Data were collected by watching drivers as they came to a stop sign to see whether or not the driver was using a mobile 'phone. (Other variables were observed as well.) It is thought that about 5% of drivers talk or text while driving, and we are interested in modeling the number found talking or texting in an observation period of one hour.
   d. There is a random sample of mothers, *all* of whom have given birth to six children. Looking at all of their births, we are interested in knowing the probabilities of 0, 1, 2, 3, 4, 5, or 6 female children. (Additional question for those in medicine: is the ***I*** condition met?)
   e. For male college students in California, the mean height is about 178 centimeters and the standard deviation about eight centimeters. We want to model the probability that a male college student is shorter than 168 centimeters.
4. ***Understanding the meaning: A bigger sample size, n = 120.*** Here we will explore what happens if we a large sample size. The context is the same: we are taking as a reasonable assumption that 20% of the 16- 35 year old population is obese, so $p = 0.2$. Now, however, $n = 120$.

- Use the software to get a graph of the the ***Binomial Distribution*** with for $n = 120$ and $p = 0.20$. Your picture should look resemble this one, but you must have the application open to answer some of the questions. First, let us choose just one of the bars (or spikes), the bar for the number 18.

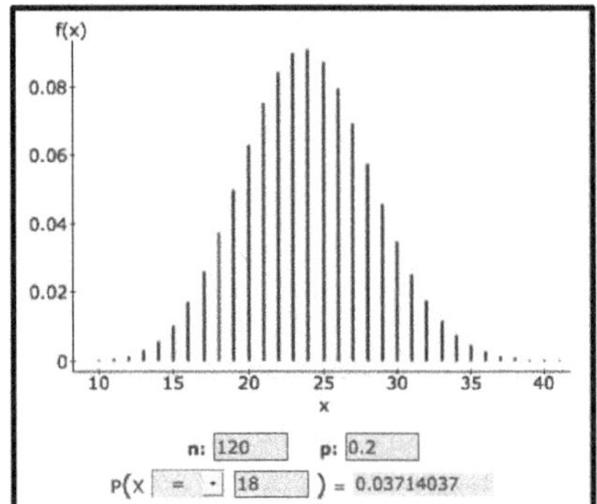

   a. What is a good name for the shape of this distribution?
   b. What is the *best* answer for the meaning of this bar in the histogram? Choose one of the responses and give a reason for your answer.
      i) It shows that there are eighteen obese people in one sample of 120.
      ii) The bar shows the probability that in a sample of 120, we have at most eighteen obese people.
      iii) The height of the bar shows the probability that in sample of 120, we get exactly eighteen obese people.

c. [*Review*] the software output shows that $P(X=18)=0.0371$. Show how 0.0371 was calculated by inserting the values of n and p in the proper places in the formula for the Binomial Model. Do not carry out the calculation at this point but show where the numbers go.

d. [*Review*] Part of the formula involves $\binom{120}{18}=\frac{120!}{18!(120-18)!}=1{,}086{,}744{,}939{,}880{,}326{,}302{,}940$.

The best answer for the meaning of this number in the formula for the probability in the Binomial Model is: (Choose one response and give a reason for your answer.)

i) The number of different ways of placing eighteen obese people in a sample of 120 (Is one of them the second person of the 120 sampled, for example?)

ii) The number of samples we would have to draw to get exactly eighteen obese people in a sample of 120

iii) The probability of getting exactly eighteen obese people in a sample of 120

iv) The number of eighteen-year-old obese people in the world

d. [*Review*] Question "b (ii)" above referred to "the probability that in a sample of 120, we have at most eighteen obese people." Find the probability that there are at most eighteen obese out of 120. Use the correct notation.

- Here is another plot of the ***Binomial Distribution*** for $n=120$, $p=0.20$. This plot (as well as the one shown on the previous page) suggests that the probability of having fewer than ten obese people where n = 120 and p = 0.20 or having fifty or more obese people where n = 120 and p = 0.20 is very unlikely. Use software to get the probabilities that fewer than ten in a Binomial Distribution where n =120 and p= 0.20 and also the probability of fifty or more where n =120 and p= 0.20. In other words, find $P(x<10)$ and $P(X\geq 50)$.

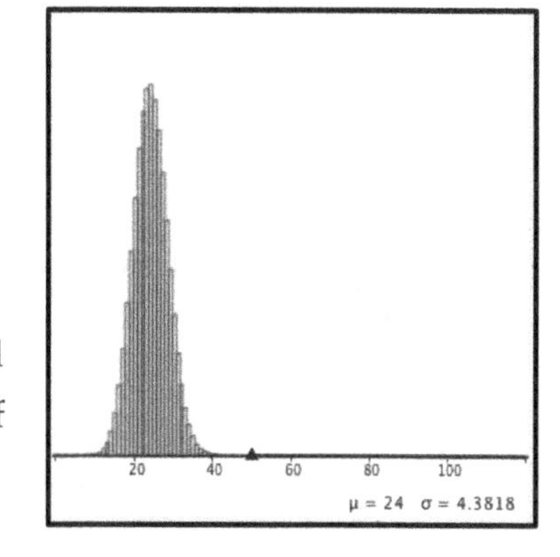

e. Write two sentences "translating" the probabilities $P(x<10)$ and $P(X\geq 50)$ found using software to the context of numbers of obese people in a random sample of 120 people.

f. If someone said that in a sample of n = 120, there are fifty obese people, decide which of the following could be true and why, judging from our Binomial Model.

i) The sample was not random. (Perhaps a voluntary response sample?)

ii) The proportion of obese people in the population is bigger than p = 0.20.

iii) The sample is very unusual.

- Use software to find $P(X\leq 15)$ and also $P(X\geq 33)$.

g. Make a sketch of the graph of the binomial distribution using a smooth curve; the spikes are not necessary) to show the value $P(X\leq 15)=0.0218$ by shading the graph. Indicate clearly what feature of the graph shows the 0.0218.

h. Now, go to the other side of the distribution and add probabilities so to confirm that $P(X\geq 33)=0.0296$. On the same as for part d, show the value 0.0296 by shading the graph.

i. From the results of parts g and h, find the probability of getting a sample of 120 in which the number of obese people is between sixteen and thirty-two (including the end points). That is, find $P(16 \le X \le 32)$. (You may check your answer software but show your calculations.)

j. On a sketch of the binomial distribution, show the answer to part "i" by shading.

k. The answer that you should have gotten for part "i" above is $P(16 \le X \le 32) = 0.9486$. The meaning of this probability is important. Choose the best answer (thinking about each) and give a reason for your choice.

   i) In a single sample of 120, we are certain to see between sixteen and thirty-two obese people.

   ii) The probability 0.9486 means that there is about a 95% chance that between sixteen and thirty-two people are obese out of every 120 in the population.

   iii) If the probability of being obese is 0.20 then the probability that we get a random sample of n = 120 with between sixteen and thirty-two obese people is 0.9486 (almost 95%).

l. For this binomial distribution with n = 120, and p = 0.20, find the mean and the standard deviation using the formulas given in the ***Notes***. (There is a box for the mean and the standard deviation.)

m. [*Review*] Using the answers to part "l," make a sketch of a Normal distribution showing with mean 24 and standard deviation 4.382, showing the values for the mean ±1, ±2 and ±3 sd.

n. [*Review*] Using the Normal distribution with the mean and standard deviation from part "l," find the probability that we will see fifteen or fewer obese people. That is, calculate $P(X \le 15)$.

o. [*Review*] Using the Normal distribution with the mean and standard deviation from part "l," find the probability that we will see thirty-three or more obese people. That is, calculate $P(X \ge 33)$.[Check the answers using software using the ***Normal Distribution*** calculator.]

p. [*Review*] Calculate, using the Normal distribution, $P(16 \le X \le 32)$.

q. Compare your answers for $P(X \le 15)$, $P(X \ge 33)$ and $P(16 \le X \le 32)$ using the Normal distribution and the answers using the binomial distribution. Are they the same? Close but not the same? Show big differences?

The comparison should show that the answers are close but not exactly the same. One reason for this is that the binomial distribution with p = 0.20 is still very slightly right skewed compared with the Normal distribution with mean at 24, even though they have the same mean. Another reason is that the binomial distribution only takes on the whole numbers (16, 17, . . .31, 32, etc.) whereas the Normal distribution takes on any number—including 16.247, 31.678. Some people adjust the fit by using (in our example) 15.5, and 32.5 instead of 16 and 33.

5. ***The shape of the Binomial Distribution Model.*** Open software with a ***Binomial Distribution*** calculator; try out different values of n—small and big—and values of $0 \le p \le 1$. Record what values you tried.

   a. Under what conditions (values of n and p) is the binomial distribution: i) right-skewed, ii) symmetric in shape, and iii) left-skewed?

   b. Under what conditions does the shape look most Normal?

About Exercise 6: Read this first, please.

The point of Exercise 6 is to see how we far we can trust results from a sample. How do we measure "trust"? What do is to calculate the probability that the sample misleads us seriously. On the other hand, we can calculate the probability that a sample gives us information we can trust.

- First of all, we will randomly sample "by hand" just to get an idea of how random sampling works and the range of results that it gives us. We will do this with a very small sample, and we choose a small sample so that the handwork will not take forever.
- Then we will apply the ***Binomial Distribution Model*** to this process to see that it gives similar results to our hand sampling.
- After that we will apply a Binomial Distribution to a larger sample size to see what happens. With both the small and the larger sample, we calculate the probability that the sample gives us wrong information, and calculate the probability that it gives reasonable information.
- Then, we will actually sample from a large collection and see how what we have done in exercise applies.

6. ***Preterm births and their incidence.*** A premature or preterm birth is one that occurs before thirty-seven weeks of pregnancy. What proportion of all births is premature? There is quite a bit of variation among countries and regions of the world and even within North America. (See http://preemiehelp.com/about-preemies/preemie-facts-a-figures/general-preemie-statistics/incidence-of-preterm-birth-by-country and http://www.statehealthfacts.org/comparemaptable.jsp?ind=39&cat=2). For this exercise, we will use a guess for the population proportion of $p = 0.10$, or 10%.

Samples of n = 10 "by hand"

PreMature Births ID Numbers
0000
0001
0002
0003
0004
0005
⋮
0632
0633
0634
0635
0636
⋮
0995
0996
0997
0998
0999

In the first part of this exercise, we will randomly choose samples "by hand"—that is, using a ten-sided die. To simulate drawing a simple random sample, imagine numbering our population of ten thousand births like this: we will gather together all of the premature births and give them serial numbers 0000 (the very first preterm birth), the second one 0001 up to 0999 (the vary last preterm birth), making one thousand preterm births, or 10% of the ten thousand births in all. Then the first full-term birth (not premature) is numbered 1000, the second one 1001, and on up to 9999. So we have a numbered list. So all the preterm births have serial numbers beginning with 0, and all the full-term births have serial numbers beginning with 1, 2,...up to 9. Number 0863 is a premature birth, but 7295 is a full-term birth. The directions are on the next page.

Full-Term Births ID Numbers
1000
1001
1002
1003
1004
1005
⋮
2874
2875
2876
2877
2878
⋮
9,995
9,996
9,997
9,998
9,999

- Roll the ten-sided die; if "0" appears we have randomly chosen one of the preterm births (because their "serial numbers" start with 0), record a "P" for the first birth in the sample. If the die shows 1, 2, 3,... (serial numbers 1000 plus), record "F" indicating a full-term birth the first birth.
- Roll the die nine more times to get the other nine births in the first random sample and sum thethe number of preterm births in your sample and record this sum.
- Repeat the process three more times so that in all you get four samples, each of $n = 10$. Each time record the total number of preterm births in each sample.
- When finished record the numbers of preterm births in each of the four different samples as the instructor directs; this may be on-line or in class.

a. In this part of the exercise, the ten thousand births are "imaginary" (later we will look at real births). However, what we have imagined makes the sampling process a simple random sample. What about the process makes it a simple random sample?

b. Why is the sampling process not a voluntary response sample?

c. The results from this exercise will be combined. Questions to answer: what proportion of the samples of $n = 10$ have just one preterm birth? What proportion of the samples of $n = 10$ have no preterm births?

Applying the Binomial Distribution Model. Each of our samples has a fixed number $n = 10$, we have set the probability (reasonably) at $p = 0.10$ for one of the two possible outcomes (Premature or Full-Term) and we have employed a random process to find the number of successes out of $n = 10$.

d. The description just above shows that the conditions for a ***Binomial Distribution*** model ***B, I, N, E*** are met. For each of the four conditions, specify exactly how the description shows that the conditions are met.

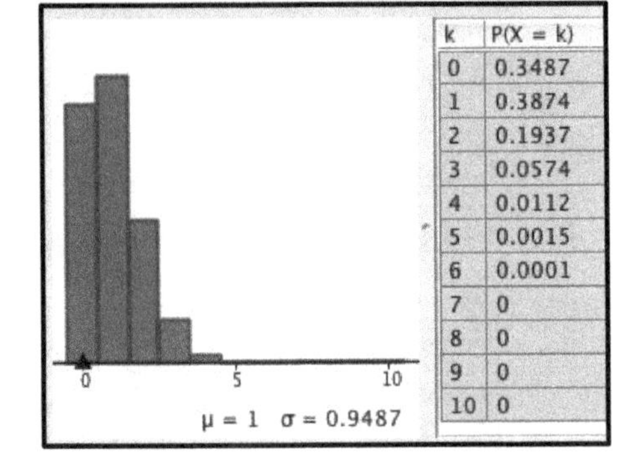

k	P(X = k)
0	0.3487
1	0.3874
2	0.1937
3	0.0574
4	0.0112
5	0.0015
6	0.0001
7	0
8	0
9	0
10	0

e. [*Review*] Shown here is the Binomial Distribution for $n = 10$, with probability $p = 0.10$. Use the formulas for the Binomial Distribution for $n = 10$ and $p = 0.10$ to confirm that $P(X = 0) = 0.3487$. [Recall that $0! = 1$ – that is "0 factorial" = 1]

f. Put into words the meaning of $P(X = 0) = 0.3487$ in the context of numbers of premature births in a sample of $n = 10$. [The answer has to say something about premature births.]

g. Thinking and discussion question. Suppose you work in a birthing center of a large hospital (a place where babies are delivered; see, e.g. http://www.mills-peninsula.org/birthcenter/). One day, of the last ten births, six of them were premature births. Given the Binomial Distribution shown above, what do you think? [There are several good answers as well as "off the wall" answers to this question. It is really meant for discussion.]

h. One meaning of the mean is "Expected value;" in this case we can interpret this as the value of premature births we "most expect to see" out of ten. Show that the mean and standard deviations given above are correct, using the formulas for the mean and standard deviation of a Binomial Distribution.

i. If the sample we got had two or more premature births we might be tempted to think that the probability of a premature birth is 0.20 (20%) or more rather than 0.10 (10%). According to the Binomial Distribution Model, what is the probability that we would get two or more premature births in a random sample of ten if the true proportion of premature births is 0.10? Use the correct notation.

A Binomial Distribution Model with a bigger sample size

- Use software to get the Binomial Distribution with $n = 100$ and $p = 0.10$. Here is a graphic of the distribution, and on the right the calculated Binomial probabilities for the numbers of premature births from zero out of 100 to 27 out of 100. [Some software may display the probabilities as spikes instead of bars, and not give a list of probabilities.]

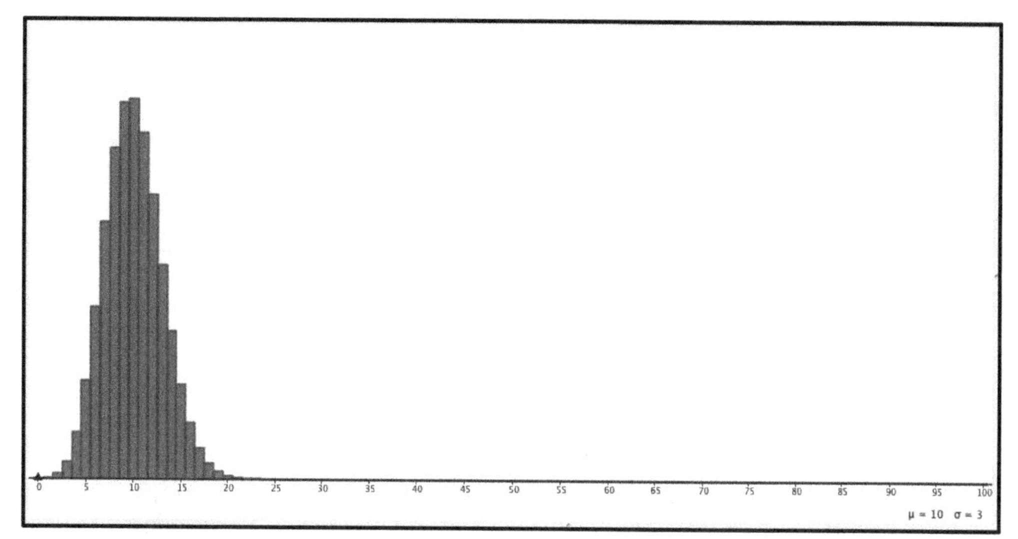

k	P(X = k)
0	0
1	0.0003
2	0.0016
3	0.0059
4	0.0159
5	0.0339
6	0.0596
7	0.0889
8	0.1148
9	0.1304
10	0.1319
11	0.1199
12	0.0988
13	0.0743
14	0.0513
15	0.0327
16	0.0193
17	0.0106
18	0.0054
19	0.0026
20	0.0012
21	0.0005
22	0.0002
23	0.0001
24	0
25	0
26	0
27	0

j. Compare the shape of this Binomial Distribution for $n = 100$ with the shape of the Binomial Distribution for $n = 10$

k. Without using the formulas or reverting to software, give the answer to five decimal places to $P(X = 50)$. Explain your answer.

L Find the probability that out of 100 births, 20 or more are premature. That is, get $P(X \geq 20)$, either by using software or by adding probabilities in the list shown here.

m. Compare the probability for $P(X \geq 20)$ with a sample size of $n = 100$ with the probability calculated for part i when we calculated with sample size of $n = 10$. Both of these probabilities can be interpreted as the probability of concluding that "20% or more of births are premature" if the true proportion is actually $p = 0.10$ (10%). What can you say about what happens to the probability of making the incorrect conclusion if the sample size is increased. Explain.

- Use software (use the ***Binomial Distribution*** calculator) to confirm that the probability of getting between five and fifteen premature births in a random sample of one hundred, that is, $P(5 \le X \le 15) = 0.9364$ with the Binomial Distribution with $n = 100$ and $p = 0.10$.). The software may also give a graphic like this.

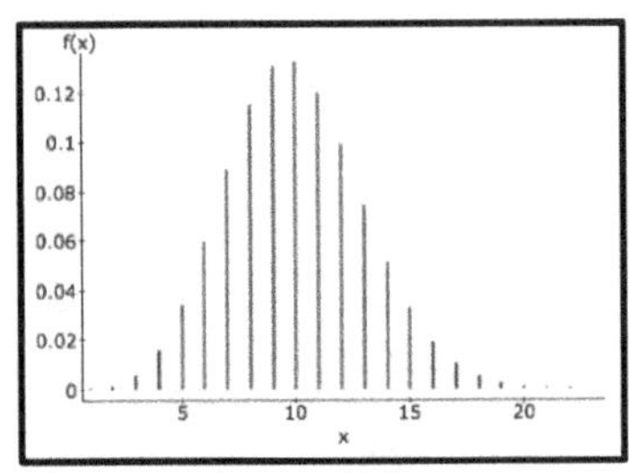

  n. Calculate the percentages (or proportions) of premature births that are implied by getting five out of 100, and then fifteen out of 100. Then complete this sentence: "If it is true that $p = 0.10$ one meaning of $P(5 \le X \le 15) = 0.9364$ is that we have a high probability (0.9364) of estimating the true proportion of premature births with an error of ______% on either side of the true 10%.

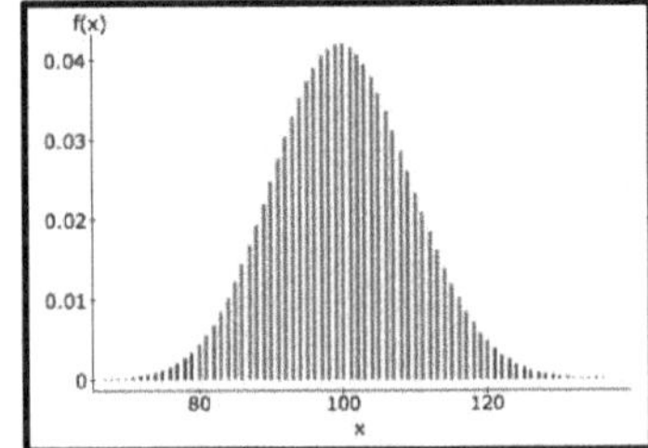

- In the software change $n = 100$ to $n = 1000$ (but keep $p = 0.10$) and find $P(80 \le X \le 120) = 0.9695$ and possibly this graphic.

  o. Calculate the percentages (or proportions) of premature births that are implied by getting 80 out of 1000, and then 120 out of 1000. Then complete this sentence: "If it is true that $p = 0.10$ one meaning of $P(80 \le X \le 120) = 0.9695$ is that we have a high probability (0.9695) of estimating the true proportion of premature births with an error of ______% on either side of the true 10%.

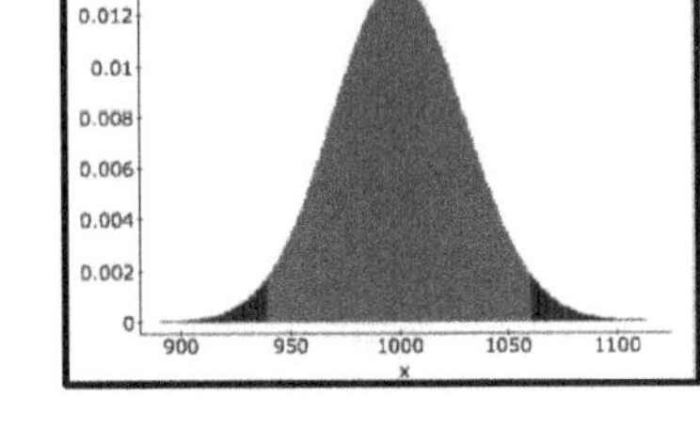

- In the software change $n = 1000$ to $n = 10000$ (but keep $p = 0.10$) and find $P(940 \le X \le 1060) = 0.9563$, and possibly this graphic.

  p. Calculate the percentages (or proportions) of premature births that are implied by getting 940 out of 10000, and then 1060 out of 10000. Then complete this sentence: "If it is true that $p = 0.10$ one meaning of $P(940 \le X \le 1060) = 0.9563$ is that we have a high probability (0.9563) of estimating the true proportion of premature births with an error of ______% on either side of the true 10%.

  q. What, briefly does this part of the exercise teach about the advantages of much data? What are the disadvantages of much data?

- The file ***DixMilleBirthsRev*** is a random sample of birth records, where n = 10,000. [These birth records are available; see http://www.cdc.gov/nchs/data_access/vitalstatsonline.htm). With software and the file ***DixMilleBirthsRev*** get a frequency table for the variable *Premature.*

  r. Notice that from this collection where $n = 10000$ the proportion of premature is about 0.1104. Is this evidence against the idea that the proportion of premature is $p = 0.10$. Discuss. Have an argument.

Frequency table results for Premature:

Premature	Frequency	Relative Frequency
Full-Term	8897	0.8897
PreMature	1103	0.1103

§3.4 Exercises: Sampling Distributions for Proportions

1. *Inference: Sample and Population* This exercise is about the proportion of students in colleges having a tattoo. Here we have shown the proportions of male and female students reporting having a tattoo for a sample at Penn State (in 1999) and samples for several semesters at one community college in Northern California.

Semester, Year and Place	Females: Proportion having a tattoo	n	Males: Proportion having a tattoo	n	Overall: Proportion having a tattoo	*n*
Both, 1999, PA	0.131	137	0.191	68	0.151	205
Spring 2008. CA	0.255	47	0.160	50	0.206	97
Autumn 2008,CA	0.254	63	0.127	63	0.190	126
Spring 2009. CA	0.372	86	0.219	64	0.307	150
Autumn 2009,CA	0.258	89	0.115	78	0.192	167
Spring 2010. CA	0.266	94	0.309	81	0.286	175
Autumn 2010,CA	0.149	101	0.146	89	0.147	190
Spring 2011. CA	0.191	94	0.218	87	0.204	181

a. ***Sample*** Study the numbers in the table and decide whether or not each of the following statements are supported by the data in the samples. Give reasons.

(i) Tattoos are more popular among students in California than at Penn State.

(ii) For the California community college students, it appears that tattoos were most popular in 2009.

(iii) Among students in California, females are more likely than males to have a tattoo.

b. ***Population*** What you see in the samples may or may not be generalizable; although it is not often done, it is helpful to think about what population the samples represent. (If we decide that the sample is a good representation of a population, we feel safe generalizing the findings to that population.) The California sample data were collected from students who had signed up for a statistics class. For each option for a "population," decide whether you think the California samples (for the information in the table above) are likely to be a representative sample of that population.

(i) all the students taking statistics at that college?
(ii) all the students at that college?
(iii) all community college students in Northern California?
(iv) all college and university students in California?
(v) all college and university students in North America?

c. The sample data were collected from all the students who had signed up for some sections of statistics at one college. Why is this sample *not* a simple random sample of all the students taking statistics at that college? (We often encounter non-random samples that may be representative.

The logic of inference For the remainder of this exercise, let us agree that the population we have in mind is "all college and university students in California." The next questions have to do with the proper use of notation. Then we tackle some logic.

Combined Class Data Spr 09

		Tattoo: n	Tattoo: y	Row Summary
Gender	f	54	32	86
	m	50	14	64
Column Summary		104	46	150

S1 = count( )

d. Calculate the overall proportion of students having a tattoo in the table for the sample for spring 2009 and assign the symbol p or the symbol $\hat{p}$. Give a reason for your choice.

e. Pete, who proclaims himself an expert on "student trends," is very happy to see the result of your calculation. He is happy because the result is in line with his idea that the *population* proportion of college and university students in California who have a tattoo is 30%, or 0.30. Suppose Pete is right; we do not know whether Pete is right, but let us suppose his idea is correct. *If* Pete really is right, what symbol should be assigned to Pete's idea, the symbol p or the symbol $\hat{p}$? Give a reason for your choice.

f. If Pete is correct that the population proportion $p = 0.30$ (so now you know the answers to questions d and e—but you still need to have the reasons right), describe the ***sampling distribution*** of sample proportions $\hat{p}$ for simple random samples of $n = 150$ drawn from the population of college and university students in California. The words "describe the sampling distribution" mean that the *shape, center*, and *spread* of the sampling distribution need to be shown.

g. Are the conditions involving $np \geq 10$ and $n(1-p) \geq 10$ met for the spring 2009 sample using Pete's idea that $p = 0.30$?

h. Use the information you have put down in your answer to part f to calculate the interval of ***reasonably likely*** $\hat{p}$ s if Pete is right and $p = 0.30$.

i. Will the $\hat{p}$ we calculated be ***reasonably likely*** or ***rare*** Give a reason for your answer.

j. Pete has an identical twin brother named Repete, but, amazingly, they disagree on nearly everything. Repete thinks that the population proportion of college and university students who have a tattoo is $p = 0.20$. Repete thinks that only 20% of all college and university students in California have a tattoo. Are the ***conditions*** for the sampling distribution of sample proportions $\hat{p}$ met if $n = 150$ and Repete is right and $p = 0.20$? Calculate.

k. Calculate the interval of ***reasonably likely*** $\hat{p}$ s if Repete is right and $p = 0.20$.

l. If $p = 0.20$ (Repete's idea) then is $\hat{p} = 0.3067$ ***reasonably likely*** or ***rare***? Give a reason.

m. If we had a $\hat{p} = 0.3067$ from an SRS of $n = 150$ from the population of all California college and university students, would Pete have good evidence that his brother Repete is wrong? What should his argument be? (*Hint:* Think *reasonably likely* or *rare*. You may wish to discuss this question in class; this question actually introduces a way of arguing that we will use.)

n. Suppose instead of forty-six students having a tattoo (which leads to $\hat{p} = 0.3067$), we actually had just twenty-nine out of the $n = 150$ who had a tattoo. Calculate the $\hat{p}$ and then decide whether Pete or Repete would have the stronger evidence for his case. Explain why one or the other has the stronger evidence.

o. Now suppose there were thirty-eight students out of the $n = 150$ who had a tattoo. Pete and Repete's mother, Maria (who knows statistics), says to her sons: "This result will *not* settle your argument." Explain, using the ideas of reasonably likely and rare to show why Maria is right.

2. ***Inference on tattoos, cont.*** All of the questions in Ex. 1 used a sample size of $n = 150$. Suppose that our sample size is only $n = 50$. We still work with Pete's idea that $p = 0.30$. The sampling distribution of sample proportions may be different with a smaller sample size or not.

   a. Determine whether the conditions involving $np \geq 10$ and $n(1-p) \geq 10$ are met.
   b. For $n = 50$, find the mean of the sampling distribution $\mu_{\hat{p}}$ and what is the $\sigma_{\hat{p}}$.
   c. Make a sketch of the sampling distribution showing the mean and the standard deviation and the interval of reasonably likely $\hat{p}$ s. Use a sketch of a Normal distribution.
   d. Will a $\hat{p}$ = 0.20 be reasonably likely or rare if $p = 0.30$ and n =50? Give a reason.

3. ***San Mateo Real Estate, Simulation*** In 2005–2006, the housing market was booming, and one result of the boom was that nearly 60% of the houses sold for over their list price. In contrast, 2007–2008 was the beginning of a depressed real estate market, although in San mateo County, the mean prices were actually higher. However, one indication of the start of the downturn was that the proportion of houses sold over list prices was lower. This exercise on simulating a sampling distribution looks at the variable *SoldOverList,* which has two categories indicating whether or not a house sold over the list price: "Not Over List Price", and "Sold Over List." We are interested in the proportion in the second category, which for 2005-2006 was at 59%.

   a. What are the cases and what is the population for this exercise? (It may seem that the answers to "cases" and "population" should be the same, and they are almost the same.)

Frequency table results for SoldOverList:

SoldOverList	Frequency	Relative Frequency
Not Over List Price	2650	0.671396
Over List Price	1297	0.328604

- With the file ***SanMateoRealEstateY0708*** and using the ***Simulating Sampling Distributions for Proportions*** facility, run a simulation for the variable *SoldOverList* looking at the category "Sold Over List." Use a sample size of $n = 100$ and run the simulation something like 9000 times. The output of the simulation will vary according to the software used, but will usually have three panels, which may be arranged horizontally or vertically. The three panels have information on:

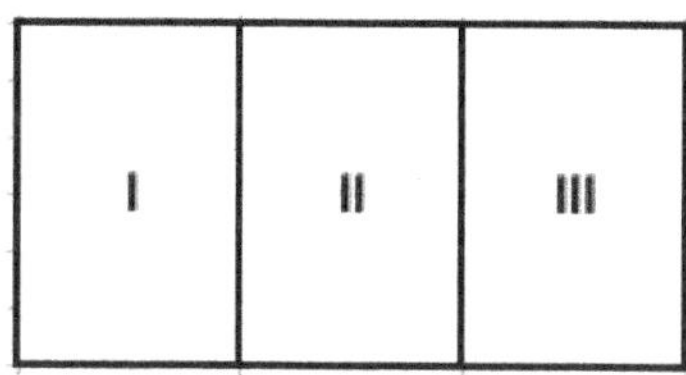

I
II
III

  I The population from which samples are drawn,

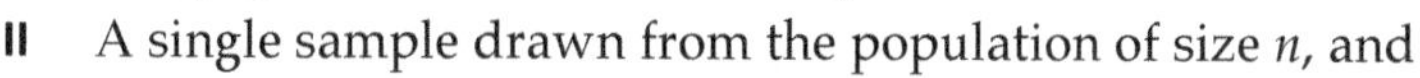

  II A single sample drawn from the population of size n, and

  III The simulated sampling distribution of the statistic (or measure) calculated each time a sample of size n is drawn. The simulated sampling distribution results from drawing a sample and calculating a statistic a large number of times

  For an example of the exact appearance of the output, consult the ***Software Supplement***.

   b. Checking the simulation: Does the proportion recorded for "Over List Price" above agree with the proportion recorded in panel I (the "population panel") in the simulation? Does the population proportion agree with the proportion recorded for a single sample in panel II? Should the sample proportion necessarily be the same?

c. Use the information in the box ***Facts about Sampling Distributions of Sample Proportions...*** to calculate the mean $\mu_{\hat{p}} = p$ and standard deviation $\sigma_{\hat{p}} = \sqrt{\frac{p(1-p)}{n}}$ of the sampling distribution of sample proportions $\hat{p}$ drawn from samples of $n = 100$ from the population of ***SanMateoRealEstateY0708***. Your simulated results for the sampling distribution in panel III should be fairly close to your calculations. Are they?

d. Why should the results for the simulation be "fairly close" to the calculations from the formulas $\mu_{\hat{p}} = p$ and $\sigma_{\hat{p}} = \sqrt{\frac{p(1-p)}{n}}$ but not exactly the same?

e. Calculate the interval of reasonably likely $\hat{p}$ s from your answer to part g.

f. Was the last sample that the software simulation collected (shown in panel II) reasonably likely or rare? Give a reason for your answer.

g. Find a value for $\hat{p}$ so that only 5% of the $\hat{p}$ s are smaller than this value. (*Hint:* Is this a "given a value—find a proportion" problem or a "given a proportion—find a value" problem?)

4. *Students and Languages: Proportions and Means.* In the combined class data for spring '09, eighty-two students of the 146 who answered the question said that they speak two or more languages. Another way of looking at the same data (the variable *Language*) is to calculate the *mean* number of languages spoken. For the spring '09 students, the mean number of languages is 1.70.

Note: parts a through h are about the sampling distribution of sample proportions.

a. When we calculate a proportion of students who speak two or more languages, are we treating the variable *Language* as a *categorical* or a *quantitative* variable? When we calculate the mean number of languages spoken, are we treating the variable *Language* as a *categorical* or a *quantitative* variable? (Notice that although *Language* is a quantitative variable, it is helpful to analyze it using "categorical" techniques.)

b. Calculate the proportion who speak two or more languages and express this proportion in two ways: first, using the P( ) notation, using X to indicate the number of languages spoken, and then secondly, using the notation for a sample proportion that was introduced in this section.

c. We do not know the *population* proportion of California college and university students who speak two or more languages; we have to "guesstimate," although our sample data at least gives the ballpark in which to guess. So let us guess that one half of the population of all college and university students in California speak two or more languages. Express this guess for the population proportion using the correct symbol for population proportion.

d. In the **Notes** consult ***Facts about sampling distributions...*** do some calculations to determine that $n = 146$ is big enough for the sampling distributions of $\hat{p}$ to be approximately Normal.

e. Again using the ***Facts about sampling distributions...*** and *also making certain to use the correct notation*, describe, with numbers and symbols and words, the center and the spread of the sampling distribution of $\hat{p}$ from samples of $n = 146$.

f. Use your answer to part e to find the interval of reasonably likely $\hat{p}$ s for samples of $n = 146$ and $p = 0.50$.

g. Make a sketch of the sampling distribution showing the mean and the standard deviation and the interval of reasonably likely $\hat{p}$ s. Use a sketch something like the one shown here but with the horizontal axis labeled with the units that are used for the proportions in the context here.

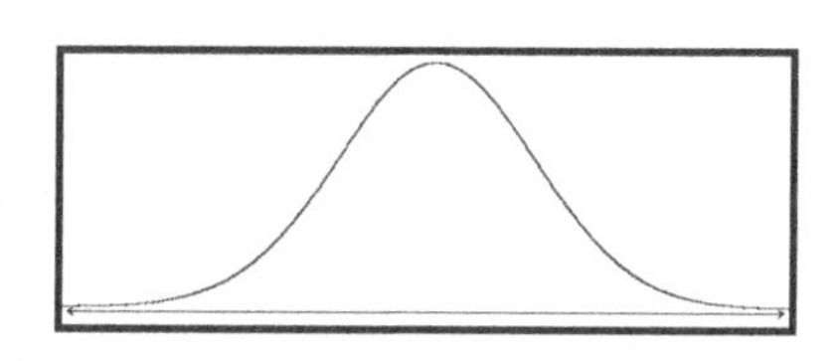

h. Is the sample proportion found from the spring '09 data a reasonably likely or rare $\hat{p}$ if $p = .50$?

Note: parts i through p are about the sampling distribution of sample means.

i. Now consider the *mean* number of languages spoken, which was given above as 1.7 for the spring '09 students. What symbol should be used for this number? Give a reason for your choice.

j. Again, we do not really know the population mean number of languages spoken or the population standard deviation for the variable *Language* for the population of all California college and university students. We can only guess, although our data gives some guidance. Let us guess that the population mean is 1.5 and that the population standard deviation is 1. Use the correct symbols for these numbers. (Notice: we can use the symbols even though we do not really know the values; the symbols are a way of *naming* what we are working with.)

k. Using the ***Facts about sampling distributions of sample means...*** in §3.2 and *also making certain to use the correct notation*, describe, with numbers and symbols and words, the center and the spread of the sampling distribution of $\bar{x}$ from samples of $n = 146$ from California college and university students.

l. Find the interval of reasonably likely $\bar{x}$ s for samples of $n = 146$ where $\mu = 1.5$ and $\sigma = 1$.

m. Make a sketch of the sampling distribution showing the mean and the standard deviation and the interval of reasonably likely $\bar{x}$ s. Use a sketch something like the one shown here but with the horizontal axis labeled with the units that are used for the proportions in the context here.

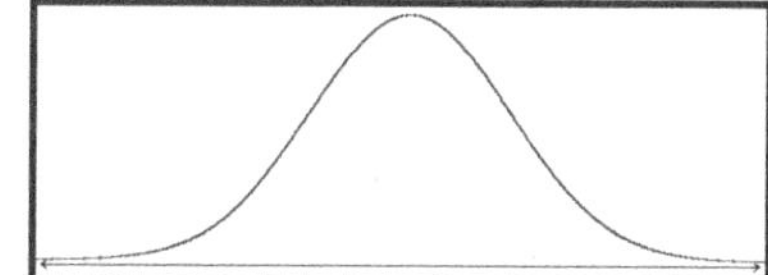

n. Is the $\bar{x} = 1.70$ from our sample of $n = 146$ reasonably likely or rare if $\mu = 1.5$ and $\sigma = 1$?

o. On your sketch of the sampling distribution, indicate our $\bar{x} = 1.70$ and calculate the probability of getting an $\bar{x}$ this large or larger with random sampling. ("With random sampling" means: if our Normal sampling distribution applies.)

p. You should have found that the probability that we would see an $\bar{x} = 1.70$ *if* $\mu = 1.5$ and $\sigma = 1$. You should have $P(\bar{x} \geq 1.7) = 0.0078$. Because we have a *rare* outcome, there are several possibilities:

- $\mu = 1.5$ and $\sigma = 1$ are wrong. Perhaps μ is actually bigger than 1.5.
- We were unlucky in our sampling; we just got one of the weird samples.
- We have a biased sample, since it is not a simple random sample of all California colleges and university students. Perhaps $\mu = 1.5$ is correct, but our college has a bigger mean.

Choose one or more of these possibilities and make your case.

5. ***Cell Phone Usage by Drivers*** It is now illegal in California to talk on a cell telephone while driving. Our question is: what proportion of drivers breaks the law at a given time? The state of Victoria in Australia has a similar law to the one in California. A small team of researchers in Melbourne, Australia, collected data by *observing* whether drivers were talking on their cell phones [D. Taylor, C. MacBean, A. Das and R. Rosli, "Handheld mobile telephone use among Melbourne drivers," *Medical Journal of Australia* (2007) 187: 432-434].

   Here is what they did: Four observers with just a clipboard stood on a corner of a road or a freeway ramp exit and observed drivers. There were three kinds of locations: one freeway ramp exit, one city street, and one urban area road. They categorized drivers in age/gender categories and, most importantly, as to whether or not the driver was using a handheld cell phone. They observed on three consecutive Tuesdays in October 2006 for one-hour periods (10.00–11.00, 14.00–15.00, and 17.00–18.00).

   They had done a previous study and were able to guess that about 1.5% of the drivers would be using cell phones. So their guess was that $p = 0.015$.

   a. Look again at the ***conditions*** for the sampling distribution of the sample proportion $\hat{p}$ which are that $np \geq 10$ and $n(1-p) \geq 10$. For the freeway exit they observed n = 8,640 drivers. Was this enough?

   b. Find the mean and the standard deviation of the *sampling distribution* of $\hat{p}$ if $p = 0.015$ and n = 8640.

   c. Calculate the interval of reasonably likely $\hat{p}$ s if $p = 0.015$ and $n = 8640$.

   d. Is the sample used by Taylor, et. al. a random sample? Give a reason for your answer. (Here is a situation in which getting a random sample is probably impossible, but getting some kind of sample is better than none.)

6. ***Premature Births Again*** In §3.3, Exercise 6 was about the Binomial Distribution and was centered around the idea that the population proportion of premature births was p = 0.10. This exercise uses the data file ***DixMilleBirthsRev*** and the variable *Premature* and treats these data as a population from which we take samples.

   - With the file ***DixMilleBirthsRev*** use software to get a frequency table of the variable Premature. The results should agree with the table shown here.

Frequency table results for Premature:

Premature	Frequency	Relative Frequency
Full-Term	8897	0.8897
PreMature	1103	0.1103

   a. Since we are treating these data as a *population* and not a *sample*, which symbol should be used for the proportion 0.1103? Should it be p or should it be $\hat{p}$?

   b. We intend to draw samples of size n = 100, and then use Normal Distribution calculations. In the last section (§3.3) we saw that with sufficiently large sample sizes for the p being used, the ***Binomial Distribution***. In this section, we are given a rule-of-thumb as one of the ***conditions*** for the Binomial Distribution being "Normal enough". Use the rule-of-thumb of $np \geq 10$ and $n(1-p) \geq 10$ to make a calculation that shows that n = 100 is large enough (albeit "just" large enough.)

- With the file ***DixMilleBirthsRev*** and using the ***Simulating Sampling Distributions for Proportions*** facility, run a simulation for the variable *Premature* looking at the category "Premature." Use a sample size of $n = 100$ and run the simulation something like 15000 times. The output of the simulation will vary according to the software used, but will usually have three panels. which may be arranged horizontally or vertically. See exercise 3 about the structure of the simulation output, as well as the ***Software Supplement.*** Shown here is one possible output of a simulation.

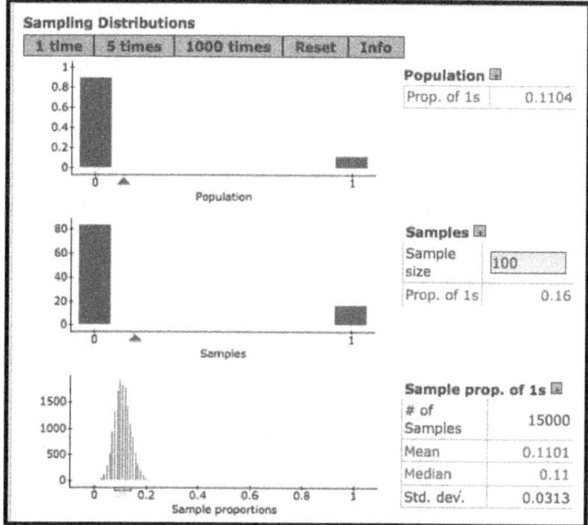

c. Use the formulas given in this section (see ***Facts about sampling distributions...***) to get the mean and standard deviation of the sampling distribution for samples of size $n = 100$ using the population proportion of premature births of $p = 0.1103$.

d. Compare the results of the calculations made in part c to the results of the mean and standard deviation of the simulated sampling distribution. The two results should be close, but probably not exactly the same, since the simulation is not the entire sampling distribution. Do the results check? (If not . . .?)

e. Use the calculations of part c to find the ***95% Reasonably Likely Interval*** for sample proportions ($\hat{p}$ s). (The answer should be $0.0489 \le \hat{p} \le 0.1717$)

f. Put into words a good meaning for the calculation made in part e, and also determine whether the $\hat{p} = 0.16$ shown above as one sample of the simulation falls in the ***95% Reasonably Likely Interval***. That is, did that sample yield a $\hat{p}$ that was "reasonably likely" or "rare"?

g. In the one sample shown in the simulation you made, was the single sample $\hat{p}$ shown "reasonably likely" or "rare"? Give a reason for the answer.

h. If thirty students in a class each draw a single random sample from this population of the file ***DixMilleBirthsRev*** and calculate the sample proportion of premature births, how many of those students would you expect to have a "rare" result? Give a reason for the number.

i. Repeat the simulation and also the calculations done in part c but with a sample size of $n = 400$ instead of $n = 100$. What has changed in the calculations?

j. Re-calculate the ***95% Reasonably Likely Interval*** for sample proportions ($\hat{p}$ s), this time with $n = 400$ but still with $p = 0.1103$. Compare the interval with the one calculated in part e. Is it wider or narrower?

k. The answer to part j should be $0.0796 \le \hat{p} \le 0.1410$, making the interval narrower. Which of the following statements follow:

   (i) We are less likely to see "wild" sample proportions of premature births that are far from the population proportion with a sample of $n = 400$ than with a sample of $n = 100$.

   (ii) The sample proportion of premature births must be between $0.0796 \le \hat{p} \le 0.1410$.

   (iii) With a bigger sample size, the range of possible sample proportions likely is smaller.

7. ***Recently Built Houses being sold in San Mateo County*** This exercises use the collection of houses being sold in San Mateo County from June 2005 to June 2006. The categorical variable *RecentHouses* distinguishes houses being sold that were built less than ten years prior to 2005-2006 ("Recently built houses") from houses built ten or more years prior to 2005-2006 ("Older houses"). For this exercise we are regarding the data in the file ***SanMateoRealEstateY0506*** as a *population.* One of the main tasks will be to describe the sampling distribution of sample proportions. Describing a distribution means to specify its shape, center and spread according to the ***Facts about Sampling Distributions of sample Proportions*** ... . Here are some summary statistics about the population of the variable *RecentHouses.*

Frequency table results for RecentHouses:

RecentHouses	Frequency	Relative Frequency
Older Houses	5001	0.91159315
Recently Built Houses	485	0.088406854

a. What is the symbol that we should use for the 0.0884, the proportion of "Recently Built Houses"? (*Hint:* Is this the population proportion or is this a sample proportion?)

b. In describing the sampling distribution of a sample proportion for a random sample of $n = 160$, what symbol should be used for a sample proportion?

c. What will be the mean of the sampling distribution of sample proportions calculated from samples of size $n = 160$ drawn from the population? Refer to the box in the ***Notes*** entitled ***Facts about Sampling Distributions of Sample Proportions...***

d. What will be the standard deviation of this *sampling distribution*? Do a calculation.

e. Will the shape of the sampling distribution be approximately Normal? Check to see if the conditions are met for the sampling distribution to be approximately Normal.

f. Make a sketch of the sampling distribution showing the mean and the standard deviation. Use a sketch something like the one shown here but with the horizontal axis labeled with the units that are used for the proportions in the context here.

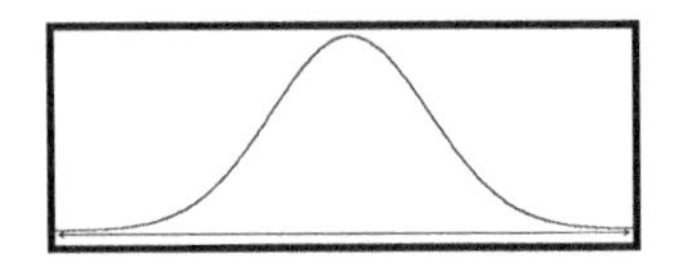

g. Use your sampling distribution to find the interval of values of the sample proportion that are ***reasonably likely*** with a sample size of $n = 160$. Refer to the box in the ***Notes*** entitled ***Reasonably Likely and Rare...***

- Use software to draw a random sample of $n = 160$ from the population ***SanMateoRealEstateY0506***. For the sample, find the sample proportion of "Recently Built Houses." Consult the ***Software Supplement*** under Sampling (to get the sample) and Frequency Tables (for the proportion).

h. The answer to part g is $0.0444 \leq \hat{p} \leq 0.1324$. Is the sample proportion in the random sample you drew "reasonably likely" or "rare"? Give a reason for the answer.

i. On the graph that was made to answer part f, indicate the *p*-hat that was found in your random sample. If your *p*-hat is bigger than $p = 0.0884$, find the probability that a random sample will give you a *p*-hat bigger than the one you got. If your If your *p*-hat is smaller than $p = 0.0884$, find the probability that a random sample will give you a *p*-hat smaller than the one you got. Use Normal Distribution calculations.

j. On the graphic, using shading, show the probability that is the answer to part i.

§4.1 Exercises on Confidence Intervals for Proportions

1. ***What is the probability of a delayed flight to LA?*** The government agency Bureau of Transportation Statistics (see www.bts.gov) collects data on every commercial flight in the United States. The file ***OntimeCombLAXSample*** is a random sample of *flights* from the San Francisco Bay Area to Los Angeles International (LAX) for 2009. For this exercise, we will be analyzing the variable *DelayOver15* whose categories are: "Departure Delay 15 min. or less" and "Departure Delay Over 15 min." Our goal will be an *estimate* of the probability that a passenger will be delayed more than fifteen minutes on a flight from the San Francisco Bay Area to LAX.

Sample of OntimeCombLAX

	Carrier_...	Date_M...	Flight_N...	Tail_Nu...	Destina...	Schedul...	Actual_...	Schedul...	ActualD...	Departu...	Wheels...	Taxiout...	DelayCa...	DelayWe...	DelayOver15
389	UA	03/03/2009	1163	N827UA	LAX	14:25	15:03	90	83	38	15:21	18	0	0	Departure Delay Over 15...
390	WN	05/24/2007	2647	N323SW	LAX	11:00	11:15	75	79	15	11:23	8	N/A	N/A	Departure Delay 15 min o...
391	WN	05/16/2009	789	N616SW	LAX	19:30	20:42	85	78	72	20:52	10	0	0	Departure Delay Over 15...
392	WN	03/16/2008	1228	N684WN	LAX	17:30	18:39	75	71	69	18:48	9	N/A	N/A	Departure Delay Over 15...
393	WN	05/03/2008	1478	N514SW	LAX	17:15	17:16	85	78	1	17:33	17	N/A	N/A	Departure Delay 15 min o...

- Use the file ***OntimeCombLAXSample*** and get a spreadsheet showing the data. It should resemble the spreadsheet above.

Frequency table results for DelayOver15:

DelayOver15	Frequency	Relative Frequency
Departure Delay 15 min or Less	332	0.78117647
Departure Delay Over 15 min	93	0.21882353

a. What are the cases for this analysis? Are they airlines, airplanes, airports, delays, flights, passengers, or times?

b. Is the variable *DelayOver15* categorical or quantitative?

c. [Review] Use the probability notation of §1.2 with the "symbol" *Over* to show the probability *from this sample* that a flight will be delayed over fifteen minutes.

d. In calculating an confidence interval estimate of the population proportion we will not use the probability notation of §1.2. What symbol should be used for the proportion of *all* flights from the San Francisco Bay Area to LAX, p or $\hat{p}$? Give a reason for your answer.

e. In calculating an estimate of the population proportion, what symbol should be used for the proportion that you calculated in part c, p or $\hat{p}$? Give a reason for your answer.

f. Using the formula for a 95% confidence interval, calculate and estimate for the proportion (or probability) of flights delayed over fifteen minutes.

- With the file ***OntimeCombLAXSample,*** use software to get a ***Confidence Interval for One Proportion***. Check the hand-calculated answer using the software result.

g. Give an interpretation (using the pattern in the ***Notes***) of the confidence interval you have calculated in the context of the problem.

h. Think about your results; if you flew to LAX from the SF Bay Area five times, would you be surprised if all five flights were delayed over fifteen minutes? Would you be surprised if one of them was delayed? Explain in the light of your estimate.

i. What would happen to your confidence interval if you increased the *level of confidence* from 95% to 99%? (You can actually see by changing the level of confidence in the software.) Explain in terms of the formula for the confidence interval.

2. *The President's Approval Rating as of May 2014*

- Log on to http://www.gallup.com/poll/171263/obama-job-approval-average-steady.aspx?utm_source=tagrss&utm_medium=rss&utm_campaign=syndication .

The president's approval rating was 44% at the time of the release of the post on the Gallup website.

a. What symbol should be used for the 44% (or 0.44), p or $\hat{p}$? Give a reason for your answer.

In that post (by scrolling down) the methodology of the survey is explained:

> **Survey Methods**
>
> Results for this Gallup poll are based on telephone interviews conducted May 1-31, 2014, on the Gallup Daily tracking survey, with a random sample of 15,724 adults, aged 18 and older, living in all 50 U.S. states and the District of Columbia.
>
> For results based on the total sample of national adults, the margin of sampling error is ±1 percentage point at the 95% confidence level.
>
> Interviews are conducted with respondents on landline telephones and cellular phones, with interviews conducted in Spanish for respondents who are primarily Spanish-speaking. Each sample of national adults includes a minimum quota of 50% cellphone respondents and 50% landline respondents, with additional minimum quotas by time zone within region. Landline and cellular telephone numbers are selected using random-digit-dial methods. Landline respondents are chosen at random within each household on the basis of which member had the most recent birthday.

b. From the information given for the sample size and the sample proportion, calculate the margin of error for a 95% confidence interval. Is it at least as small as what Gallup say?

c. Software gives the results as shown. Show how the lower and upper limits of the confidence interval were calculated, using the formula.

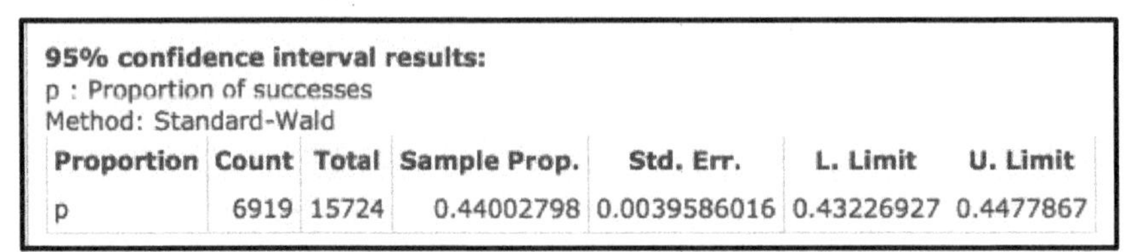

95% confidence interval results:
p : Proportion of successes
Method: Standard-Wald

Proportion	Count	Total	Sample Prop.	Std. Err.	L. Limit	U. Limit
p	6919	15724	0.44002798	0.0039586016	0.43226927	0.4477867

d. Indicate in the calculation which part of the calculation is the margin of error.

e. Express your confidence interval using mathematical notation; that is, *L.Limit* <   < *U.Limit* using the numbers given above, and the correct symbol between. What symbol should be placed between the inequality symbols, p or $\hat{p}$? Give a reason for your answer.

f. Give a valid interpretation of the 95% confidence interval that you have calculated.

g. Will a 99% confidence interval be wider or narrower? Give a reason for your answer. (You can answer by actually doing a calculation or by reasoning.)

h. What is the term by which the numbers 90%, 95%, 99% is known in the context of confidence intervals?

i. Complete this sentence: "In general, if the sample proportion is the same and the level of confidence is the same then with a bigger sample size, we will have a ______________ margin of error." Give a reason for the answer.

Special Exercise 3: Confidence Intervals on Real Estate

We are interested in the proportion (or the probability) that a house being sold in San Mateo County is a *Ranch* house, and we will use our new found ability to estimate using confidence intervals to do this.

- Use software with the file ***SanMateoRealEstateY0506*** to get a ***random sample*** of size $n = 64$ that includes the variable *Style2.*
- Get a frequency table for the variable *Style2* for the sample just drawn. It should resemble this frequency table, although the numbers will be different. (The numbers should be about the sizes of the numbers here; check that they sum to 64, if the sum is not shown. If the numbers are in the thousands, it means the population and not the sample was chosen.)

Frequency table results for Sample(Style2):

Sample(Style2)	Frequency	Relative Frequency
Contemporary	10	0.15625
Other	27	0.421875
Ranch	20	0.3125
Traditional	7	0.109375

We are interested in the proportion (or the probability) that a house is a *Ranch* house.

1. If we were in the "world of probability calculations" (§1.2) and if we use R for "ranch house," calculate the probability that a house is a ranch house" and express the probability using the correct notation. (Remember to use your own sample data and not the data shown in the table above.)
2. However, we are not in the world of probability calculations (though that world is just next door); we are in the world of making confidence intervals, so we will use either p or $\hat{p}$. Which one should we use? (*Hint:* Are the numbers from a sample or a population?) Give a reason for your choice.
3. Using the confidence interval formula, calculate a 95% confidence interval for p using your $\hat{p}$. (The software instructions just below will allow you to check your results.)

- Check the results with the file ***SanMateoRealEstateY0506*** using software to get a ***Confidence Interval for One Proportion***.

4. a. In your calculation, what is the value of the "margin of error"?
   b. Subtract the CI lower limit *from* the CI upper limit and divide by 2; does this number agree with your answer to a?
   c. Add the CI lower limit to the CI upper limit and divide by 2; what is the symbol for the number you get?
   d. True or false, and explain: "The margin of error is just half the width of the confidence interval."
5. Give an interpretation of your confidence interval as outlined in the boxes in the ***Notes,*** making certain that what you say is the context of the data, which are "houses sold in San Mateo County in 2005–2006."
6. Compare your confidence interval with the confidence interval calculated by someone else with a different sample. Do you expect their confidence interval to be exactly the same as yours (that is, assuming that both of you did everything correctly)? Similar to yours? Why or why not?

- Find a ***Swift Diagram*** with the reasonably likely intervals shown on the last page of these exercises.

•7. a. On the x axis of the ***Swift Diagram,*** plot your $\hat{p}$. Draw a light vertical line for your $\hat{p}$. (That is why there is a scale at the top of the plot.)
   b. On this light vertical line, plot the lower limit and the upper limit for your estimate of p as it is expressed in the confidence interval. (See the ***Swift Diagram*** in the ***Notes.***)

c. Darken in the part of the line that is your confidence interval.

- Using software, get a frequency table for the variable *Style2*. Use the file ***SanMateoRealEstateY0506***.

8. a. The frequency table just found gives you the *population* number for different styles of houses. Calculate the population proportion of "Ranch" houses and give it the correct symbol. Also indicate this *population proportion* on the ***Swift Diagram*** by drawing a horizontal line. (Why should the line be horizontal and not vertical?)

   b. Did your confidence interval capture (or include) the true population proportion p?

A Bigger Sample Size We will see the effects of getting a bigger sample size on the confidence interval. Instead of a sample size of $n = 64$, we will have a sample size of $n = 144$, but we are interested in the proportion of ranch houses in the population p, and we want to estimate this p using our $\hat{p}$ of ranch houses.

9. Look at the formula for the confidence interval.

   a. If you increase the sample size, what will happen to the size of the margin of error? Explain your answer by referring to what the formula does.

   b. If you increase the sample size, what should happen to the width of the confidence interval? Explain your answer.

- Use software with the file ***SanMateoRealEstateY0506*** to get a ***random sample*** of size $n = 144$ that includes the variable *Style2*.
- Get a frequency table for the variable *Style2* for the sample just drawn. The frequency table should resemble the one shown here.

Frequency table results for Sample(Style2):

Sample(Style2)	Frequency	Relative Frequency
Contemporary	27	0.1875
Other	49	0.34027778
Ranch	46	0.31944444
Traditional	22	0.15277778

10. a. Use a calculator to calculate the 95% confidence interval for the population proportion of ranch houses using the sample proportion $\hat{p}$ found from your Summary Table, and express your answer using the proper notation.

    b. What is the margin of error for your confidence interval for p using $n = 144$?

    c. Compare the size of your margin of error for the CI based on $n = 144$ with the margin of error for the CI based on $n = 64$. Does the comparison follow what you predicted in answering question 9a?

    d. Give a good interpretation of your new CI.

More Confident What happens when we increase the level of confidence from 95% to 99%?

11. a. What number in the formula will be different for the calculation of the 99% CI compared with the 95%?

    b. Will the margin of error be bigger or smaller with 99%? Explain.

    c. Do the calculation and express the answer using proper notation.

    d. Compare the size of your margin of error for the 99% CI with the 95% CI. Does the comparison follow what you predicted in answering question 11b?

- Use software (***Confidence Interval for One Proportion***) to check answers to questions 10 and 11.

Special Exercise 4: Internet in Australia?

Australian High School Students The data for this exercise are from a sample of high school students in the Australian states of Queensland and Victoria. The sample size is $n = 200$.

CAS Australia Vic Qld

	Qtime	Sex	BrthMnth	BrthYear	BrthPlce	BedRms	PeoplHme	InetAxs
154	19	Male	July	1990	Victoria	5	6	Yes - br...
155	42	Male	May	1989	Victoria	4	4	Yes - br...
156	17	Male	July	1992	Queensl...	3	4	Yes - br...
157	23	Male	January	1992	Queensl...	4	4	Yes - br...
158	20	Female	September	1990	Victoria	3	3	Yes - dia...
159	18	Male	March	1989	South Au...	4	2	Yes - dia...
160	16	Male	January	1992	Victoria	5	5	Yes - br...

1. Based on the description just above, write down what the cases are and what population you could reasonably say the sample represents. (It is reasonable to ask questions about how the data for the Censuss @ School were collected.)

Here are the results by year of school for the type of Internet connection the students reported.

CAS Australia Vic Qld

		Year_of_High_School			Row Summary
		First Year	Second Year	Third or Fourth Year	
InetAxs	No - Internet connection	15	9	1	25
	Yes - broadband connection	69	37	14	120
	Yes - dial-up connection	27	15	10	52
	Yes - other (include Internet access through mobile phone etc)	1	1	1	3
	Column Summary	112	62	26	200

S1 = count ( )

2. From the table above, you can see that sixty-nine out of the 112 first-year students said that they had a broadband Internet connection at home.
   a. Calculate a proportion from the "69 out of the 112" and give this proportion its correct symbol that we will use in calculating a confidence interval. Notice that now for the sample of just first-year students, n is smaller than it was for the entire sample. (So, now, $n = 112$; this is the n you will use for the other parts of the question.)
   b. Determine whether the conditions necessary for calculating a confidence interval are met for the proportion of first-year students who have a broadband Internet connection at home. (See the box in the ***Notes.***)
   c. Using the formula for a confidence interval, calculate a 95% confidence interval for the proportion of first-year students in Queensland and Victoria who have a broadband Internet connection. (Ans: 0.616 ± 0.090 or $0.526 < p < 0.706$)
   d. One part of the formula is $\sqrt{\frac{\hat{p}(1-\hat{p})}{n}}$. Confused Conrad puts this into his calculator: $\sqrt{\frac{69(1-69)}{112}}$ and he finds that his calculator refuses to give him an answer. Tell Conrad what his mistake is, and why his calculator is being troublesome.
   e. Give a correct interpretation for your confidence interval.
3. For the second-year students, the confidence interval for the proportion of students who have broadband Internet connection comes out as 0.597 ± 0.122; this is the interval $0.475 < p < 0.719$.
   a. The margin of error (the number to the right of the $\pm$) is 0.122. Show that you get this number by getting half the width of the confidence interval.

b. What happens to the margin of error when you increase the sample size? Explain by referring to the formula.

c. Suppose you wanted the margin of error to be as small as 0.06 (you want that amount of accuracy) for the second-year students. Calculate the sample size of second-year students you would have to have if the sample proportion $\hat{p}$ is the same.

d. Here are some bad interpretations for the confidence interval in this question, which is $0.475 < p < 0.719$. For each bad interpretation, say why it is incorrect.

> "The confidence interval is 0.475 to 0.719."

> "This confidence interval shows that 95% of the students have Internet connection between 47.5% and 71.9% of the time."

> "We can be 95% confident that the sample percentage of second-year students who have broadband Internet connection in Victoria and Queensland states is between 47.5% and 71.9%." (*Hint:* The problem is not that the interpretation uses percentages; that is acceptable: percentages are just another way of expressing proportions.)

e. Give a correct interpretation for the confidence interval $0.475 < p < 0.719$ or 0.597 ± 0.122.

4. ***Thinking about Confidence Intervals.*** You have all the information you need to answer the questions below; you just need to think and perhaps think in reverse.

Large parts of Australia are dry. One of the questions all two hundred high school students were asked was: "Do you agree or disagree that Australia will always have plenty of water?' The confidence interval for the "disagree" answer to this question (so that the students showed some concern about the supply of water) was $0.652 < p < 0.778$.

a. What was the $\hat{p}$? (Think of how the interval was calculated.)

b. What was the margin of error?

c. Approximately what number of students disagreed?

5. Suppose we wanted confidence intervals for the proportion of Australian students who do *not* have Internet connection.

a. For the first-year students, are the conditions met for calculating a confidence interval? Explain.

b. For the second-year students, are the conditions met for calculating a confidence interval? Explain.

6. The Australian Bureau of Statistics administers the Census @ School Project, and a great many schools in Australia participate. When a school participates, all of the students in the school answer the questionnaire, which is administered online. However, school participation is voluntary, so not all schools in Australia are part of the project.

a. Are the data from a simple random sample of Australian high school students? Give reasons for your answer. If you think that the data area not SRS, what label would you give the sampling process? (Check back to §3.1 for the various types of sampling.)

b. If the data are not from an SRS, what implication does that have for the calculation of confidence intervals?

Special Exercise 3

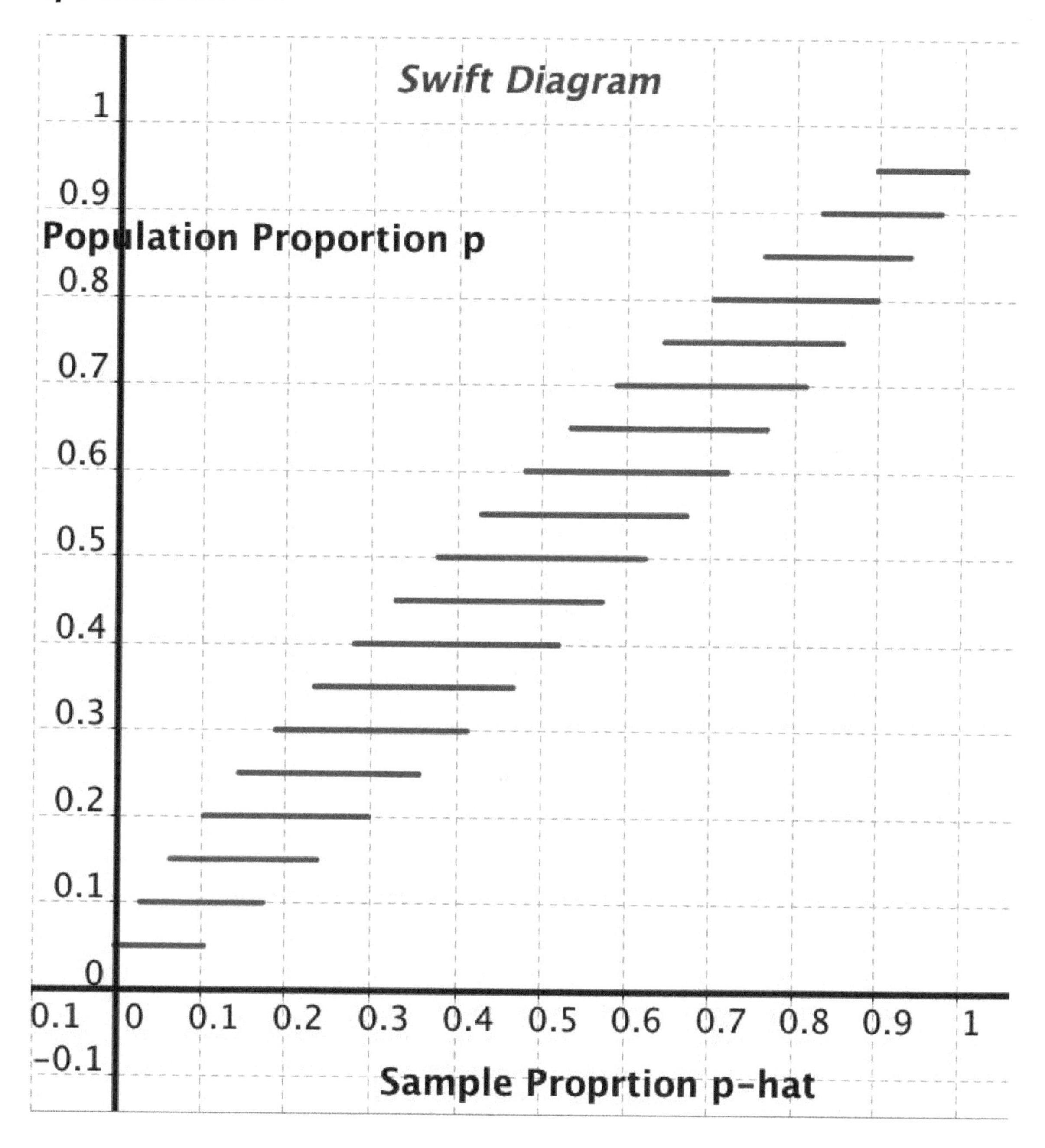

§4.2 Exercises on Hypothesis Testing for Proportions

1. Shoes on Beaches in the Netherlands. Often hypothesis testing is used just to test the idea of whether there is something more than random variability happening. Here is an example.

There is a light-hearted discussion among biologists who study seabirds in Northern Europe about the proportion of left and right shoes found on beaches. Martin Heubeck, who works in the Shetland Islands in Scotland, relates that at a seabird conference in Glasgow,

> ...the conversation drifted around to the amazing variety and number of shoes found during beached bird surveys. Mardik Leopold, normally an amiable and respected seabird biologist, claimed that it was a little known fact that due to some physical process or other, more left than right shoes wash ashore on beaches, at least in the Netherlands. [Shetland Bird Club Newsletter 107 (Spring 1997)]

Here are Leopold's sample data for the Dutch island of Texel.

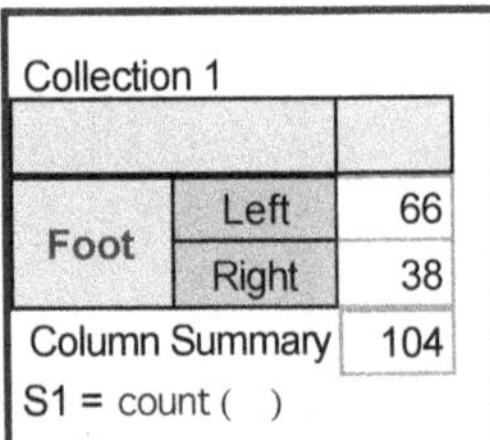

Collection 1

Foot	Left	66
	Right	38
Column Summary		104
S1 = count ( )		

Our statistical question is: *The proportion of left shoes should be near 50%; so does the proportion of left shoes deviate from 50% so much that we think something more than random variability accounts for the high proportion of left shoes on Texel Island?* We can analyze these data using a hypothesis test.

a. First, use the data for Texel to calculate the sample proportion of *left* shoes. (You should get 0.635, rounded to three decimal places.) Assign the correct symbol to this proportion; should the symbol be $\hat{p}$, p, or p_0? Give a reason for your choice of symbol.

b. What is the sample size? Assign the correct symbol to the sample size.

Step 1: Setting Up the Null and Alternate Hypotheses. Our interest will be in the population proportion p of *left* shoes on beaches. (That, everything will refer to *left* shoes.)

c. What is the logical proportion of left shoes in the world? That is, what is a good choice for p_0?

d. Write the null hypothesis using the notation shown in the ***Notes.***

e. The alternate hypothesis should be written as though you have no expectation that the number of left shoes should outnumber the right shoes or right outnumber the left, only that they *may* be different. Write the alternate hypothesis using the notation shown in the ***Notes.***

How did the shoes get on the beaches? Were they left there by beach-goers or lost from ships passing by? In a sense, it does not matter since whatever it is, the p_0 should be probably 0.50, since the population proportion of left shows should be near half the shoes worn.

Step 2: Checking the Conditions. There are three conditions to be checked, some of them difficult.

f. Simple Random Sample. State why the shoes found on beaches do *not* conform to the definition of a deliberately chosen simple random sample. (That leaves unclear whether how the shoes got there is a "random process" or not; that is what is being tested.)

g. Check the conditions $np_0 \geq 10$ and $n(1-p_0) \geq 10$. There is an easy way of doing this, which you may only discover after you have gone through the calculations. If you see the easy way, state why it works—perhaps using fractions.

h. The population is required to be ten times the size of the sample. Whatever the source of the shoes on the beaches (beach-goers or washed-overboard shoes), do you think that there are more than 1,040 shoes in the population? If so, the condition is satisfied.

Step 3: Calculating the Test Statistic

i. Calculate the test statistic using the formula given in the ***Notes.***

j. Draw a sketch of a Normal distribution (see below) and show on it the p_0. In your calculation of the test statistics, the denominator should have $\sqrt{\frac{p_0(1-p_0)}{n}} = \sqrt{\frac{0.50(1-.50)}{104}} \approx 0.049$. Use this to find the locations of 0.500 ± 0.049, $0.500 \pm 2*0.049$ and $0.500 \pm 3*0.049$ on a sketch of the sampling distribution we are using in this hypothesis test. Locate $\hat{p} = 0.635$ on your sketch.

k. From the value of the *test statistic,* is the $\hat{p}$ *rare* or *reasonably likely* given the p_0 we are using? Give a reason for your answer. (You should also be able to see this from your sketch.)

Step 4: Calculating the p-value. (We will do this in two sub-steps: the answer to part *l* is half the p-value.)

l. Your test statistic (your answer to part *i*) should be a positive number. Use the Normal Distribution Chart to get the area (i.e., the probability) in the right tail. (Will you subtract a probability from 1?)

m. Your answer to part *l* (just above) is only one-half the *p*-value. Use your result to get the *p*-value.

n. On your sketch of the Normal distribution, show the *p*-value by shading tails.

Check your calculations using software.

- Open the file ***TexelShoes1997*** and use software with the variable *Foot* to get a ***Hypothesis Test for One Proportion.***

o. Check the calculated values for the $\hat{p}$, for the z, and for the *p*-value. If in accord, answer: "OK."

p. How should the very small value of the *p*-value shown on the sketch? Get this correct! (The *p*-value is ***not*** shown by a point on the horizontal axis!)

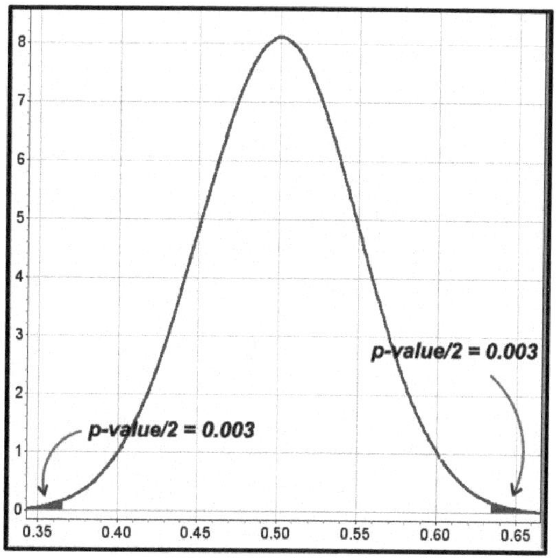

p-value/2 = 0.003

p-value/2 = 0.003

Step 5: Interpreting the result.

q. Is your test statistically significant? Give a reason for your answer.

r. With this test, can you reject the null hypothesis? Give a reason for your answer.

s. Give an interpretation in the context of the statistical question posed at the beginning. That is, do the results of our hypothesis test show that "something more" than random variability "accounts for the high proportion of left shoes" on Texel Island beaches?

(***Note:*** Your answers to parts q and r should be "yes." No one knows why there should be a significantly higher proportion than 50% of left shoes on Texel beaches. Do we have just a "weird" sample? It could be, or perhaps the shoe "wash up" process favored left shoes for some reason. Or did the shoe collection process favor left shoes?)

2. ***Shoes on Beaches in Scotland.*** Dr. Martin Heubeck (of Scotland) responded to his Dutch colleagues by collecting some data on shoes on Scottish beaches. Here is how he relates it:

> The gauntlet having been flung, I decided to record the same on 'my' beaches during the end of the February beached bird survey. I've done some crazy things in my time, but picking up shoes and wellingtons in Force 9 sleet takes some beating and I just hope few people saw what I was doing! [*Shetland Bird Club Newsletter **107** (Spring 1997)*]

Collection 1		
Foot	Left	87
	Right	124
Column Summary		211
S1 = count ( )		

Actually, Dr. Heubeck managed to collect sample data for both February and March. Here is the frequency table for all the shoes he collected.

a. Calculate the proportion of *left* shoes collected and assign the correct symbol (the one that we will use in hypothesis testing) to it.

b. Compare the *sample* results for Scotland to the *sample* results for Texel Island. What are the differences? Where do right shoes "go"? Left shoes?

c. To do the same kind of hypothesis test as we did for the Texel Island data, we need to set up the null and alternate hypotheses. Set up the hypotheses using the correct notation.

d. Check the sample size conditions for the test (i.e., np_0 and $n(1-p_0)$).

e. Calculate the test statistic for this test, showing your work.

f. Judging from your test statistic, is the sample proportion $\hat{p}$ reasonably likely or rare given the p_0 we are using? Give a reason for your answer. (Consult the ***Notes.***)

g. Before calculating the *p*-value, you should be able to predict whether the *p*-value will be smaller or bigger than 0.05. Predict and give a reason for your prediction.

h. Make a sketch of the sampling distribution, showing the location and value of p_0 and also the approximate location and value of $\hat{p}$.

- Open the file **ScottishShoes** and use software with the variable *Foot* to get a ***Hypothesis Test for One Proportion*** to check the numerical answers to the calculation.

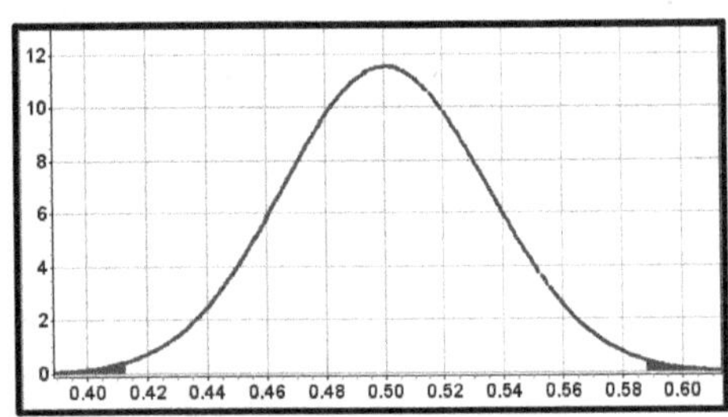

i. The sketch you made should resemble this picture, with the shading on the left starting at the position of the $\hat{p} = 0.412$ and then repeated on the right hand side. This picture shows the *p*-value; how does it show the *p*-value? [*Hint:* Compare the Texel graph.]

j. One way of understanding the difficult idea of *p*-value is to think of it as: "The probability of getting results *as extreme* as the data we actually have, if H_0 is true." (And so, a small *p*-value says that the results we have are rare or extreme, and a large *p*-value says that our data are not extreme.) How does putting a *p*-value in the "tails" of a distribution show "extremeness"?

k. Can you reject the null hypothesis? Give a reason for your answer.

l. Give a good interpretation of your test in the context of the data.

m. Confused Conrad's answer to part l reads like this: "The test is not statistically significant because the *p*-value is so small. Only 1.1% of shoes left on Scottish beaches are left shoes." Poor Conrad; he is really confused! Identify as many of his errors as you can.

3. ***Right-Handers*** The traditional figure given for the percentage of left-handers is from 10%–12%, (See the paper cited in the ***Notes*** (Medland, et.al, 2004) and other research suggests that another 4% or 5% of people are "partially" left-handed. That is, they do some things with their left hand and some things with their right hands. In the sample data from the file ***CombinedClassDataY09*** about 5% of the students identified themselves as "ambidextrous." These students may be among those who do some things left-handed and other things right-handed. If about 15% of the population is either left-handed or ambidextrous, that leaves 85% who are "purely" right-handed. We will use our data to test whether the idea that 85% of the population of CSM students is right-handed or whether it is some other number.

Combined Class Data 09

	DominantHand			Row Summary
	Right	Left	Ambidextrous	
	274 0.864353	27 0.0851735	16 0.0504732	317 1

S1 = count( )
S2 = rowProportion

a. [Review] Using the table calculate $P(R)$, where R stands for "right-handed," confirm that the "not" formula from §1.2 $P(Not\ A) = 1 - P(A)$ can be used to give you the proportion of left-handed or ambidextrous combined.

b. Using the *notation for hypothesis tests,* what symbol should be used for the fraction $274/317$: $P(R)$, $\hat{p}$, p or p_0? Give a reason for your answer and say briefly why the others are wrong.

c. What number are we using for the p_0 for this test? Why is it not 0.864353?

d. ***Step 1:*** Using the information in the answers to parts b and c, set up the null and alternate hypotheses for this test. Use the correct notation.

e. ***Step 2:*** We have already discussed in the ***Notes*** that our sample is not a simple random sample. Are the other conditions for a hypothesis test met? Show simple calculations to make your point.

f. Describe the sampling distribution of $\hat{p}$ s for the p_0 we have chosen. Your description should include shape, center, and spread.

g. Use the answers to part f to make a sketch of the sampling distribution, showing the mean and the standard deviation.

h. ***Step 3:*** Calculate the test statistic using the correct formula and using the correct notation.

i. Judging from the test statistic, is the $\hat{p}$ reasonably likely or rare? Give a reason for your answer.

j. Judging from the test statistic, will the p-value be smaller than 0.05 or bigger than 0.05? Give a reason for your answer, perhaps accompanied by a sketch.

k. ***Step 4:*** Find the p-value, using the correct notation. Use the Normal chart.

- Open the file ***CombinedClassDataY09*** and using the variable *DominantHand* get a ***Hypothesis Test for One Proportion*** to check the numerical answers to the calculation done by hand.

l. Is your test statistically significant? Give a reason for your answer.

m. Can you reject the null hypothesis? Give a reason for your answer.

n. ***Step 5:*** Give a good interpretation of your test in the context of the data.

o. Here is a picture of the sampling distribution (with mean at $p_0 = 0.85$) that was used in the hypothesis test. In this graph, what does the amount of shading indicate? Is it the z, the $\hat{p}$ or the p-value?

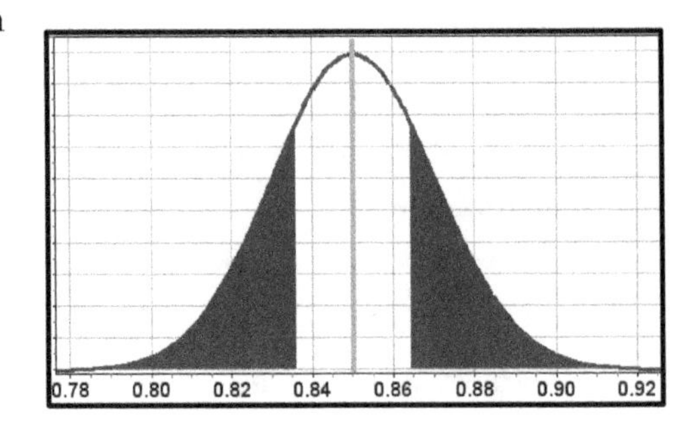

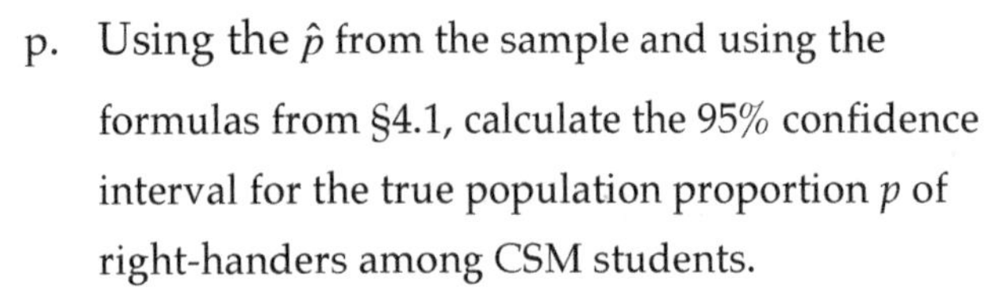

p. Using the $\hat{p}$ from the sample and using the formulas from §4.1, calculate the 95% confidence interval for the true population proportion p of right-handers among CSM students.

q. Does your confidence interval include the p_0 we have been testing?

r. True or false, and explain why. If a hypothesis test is *not* statistically significant then the confidence interval based upon a $\hat{p}$ will include the p_0 for the test.

s. True or false, and explain why. If a hypothesis test (such as we have been doing) *is* statistically significant then the confidence interval based upon a $\hat{p}$ will *not* capture the p_0 for the test.

4. ***The King of Rufutania's Seventieth Birthday and Palace Intrigue.*** The king of Rufutania is a megalomaniac, jealously guarding his standing and power and always on the lookout for unrest in his kingdom. His picture appears everywhere as well as on the coins of his realm, which all bear his image on the head side. To celebrate his seventieth birthday, the king directs the director of the Royal Mint to make coins that come up heads exactly 70% of the time when flipped. The questions are on the next page.

a. The king, always suspicious, suspects that the director of the Royal Mint has not followed his directions. The king summons the director to bring twenty coins, chosen at random, which the king will flip. When he was a youth, the king had taken statistics (some of which he has forgotten, however) and agrees to abide by a hypothesis test based on the flipping of the twenty coins. What are the null and alternate hypotheses? What is the p_0?

b. Is the sample size large enough for the test to be valid?

c. At the answer to part b, the king is furious, but the director of the Royal Mint, scrambling for his life, suggests they do a test with forty coins. Now is the sample size large enough? Explain.

d. They do the coin flipping, and 25 of the 40 flips come up heads. Will the director of the Royal Mint lose his head? (He will lose his head if the king is convinced that the coins come up heads less than 70% of the time.) Explain with a hypothesis test. Follow the five steps. Be complete.

e. The enemy of the director of the Royal Mint arranges to collect $n = 400$ coins, and, surprisingly, 250 of them turn up heads when flipped, yielding the same sample proportion $\hat{p}$ as with $n = 40$. Should the director of the Royal Mint be very worried? Explain why or why not. You may well have to do the hypothesis test again.

f. What is the (statistical) moral of this story?

5. ***Pete Repete Reprise*** Recall the question about Pete and Repete and their argument about the proportions of college students having tattoos. Recall that:

- Pete's idea was that the population proportion is $p = 0.30$, (so for Pete, $p_0 = 0.30$)
- Repete's idea was that the population proportion is $p = 0.20$, (and so, $p_0 = 0.20$)

Here are data for many semesters, and we can use these data to evaluate their ideas:

	Females			Males			Overall		
Semester, Year and Place	Proportion having a tattoo	Number having a tattoo	n	Proportion having a tattoo	Number having a tattoo	n	Proportion having a tattoo	Number having a tattoo	n
Both, 1999, PA	0.131	18	137	0.191	13	68	0.151	31	205
Spring 2008. CA	0.255	12	47	0.160	8	50	0.206	20	97
Autumn 2008,CA	0.254	16	63	0.127	8	63	0.190	24	126
Spring 2009. CA	0.372	32	86	0.219	14	64	0.307	46	150
Autumn 2009,CA	0.258	23	89	0.115	9	78	0.192	32	167
Spring 2010. CA	0.266	25	94	0.309	25	81	0.286	50	175
Autumn 2010,CA	0.149	15	101	0.146	13	89	0.147	28	190
Spring 2011. CA	0.191	18	94	0.218	19	87	0.204	37	181

- This exercise may be done by hand (and calculator) or using software. If using software, use the ***Hypothesis Test for One Proportion*** for the first part and use the facility that allows the sample size, the $\hat{p}$ and the p_0 to be entered. See the ***Software Supplement***.
- Use the information given above for Spring 2008, CA that shows that 20 out of the 97 students had tattoos. Use these data to test Pete's idea that $p_0 = 0.30$. Use a two-sided hypothesis test.

a. Is the test statistic in the reasonably likely region or the rare region?

b. What is the *p*-value?

c. So does Pete lose this argument? Why?

d. Repeat (!!) the test of Pete's idea that $p_0 = 0.30$ but now using the overall data for Spring 2010 showing that 50 out of 175 students had tattoos. What is the test statistic and what is the *p-value*? Does Pete lose this argument? Express the outcome in terms of the null hypothesis.

e. With these same data (Spring 2010), test Repete's idea that $p_0 = 0.20$. (What do you have to change?) Does Repete lose? Explain why or why not.

- With the Spring 2010 data shown above, get a ***Confidence Interval for One Proportion***, either by hand calculation or by using software.

f. The confidence interval gives plausible values for the population proportion *p*. Explain how it supports Pete's idea and does not support Repete's idea.

g. Get the confidence interval for the Autumn 2010 overall data that shows that 28 out of 190 students had tattoos. What can you say about Pete's and Repete's ideas? Do your results support either one? Why?

6. ***Born on Sunday?*** Are babies more likely or less likely to be born on some days of the week rather than others? Are they less likely to be born on weekends? Or more likely? Specifically, are babies less or more likely to be born on *Sundays*? We can answer this question, because we have a random sample of all births in the USA for 2006. For this exercise we will work with a small random sample ($n = 210$) of births. Our statistical question is:

"Do we have evidence that for the population of babies born in the USA, the proportion of Sunday births is different from what we expect if births for each day of the week are equally likely?"

TwoTenBirthSample		
SundayBirth	Not Sunday	191
	Sunday	19
	Column Summary	210
S1 = count ( )		

- Open the file ***TwoTenBirthSample*** and use software to get the frequencies of Sunday births and "not Sunday" births recorded in the variable *SundayBirth*. The table should be similar to the one shown here.
    a. First, use the data shown here to calculate the sample proportion of *Sunday* births in our small sample of $n = 210$. Assign the correct symbol to this proportion; should the symbol be $\hat{p}$, p or p_0 ? Give a reason for your choice of symbol.
    b. Thinking about the statistical question: If babies are *equally likely* to be born on any specific day of the week, then in a sample of $n = 210$, how many babies should be born on Sunday?
    c. If babies are *equally likely* to be born on any specific day of the week, then what *proportion* of babies should be born on Sunday? Do a calculation, and in this instance, give the answer rounded to *six* decimal places.
    d. The proportion that was calculated in part c is the proportion of Sunday births, if the proportions by day are equally likely, and it will be a part of both our null and alternate hypotheses. What is the best symbol for the proportion calculated in part c? Should the symbol be $\hat{p}$, p or p_0 ?
    e. Give a reason that the answer to part d should *not* be p rather than the symbol you chose.

Step 1: Setting up the Null and Alternate Hypotheses

f. So far, we have seen that we have a sample proportion $\hat{p} = \frac{19}{210} \approx 0.0905$ and we have determined that our idea for the population proportion is $p_0 = 1/7 \approx 0.142857$. Using the H_0 format and symbols shown in the ***Notes***, write the Null Hypothesis.

g. Write the Alternate Hypothesis using the H_a format shown in the Notes. So which symbol of the three: $>, <,$ or $\neq$ is the most appropriate for the alternate hypothesis? (To decide which of the symbols of $>, <,$ or $\neq$ should be used in the alternate hypothesis, read the statistical question. Our statistical question does not state that the proportion born on Sunday is to be greater than the "equally likely" proportion nor does it predict that the population proportion will be less than the "equally likely" proportion. It allows for *either* greater or less with the word "different." Note: the H_a is based upon our statistical question, not our sample data!)

Step 2: Checking the Conditions There are three conditions to be checked.

h. *Simple Random Sample*: Our sample of $n = 210$ *is* a Simple Random Sample of all of the births recorded in the year 2006. So, if our population of interest is *all* births in the USA, is the condition that our sample is a SRS met? (We must always check for randomness.)

i. Check the (second) condition that $np_0 \geq 10$ and also $n(1-p_0) \geq 10$. (This is the condition that must be met if we use a Normal Distribution rather than a Binomial Distribution.)

j. A third condition is that the size of the population should be at least ten times the size of the sample? Is that condition met if the population is all births in the USA in 2006? Give a very brief reason.

Step 3: Calculating the Test Statistic

k. Calculate the test statistic using the formula for the test statistic.

l. Does the value of the test statistic indicate that the $\hat{p}$ is *rare* or *reasonably likely*? Give a reason for your answer.

m. Draw a sketch of a Normal Distribution, (as shown) and put $p_0 \approx 0.143$ as the mean of the distribution. In your calculation of the test statistic, the denominator should have been $\sqrt{\frac{p_0(1-p_0)}{n}} = \sqrt{\frac{0.142857(1-0.142857)}{210}} \approx 0.024$. This number is the standard deviation of the sampling distribution we are using. Use the numbers for the mean and standard deviation to determine the location of *p*-hats that are one and two standard deviations from the mean; that is 0.143 ± 0.024, 0.143 ± 2(0.024). the picture has started the process. Locate the $\hat{p} \approx 0.0905$ on the sketch. [Your answer to part k should be $z \approx -2.17$. Does the *p*-hat shown in your sketch come out to be just beyond two standard deviations from the mean? Your sketch and your conclusion to part l (i.e. part "el") should agree.]

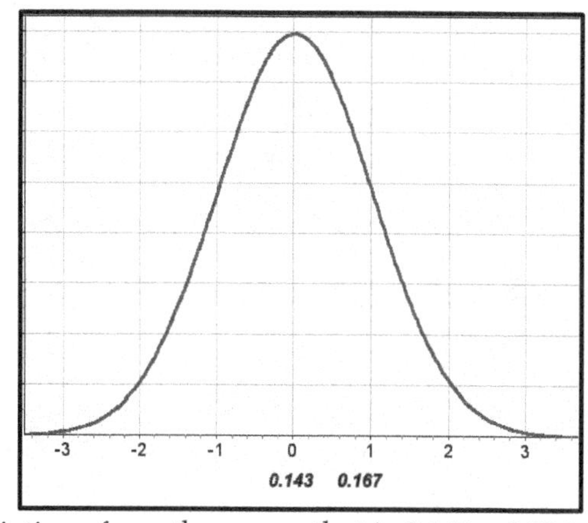

Step 4: Calculating the p-value We will calculate the p-value in two sub-steps; part "n" is the first.

n. Using the test statistic calculated in part k, the Normal Distribution Chart or software, find the probability $P(\hat{p} < 0.0905) = P(z < -2.17)$. On your sketch show this probability by appropriate shading and an arrow pointing to the shading.

o. Because our alternate hypothesis is two sided (H_a: $p \neq 0.142857$), the p-value is twice what was found in part n. Show the *p*-value by shading the graph made to answer part m.

Check your calculations using software.

- Open the file ***TwoTenBirthSample*** and using the variable *SundayBirth* use ***Hypothesis Test for One Proportion*** to do the calculations. Check the results against the hand calculations done above. (The ***Software Supplement*** uses this example.)

Here are software results for the hypothesis test.

Hypothesis test results:
Outcomes in : SundayBirth
Success : Sunday
p : Proportion of successes
H_0 : p = 0.142857
H_A : p ≠ 0.142857

Variable	Count	Total	Sample Prop.	Std. Err.	Z-Stat	P-value
SundayBirth	19	210	0.09047619	0.024147254	-2.1692242	0.0301

p. For this test, the p-value is 0.0301. How is this p-value shown on the graph below (which should resemble the one you made for part m. Is the p-value:

(i) on the "x-axis" with 0.03 just to the right of zero, OR is it (ii) by the slivers of shading to the left of 0.10 and to the right of 0.20.

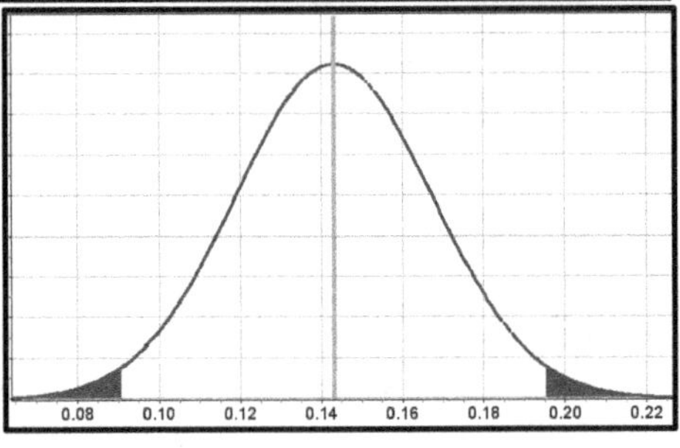

Step 5 Interpreting the Result

q. Is the test statistically significant? Give a reason for the answer.

r. Do we have evidence against the Null Hypothesis, or is our test consistent with the H_0?

s. Give an interpretation of the result in the context of the statistical question. That is, do our results indicate that babies are just as likely to be born on Sunday as any other day, or not?

t. Using $\hat{p} \approx 0.0905$ and n = 210 find (either by calculating or by using software) a 95% Confidence Interval for the proportion of Sunday births p.

7. Births on Monday? The exercise just above showed evidence that the proportion of births on Sunday is different from the proportion one would expect if each day's births were equally likely. Our small sample of n = 210 shows evidence that in the population of all births, Sunday births are less numerous than expected. What about Monday? We can do the same kind of test for that day of the week. Having seen that the proportion for Sunday is lower, perhaps we may think that the proportion for Monday is higher than the $p_0 = 1/7 \approx 0.142857$ expected if the proportions for each day of the week are equal. Our statistical question is:

Do we have evidence that the proportion of births on Monday is higher than one-seventh of all births?

- Open the file ***TwoTenBirthSample*** and use software to get the frequencies of Monday births and "not Monday" births recorded in the variable *MondayBirth*. The table should be similar to the one shown here.

TwoTenBirthSample

MondayBirth	Monday	33
	Not Monday	177
	Column Summary	210

S1 = count ( )

a. Calculate the proportion of Monday births in the sample and assign the correct symbol. Is the proportion higher than the expected proportion?

b. Set up the Null and Alternate Hypotheses for the statistical question written in italics just above. [Should the alternate hypothesis be one-sided or two sided? Be prepared to give a reason for the choice. Notice the language in the question mentions "higher."]

c. Although we checked the conditions for these data in Exercise 6, give reasons why each of the three conditions [(i) Random sampling, (ii) $np_0 \geq 10$ and $n(1-p_0) \geq 10$ and (iii) Population large compared with the sample.] are met.

d. Calculate the test statistic for the hypothesis test.

e. Judging from the test statistic, do you think that this hypothesis test will be statistically significant? Give a reason for your answer.

f. Use the Normal Distribution Chart to find the p-value.

g. Give an interpretation of the hypothesis test in the context of the statistical question asked.

h. Make a sketch of the sampling distribution used, and show (i) the mean of the sampling distribution, (ii) the p-hat on the x-axis, and (iii) by shading, and labeled, the p-value.

Check your calculations using software.

- Open the file ***TwoTenBirthSample*** and using the variable *MondayBirth* use ***Hypothesis Test for One Proportion*** to do the calculations. Check the results against the hand calculations done above.

i. Your sampling distribution sketch should resemble this graphic. Explain why the shading is only on the right side, and not on both sides.

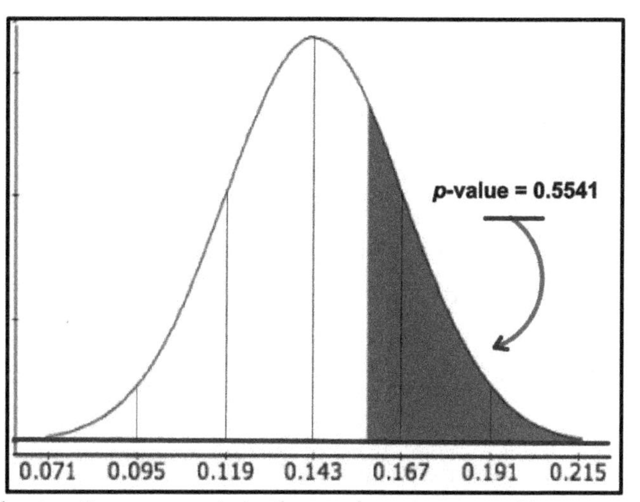

j. The p-value for this test is 0.28. Confused Conrad, to attempting to interpret the p-value, writes: "The test is statistically significant because this is a big p-value and shows that 28% of the children are born on Mondays, which is a lot" Correct all of CC's mistakes.

k. We have failed to reject the Null hypothesis. Our test statistic shows that given our H_0 our sample proportion is reasonably likely; our p-value indicates that there is a fairly high probability that we would get our p-hat or one greater if the H_0 is true. Have we therefore proved the Null Hypothesis to be true? Why or why not? [This question may be discussed in class in general.]

§4.3 Exercises on Two Proportions Inference

Special Exercise 1: Smoking and Premature Births

Here are some sample data for births showing the relationship between birth status (whether the birth was "premature" or "full-term") and the smoking status of the mother (smoker or not.)

EnglishSmokingPregnancyData.csv

		BirthStatus		Row Summary
		Full-Term	PreMature	
SmokingStatus	Non-Smoker	4036	365	4401
	Smoker	465	49	514
	Column Summary	4501	414	4915

S1 = count ( )

The data are from England, but we do not know whether they are a random sample of all English births or not; they may have come from just one hospital or one county's records. Also, we do not know when the data were collected. Our statistical question is:

Does mothers' smoking affect the likelihood of a premature birth?

Step 0: Decide what we are studying. We need to decide exactly the definition of the proportions we will study. The question "Does mothers' smoking affect the likelihood of a premature birth?" leads us to decide to think about:

p_S=The population proportion of premature births for mothers who are smokers (or we could say: "The probability of a premature birth for mothers who are smokers")

p_N=The population proportion of premature births for mothers who are *non*-smokers (or we could say: "The probability of a premature birth for mothers who are *non*-smokers")

1. a. What are the cases for these data? (There are two possible good answers and lots of wrong ones.)
   b. Which variable should be the explanatory variable and which variable the response variable?
   c. In our statistical question (not in life!) is a "success" a *premature birth* or a *full-term* birth?
   d. Explain why it is that there are no "hats" in the notation used above.

Step 1: Write Null and Alternate Hypothesis.

2. Write the null hypothesis using the notation introduced above.
3. The form of the alternate hypothesis depends upon how you think smoking affects the proportion (i.e., the probability) of a premature birth.
   a. Do you think that mothers' smoking makes the proportion (probability) of a premature birth higher for all mothers in the population? Lower? Or have you no idea how smoking affects premature births?
   b. Does your answer in part a lead to a one-sided test or a two-sided test? Give a reason for your answer.
   c. If your answer to part b is one-sided (you may well have said $H_a : p_S > p_N$ where p_S is the probability of a premature birth for a smoking mother and p_N is the probability of a premature birth for a non-smoking mother), we will come back to that (perhaps as homework). Just for practice, we are going to test the two-sided alternative. Write the alternate hypothesis for a *two-sided* test.

Step 2: Check the Conditions.

4. Are we sure that we have independent random samples? Or are we not certain? (We will proceed with the test even if we are not certain, but then we will have some doubt about our conclusions.)
5. a. Calculate $\hat{p}_N$ $\hat{p}_S$ and $\hat{p}$, our sample estimate of the overall proportion of premature births.

   b. Determine whether all of $n_S\hat{p}$, $n_S(1-\hat{p})$, $n_N\hat{p}$, $n_N(1-\hat{p})$ are all at least 5.
6. One of the conditions that is usually not a problem is that the population from which the sample was drawn should be at least ten times the size of the sample. Here we have a sample size of 4,915. Are there likely to be at least ten times this many births in England in a year?

Step 3: Compute the Test Statistic.

7. Look up the formula in the ***Notes*** for the test statistic. Write down the formula.
8. Using the numbers given, calculate the test statistic. You should have $\hat{p}_S = 0.0953$, $\hat{p}_N = 0.0829$, and $\hat{p} = 0.0842$, and you should come out with $z = 0.95744$ for the test statistic.
9. a. Judging from the value of the test statistic, does the test look like it is *statistically significant* or *not statistically significant?* Give a reason for your answer.j

   b. Judging from the value of the test statistic, will we reject or *not* reject the null hypothesis?

Step 4: Compute a p-value.

10. We have enough information to say whether the p value will be bigger or smaller than $\alpha = 0.05$ even before we find the p-value. Will the p-value be bigger or smaller than $\alpha = 0.05$? And why?
11. Make a sketch of a Normal distribution (the sampling distribution) showing the mean, the test statistic, and by shading the p-value. Consult the pictures in the ***Notes.***
12. a. Calculate the p-value using the Normal Distribution Chart.

    b. Is the value you got consistent with your answer to question 9?

    c. Is the value you got consistent with your answers to question 10? (If it is not then you either made a mistake or your answer to question 9 is not correct.)

Step 5: Interpretation. Consult the guidelines in the ***Notes for §4.2.*** An interpretation should state the evidence in the context of the problem—that is, in this case, smoking and premature births.

13. Write your interpretation for this hypothesis test in the context of the data.

Confidence Interval: Difference in Proportion of Premature Births between Smoking and Non-Smoking Mothers.

14. a. Using the formula for a confidence interval for two proportions,

    $$\hat{p}_1 - \hat{p}_2 \pm z^* \sqrt{\frac{\hat{p}_1(1-\hat{p}_1)}{n_1} + \frac{\hat{p}_2(1-\hat{p}_2)}{n_2}}$$, calculate a 95% confidence interval for the difference of the proportions of premature births between births to smoking mothers and non-smoking mothers.

    b. Does your confidence interval include zero?

    c. Write a conclusion (in context) about your confidence interval that takes into account your answer to part b. (This may be difficult to put in elegant English; ask for help.)

- Check the numerical values of the calculations by using software for a ***Hypothesis Test for Comparing two Proportions*** and a ***Confidence Interval for Comparing Two Proportions*** using the numbers in the table at the beginning of the exercise. Consult the ***Software Supplement.***

A one-tailed test.

15. You were asked to do a two-tailed test, but you may think that the proportion of premature births to smoking mothers should be higher than to non-smoking mothers.
    a. Write up the null and alternate hypotheses for this one-tailed test, using the correct notation.
    b. Which parts of the significance test will be different and which parts will be exactly the same for a one-tailed test as for a two-sided test? If they are the same, write "same"; otherwise, show what you would get for the one-tailed test.

       (i) Conditions? (ii) Test Statistic? (iii) Picture? (iv) p -value?

2. ***Politics and Gender*** It is generally thought that women are more liberal politically than men. Is this true for college students or is it that the proportion of students who have liberal political views is essentially the same, independent of gender.

- Use the file ***CombinedClassDataY09*** and use software to get a ***Two-way table*** that is equivalent to the one shown. We will consider these data to be a sample of *all* the students at the college. However, the sample may not be a random sample of all students. (See Exercise 1 in §3.4)

Combined Class Data 09

		PolticalView			Row Summary
		Liberal	Moderate	Conservative	
Gender	F	95	69	11	175
	M	70	61	11	142
Column Summary		165	130	22	317

S1 = count( )

Our statistical question is: *In the population from which these data are drawn, are female students more likely to have liberal political views than male students?*

a. Our interest is in the proportion of male students whose political view are "liberal" and the proportion of female students whose political views are "liberal." What symbols should be used to designate the *population proportions*? (Should the symbols have "hats" or not? Notice that we are really only interested in "Liberal" views as contrasted to all other political views.)
b. Set up the null and alternate hypotheses for the test that we are doing, using the correct notation. Does the statistical question imply a one- or a two-sided test? Why?
c. Calculate the sample proportion of males who have liberal political views and the sample proportion of females who have liberal political views and assign the correct symbols to these proportions.
d. Check whether the conditions for using the Normal distribution are met for this problem.
e. Calculate the test statistic.
f. From the value of the test statistic, does it appear that that the hypothesis is *statistically significant* or *not?* Give a reason for your answer.
g. Make a drawing of the Normal sampling distribution and show the mean and the location of your test statistic.
h. Considering the test statistic that you got, should the p-value be bigger or smaller than $\alpha = 0.05$? Give a reason for your answer, perhaps aided by your drawing.
i. Use the Normal Distribution Chart or software to find the p-value. Show your work using correct notation.

j. Show the *p*-value on your drawing by shading (part g) in an area.

- Use software with the file ***CombinedClassDataY09*** to check the calculations for the hypothesis test. Consult the ***Software Supplement*** under ***Hypothesis Test for Comparing Two Proportions***.

k. Did your calculated numbers agree with the numbers from software? Did your graphic resemble this one?

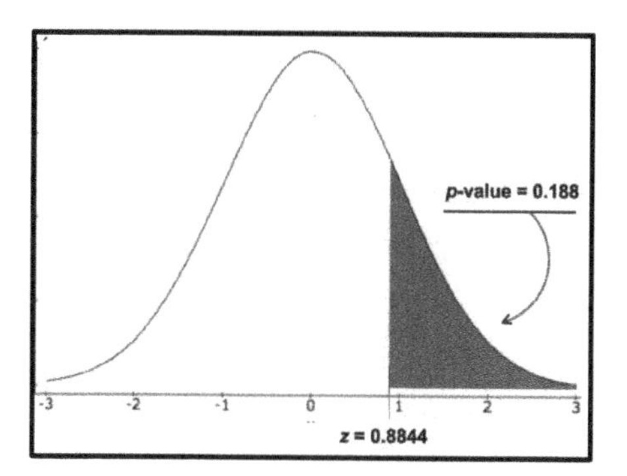

l. Put together all of your evidence to answer the question: "Can we say that in the population of *all* students at the college, female students are more likely to have liberal political view than male students?" Consider also the fact that the sample is not a random one.

3. *Tattoos and Politics* What is the relationship between having a tattoo and political views? The data are the same student data as in Exercise 2: They are students taking statistics at a community college, so the sample may be representative of the students at the college. However, it is not a random sample. Our statistical question is:

Are students in this college who have tattoos more likely to have liberal political views?

- Use the file ***CombinedClassDataY09*** and use software to get a ***Two-way table*** that is equivalent to the one shown showing the relationship between having a tattoo and political views in the sample of students for the year 2009. The two-way table should resemble the one here, although the software may put the categories in alphabetical order.

CombinedClassDataY09

		PolticalView			Row Summary
		Liberal	Moderate	Conservative	
Tattoo	N	114	108	17	239
	Y	51	22	5	78
Column Summary		165	130	22	317

S1 = count ( )

a. Our interest is to compare and generalize about the proportion of students who have tattoos whose political view are "liberal" with the proportion of students who do *not* have tattoos whose political views are "liberal." If we were using the conditional notation of §1.2, should we be calculating $P(L \mid N), P(L \mid Y)$ or should we be calculating $P(N \mid L), P(Y \mid L)$? (Think: which variable are we taking as the explanatory variable according to the way the statistical question is phrased?)

b. Since we are not in §1.2 and are dealing with hypothesis tests, what symbols should be used to designate the *population proportions*? (What subscripts should the symbols have? Should the symbols have "hats"?)

c. Set up the null and alternate hypotheses for the test that we are doing, using the correct notation. Should this be a one sided or a two sided hypothesis test, given the way the statistical question is phrased?

d. Assign the correct symbols to the sample proportion of students who have tattoos who have liberal political views and the sample proportion of students without tattoos who have liberal political views. The actual calculations are the ones that were done in part a.

e. Check whether the conditions for using the Normal distribution are met for this problem. Explain how you checked.

f. Calculate the test statistic for the hypothesis test.

g. From the value of the test statistic, does it appear that that the hypothesis is *statistically significant* or *not?* Give a reason for your answer.

h. Make a drawing of the Normal sampling distribution and show the mean and the location of your test statistic.

i. Considering the test statistic that you got, should the *p*-value be bigger or smaller than $\alpha = 0.05$? Give a reason for your answer, perhaps aided by your drawing.

j. Use the Normal Distribution Chart or software to calculate the *p*-value.

k. Show the *p*-value on your drawing by shading in an area.

- Use software with the file ***CombinedClassDataY09*** to check the calculations for the hypothesis test. Consult the ***Software Supplement*** under ***Hypothesis Test for Comparing Two Proportions*** for help.

l. Here are two possible answers to using software to get results. Is one correct and one wrong? Or are both correct? What accounts for the differences, especially in the direction of the inequality?

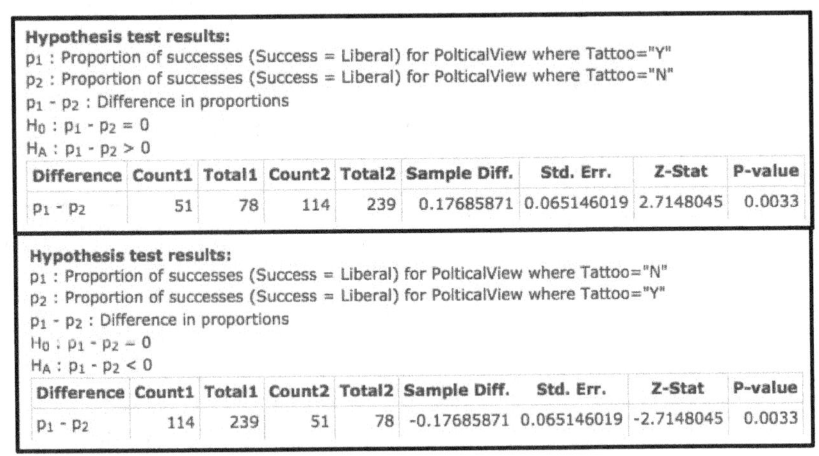

Hypothesis test results:
p_1 : Proportion of successes (Success = Liberal) for PolticalView where Tattoo="Y"
p_2 : Proportion of successes (Success = Liberal) for PolticalView where Tattoo="N"
$p_1 - p_2$: Difference in proportions
$H_0 : p_1 - p_2 = 0$
$H_A : p_1 - p_2 > 0$

Difference	Count1	Total1	Count2	Total2	Sample Diff.	Std. Err.	Z-Stat	P-value
$p_1 - p_2$	51	78	114	239	0.17685871	0.065146019	2.7148045	0.0033

Hypothesis test results:
p_1 : Proportion of successes (Success = Liberal) for PolticalView where Tattoo="N"
p_2 : Proportion of successes (Success = Liberal) for PolticalView where Tattoo="Y"
$p_1 - p_2$: Difference in proportions
$H_0 : p_1 - p_2 = 0$
$H_A : p_1 - p_2 < 0$

Difference	Count1	Total1	Count2	Total2	Sample Diff.	Std. Err.	Z-Stat	P-value
$p_1 - p_2$	114	239	51	78	-0.17685871	0.065146019	-2.7148045	0.0033

m. Put together all of your evidence to answer the question: "Can we from our data say that in the population of *all* students at the college, those with tattoos are also more likely to have liberal political views?" Consider also the fact that the sample is not a random sample.

n. Calculate a 95% confidence interval to show how much more likely the students with tattoos are to have liberal political views than the students without tattoos.

o. Give a good interpretation of your confidence interval.

- Use software with the file ***CombinedClassDataY09*** to check the calculations for the confidence interval. Consult the ***Software Supplement*** under ***Confidence Interval for Comparing Two Proportions*** for help. You should have something like the following.

p. What would happen to the lower and upper limits of the confidence interval if the order of Tattoo ="Y" and Tattoo = "N" were reversed?

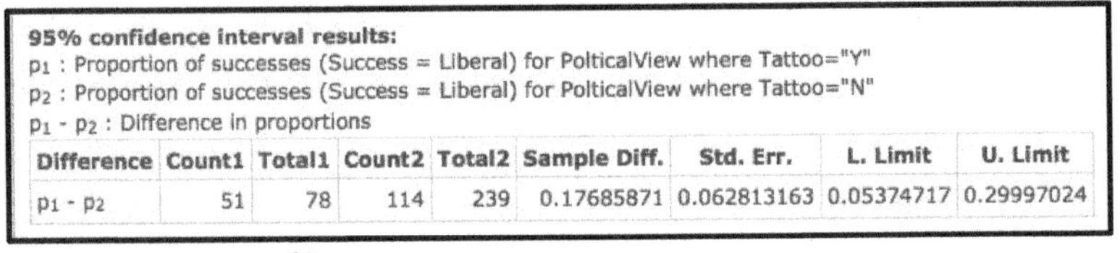

95% confidence interval results:
p_1 : Proportion of successes (Success = Liberal) for PolticalView where Tattoo="Y"
p_2 : Proportion of successes (Success = Liberal) for PolticalView where Tattoo="N"
$p_1 - p_2$: Difference in proportions

Difference	Count1	Total1	Count2	Total2	Sample Diff.	Std. Err.	L. Limit	U. Limit
$p_1 - p_2$	51	78	114	239	0.17685871	0.062813163	0.05374717	0.29997024

4. *Another Smoking and Childbirth Problem.* From the same data as the example of smoking and premature births, they also recorded whether the child survived. Here are the data.

Collection 1

		Survival		Row Summary
		Died	Survived	
Smoking_Status	Non-Smoker	74	4327	4401
	Smoker	15	499	514
	Column Summary	89	4826	4915

S1 = count( )

a. Test whether smoking affects the probability of child survival using a hypothesis test. Show all the five steps clearly: set up the hypotheses, check the conditions, calculate the test statistic and the p-value, and come to a conclusion. Write an interpretation of your conclusion in the context of the data. The test you carry out may be a one-sided or two-sided test, and you may choose the significance level (that is, you may choose the alpha).

- b. Check your calculations by using software to get a ***Hypothesis Test for Comparing Two Proportions***. Consult the ***Software Supplement*** where this example is used.

5. *Better data for Smoking and Premature Births.* We have data that are actually better than the data in Exercise 1 because we are certain that the data are a simple random sample of all births in 2006. The source of the data used was not specified, except that they came from England. The file ***QuattroMilBirths*** is a simple random sample of approximately 4000 births from the entire population of births recorded in 2006 in the USA.

- Use the file ***QuattroMilBirths*** and software to get a two-way table showing the relationship of the variables *Premature* and *MotherSmokes*

QuatroMilleSample

		Premature		Row Summary
		Full-Term	PreMature	
MotherSmokes	No	1673	174	1847
	Yes	195	30	225
	Column Summary	1868	204	2072

S1 = count( )

a. What are the cases for these data?

b. We will look at the proportion of premature (rather than full-term) births; calculate from the table the proportions $\hat{p}_N$, $\hat{p}_S$ and also $\hat{p}$. (You will be able to check your answers shortly.)

c. What is the meaning of $\hat{p}$ compared with the meaning of $\hat{p}_N$ and $\hat{p}_S$?

d. Checking the conditions. Here we are fortunate to have a simple random sample. The population of all births in 2006 must be bigger than 10 x 2072, so we just need to check whether our sample sizes are big enough. All of $n_N\hat{p}, n_N(1-\hat{p}), n_S\hat{p}, n_S(1-\hat{p})$ should be greater than 5. However, we need only check: the one that is likely to be smallest. Show the calculation.

- With the file ***QuattroMilBirths*** use software to perform the ***Hypothesis Test for Comparing Two Proportions*** to compare the probability of a premature birth to mothers who smoke compared with the probability of a premature birth to mothers who do not smoke. Consult the ***Software Supplement***, where this example is shown. The output should be similar (but not necessarily the same) as the output shown here:

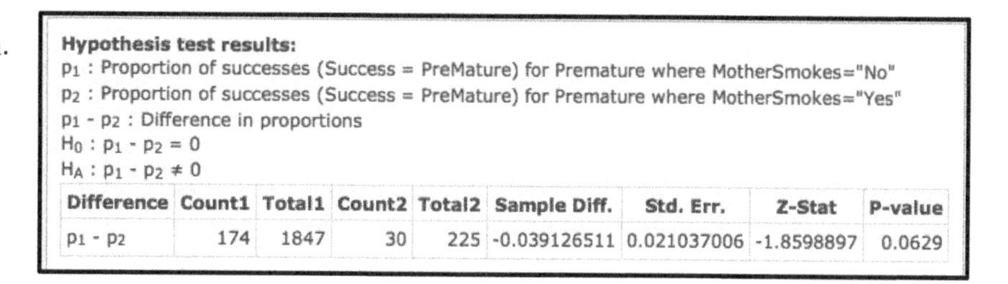

Hypothesis test results:
p_1 : Proportion of successes (Success = PreMature) for Premature where MotherSmokes="No"
p_2 : Proportion of successes (Success = PreMature) for Premature where MotherSmokes="Yes"
$p_1 - p_2$: Difference in proportions
$H_0 : p_1 - p_2 = 0$
$H_A : p_1 - p_2 \neq 0$

Difference	Count1	Total1	Count2	Total2	Sample Diff.	Std. Err.	Z-Stat	P-value
$p_1 - p_2$	174	1847	30	225	-0.039126511	0.021037006	-1.8598897	0.0629

e. Write, in symbol form, the hypotheses for the test shown in the software output. Is the test a one-sided or a two-sided test? How do you know?

f. Using your calculations from part c, show how the test statistic was calculated. Notice that the test statistic is negative.

g. Based upon the test statistic and the *p*-value, give an interpretation of the hypothesis test. Discuss with great insight whether the results are consistent with what was found in Exercise 1.

h. Here is a graphic of the sampling that was used in the hypothesis test. Make a copy of this Normal Distribution, and fill in the boxes with the relevant information using the software output.

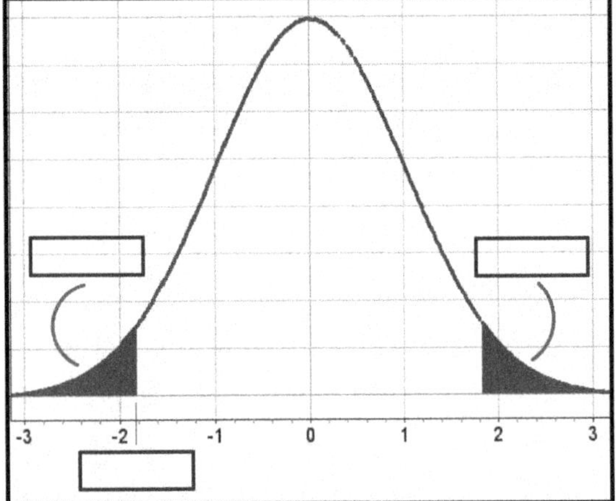

i. Suppose a one-sided test instead of a two-sided test was carried out with these null and alternate hypotheses: $\begin{matrix} H_0: P_S - P_{NS} = 0 \\ H_a: P_S - P_{NS} < 0 \end{matrix}$.

Which numbers in the output would not change and which numbers would change, and how? [Of course, you can modify the software to conduct the one-sided test to confirm your idea.]

j. Would the graphic shown in part h be different? If so, how? If not, why not?

§4.4 Exercises for Testing Independence

Special Exercise 1: Smoking and Birth Status Again

Our statistical question can be expressed in two ways:

- *Does the mother's smoking affect the likelihood of a premature birth?* *OR*
- *Is birth status independent of smoking status?*

We have seen these data before. We concluded that the probability of a premature birth is higher for births of smoking mothers than it is to births of non-smoking mothers; however, the difference of proportions was not big enough to reject the null hypothesis of no difference. Our hypothesis test was not statistically significant. Now we will analyze the same data but with a Test for Independence.

EnglishSmokingPregnancyData.csv

		BirthStatus		Row
		Full-Term	PreMature	Summary
SmokingStatus	Non-Smoker	4036	365	4401
	Smoker	465	49	514
	Column Summary	4501	414	4915

S1 = count ( )

1. Refer to the ***Notes*** for the definition of independence of two events. Call the event of a premature birth *Prem*, the event that the mother is a nonsmoker ***N***, and the event that the mother is a smoker ***S.***
   a. Use the correct conditional probability notation and show that the events Prem and S are not independent.
   b. Use the correct conditional probability notation and show that the events Prem and N are not independent.

Here is some background from the ***Notes.***

- We call the numbers in the table "counts." The count for the Smoker-Premature cell is 49.
- We compare the counts (or frequencies) actually in the table (the observed counts) with the counts the table would have if Smoking Status and Birth Status were independent.
- The expected counts (or frequencies) are the frequencies (or counts) expected if the proportions premature are the same for the smoking mothers and the non-smoking mothers.

Step 1: Setting Up Hypotheses

2. Refer to the ***Notes*** about setting up hypotheses for a test for independence. Set up the hypotheses for our statistical question above.

Step 3a and Step 2: Getting Expected Counts and Checking the Conditions How do we proceed? We begin by making the table of expected counts, which are the counts expected if there actually is no difference in proportion premature between smokers and non-smokers:

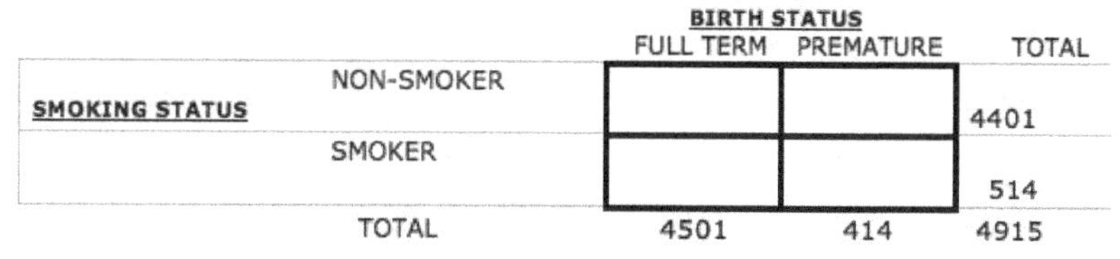

		BIRTH STATUS		
		FULL TERM	PREMATURE	TOTAL
SMOKING STATUS	NON-SMOKER			4401
	SMOKER			514
	TOTAL	4501	414	4915

The expected count for the "non-smoker premature cell" (the one with 365 in the actual table) is

$$\begin{pmatrix}\text{Overall Proportion}\\ \text{of Premature Births}\end{pmatrix}\times\begin{pmatrix}\text{Total Number}\\ \text{of NonSmokers}\end{pmatrix} \text{ which is: } \frac{414}{4915}\times 4401\text{, which is } 0.084231943\times 4401 = 370.70$$

so that 370.70 is the expected count for that cell. Notice that we do not round to the nearest whole number, even though there cannot be .70 of a premature birth.

3. On the same pattern, get the expected count for the "smoker-premature cell." The only difference in the formula is that you will 514 in place of 4401.
4. Get the other two expected counts. (Again, show your work, although there is not much.) The expected counts that you should get are in the output shown below.
5. Add the expected counts vertically and horizontally. What do you notice compared to the original observed table? What you notice should always be true if you have calculated the expected counts correctly.
6. Are the conditions met for the hypothesis test to proceed?

Step 3b: Comparing Expected and Observed: The Chi-Square Goodness of Fit Test Statistic

The formula is $\chi^2 = \sum_{\text{All cells}} \frac{(O-E)^2}{E}$ where O stands for "observed" and E stands for "expected," and the summation is for all the cells in the table.

	Full-Term	PreMature	Total
Non-Smoker	4036 (4030.3)	365 (370.7)	4401
Smoker	465 (470.7)	49 (43.3)	514
Total	4501	414	4915

Chi-Square test:

Statistic	DF	Value	P-value
Chi-square	1	0.9166947	0.3383

7. If the expected and the observed happened to be the same numbers, what would the value of χ^2 be?
8. If chi-square is small, does that support the null hypothesis or the alternative hypothesis? Give a reason for your answer. Conversely, if chi-square is large, which hypothesis does that result support? Give a reason for your answer.
9. Here is part of the calculation of the test statistic. Copy the completed parts and fill in the blank parts.

$$\chi^2 = \frac{(4036-4030.30)^2}{4030.30} + \frac{(\quad -370.70)^2}{370.70} + \frac{(\qquad)^2}{470.70} + \frac{(\qquad)^2}{\qquad} = 0.917$$

Step 4: Getting the p-value

10. Calculate the degrees of freedom for our table for Smoking Status and Birth Status using the df formula.
11. a. Read the appropriate row on the ***Chi-Square Distribution Chart*** and determine the Critical Value—the value beyond which a c2 that we have calculated would be considered rare if a = 0.05 and the df = 1.
    b. Judging by the test statistic and your answer to part a of this question, is our hypothesis test statistically significant? Give a reason for your answer.
12. a. Using the Chi-Square Distribution Chart and the procedure outlined in the ***Notes***, get an approximation to the p-value for our hypothesis test.
    b. Does your approximation agree with the exact calculated p-value?
    c. Does the p-value lead you to reject the null hypothesis or not reject the null hypothesis? Give a reason for your answer.
    d. Does your answer to 12c agree with your answer to 11b? Explain how the two answers agree.

Step 5: Interpretation

13. Write a good interpretation in the context of the data. "In the context of the data" means that you must say something about smoking and premature births.
14. We did this same hypothesis test using a Comparison of Proportions Test, where the null hypothesis was that there was no difference in proportion of premature births by the smoking status of the mother.
    a. Without doing that test again, state what the test statistic would be using $\chi^2 = 0.916$? (Hint: See: "Test of independence for a two-by-two table.")
    b. State the p value for the Comparison of Proportions Test? Give a reason for your answer.

2. *Tattoos and PoliticalView* In the exercises for §4.3, we explored the relationship between having a tattoo and having liberal political view. However, the categorical variable *PoliticalView* has three categories. Thus, it would appear that a good alternative for analyzing these variables is a Chi Square Test for Independence.

- The data are found in ***CombinedClassDataY09*** and the two-way table shown below may easily be produced using software.

CombinedClassDataY09

		PoliticalView			Row Summary
		(a) Liberal	(b) Moderate	(c) Conservative	
Tattoo	No	114 0.476987	108 0.451883	17 0.0711297	239 1
	Yes	51 0.653846	22 0.282051	5 0.0641026	78 1
Column Summary		165 0.520505	130 0.410095	22 0.0694006	317 1

S1 = count ( )
S2 = rowProportion

a. [*Review*] Show, using the definition of independence, that the *events* L (for liberal) and Y (for having a tattoo) are *not* independent. Use the notation correctly.

b. Are the probabilities that are calculated $P(L \mid N), P(L \mid Y), P(Mod \mid N), P(Mod \mid Y)$ etc., or are they $P(N \mid L), P(Y \mid L), P(N \mid Mod), P(Y \mid Mod)$ etc.? (*Hint:* Which variable is being regarded as the "explanatory"?)

c. Does it appear to you that the two variables are independent or not independent? Or (in different words) does it appear that there *is* a relationship (or *no* relationship) between having a *Tattoo* and *PoliticalView?*

d. We will test whether the two variables *Tattoo* and *PoliticalView* are independent or not with a Chi-square hypothesis test. We will let software do most of the calculations.. Set up the null and alternate hypotheses for this test, using the language specified in the ***Notes.***

- Using the file ***CombinedClassDataY09*** use software to get the results of the Chi-Square test for Independence. Choose the option that shows the Expected Counts if the software has that option. The output should resemble what is shown on the next page.

e. The expected count for the "Liberal-No" cell is 124.4 and the "Conservative-Yes" cell is 5.4. Show how these numbers were calculated.

f. [*Test-like question.*] Complete this sentence in the context of this hypothesis test: "The expected counts are the counts expected if the ____________ __________were true and the variables were ____________________."

g. Are the conditions met for the test for independence?

h. Slightly Slack Silas answers question g by saying. "We are okay because all of the counts are five or above, even though one of the counts is exactly five." SSS is making a mistake; what is it?

i. [*Test-like question*.] Show how the software calculated the test statistic by putting the numbers in the formula. You can check yourself by doing the calculation, but the important thing is that you show where the numbers go. All of the numbers that are needed are shown.

j. The degrees of freedom is shown as $df = 2$. Show how this was calculated.

Contingency table results:
Rows: Tattoo
Columns: PoliticalView

Cell format
Count Expected count

	(a) Liberal	(b) Moderate	(c) Conservative	Total
No	114 (124.4)	108 (98.01)	17 (16.59)	239
Yes	51 (40.6)	22 (31.99)	5 (5.41)	78
Total	165	130	22	317

Chi-Square test:

Statistic	DF	Value	P-value
Chi-square	2	7.7118537	0.0212

k. For $df = 2$, consult the ***Chi-Square Distributions Chart*** and or software to find the ***Critical Value*** for χ^2 for $\alpha=0.05$. Is the χ^2 for this test above that Critical Value? Does that make out result "rare" or "reasonably likely"?

- Use software with the ***Chi-Square Probability Distribution*** with $df = 2$ to show the probability of getting a value of the test statistic 7.712 or larger. The graph should resemble the one shown here, with just a "sliver" of shading under the curve.

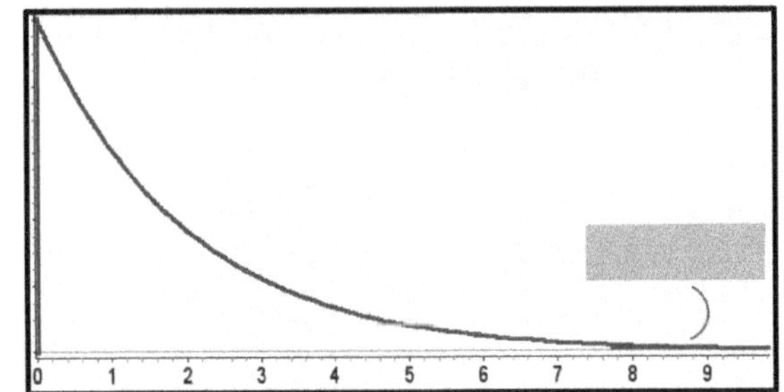

l. Where does the shading start, and what does the shading represent? Make a rough sketch of the graph, and show the answer for the "meaning of the shading" as a kind of "cartoon balloon" as shown here.

m. Is our test *statistically significant?* Give at least one reason for the answer.

n. Do we have enough evidence to reject the null hypothesis of independence? Give at least one reason for your answer.

o. Give a good interpretation of the results of this test in the context of the data.

p. A Chi-Square Test of Independence merely tells us that we can confident that there is an association between the variables. If the test is significant, it does not tell us the details of the matter. Use the contributions to the Chi-Square and the row and column percentages (shown here) to determine the details of the association between political views and having a tattoo.

Contingency table results:
Rows: Tattoo
Columns: PoliticalView

Cell format
Count (Row percent) (Column percent) Contributions to Chi-Square

	(a) Liberal	(b) Moderate	(c) Conservative	Total
No	114 (47.7%) (69.09%) (0.87)	108 (45.19%) (83.08%) (1.02)	17 (7.11%) (77.27%) (0.01)	239 (100%) (75.39%)
Yes	51 (65.38%) (30.91%) (2.66)	22 (28.21%) (16.92%) (3.12)	5 (6.41%) (22.73%) (0.03)	78 (100%) (24.61%)
Total	165 (52.05%) (100%)	130 (41.01%) (100%)	22 (6.94%) (100%)	317 (100%) (100%)

3. *Take Away Favourites* "Take away" is the term used in the UK, Australia, Hong Kong and other places rather than "Take Out." The Census @ School survey conducted in Australia asked about high school students' favorite "take away food" using the question shown here. See Exercise 4 in §2.1 for more about the Census @ School surveys in Australia; specifically, note whether the data can be regarded as a random sample of Australian high school students, or not. In the file ***CASAustralia2011Combined*** some of the categories have been combined into "Other."

17. What is your favourite type of take-away food?
Select one only.

- Chicken (e.g. BBQ chicken)
- Chips/Fries
- Fish (e.g. Fish and Chips)
- Fruit/Fruit Salad
- Hamburgers
- Kebabs/Wraps
- Noodle Dishes
- Pies/Pasties
- Pizza/Pasta
- Rice Dishes (e.g. Sushi)
- Rolls/Sandwiches
- Salads
- Other
- None

A. Our first statistical question is:

Is there a gender difference in favorite "take away" food amongst Australian high school students?

- With the file ***CASAustralia2011Combined*** get a two-way table relating the responses to the question (the variable *TakeAwayFavorite*) and *Gender*. Show the Chi-Square Test for Independence result and the expected values. The output should resemble what is below.

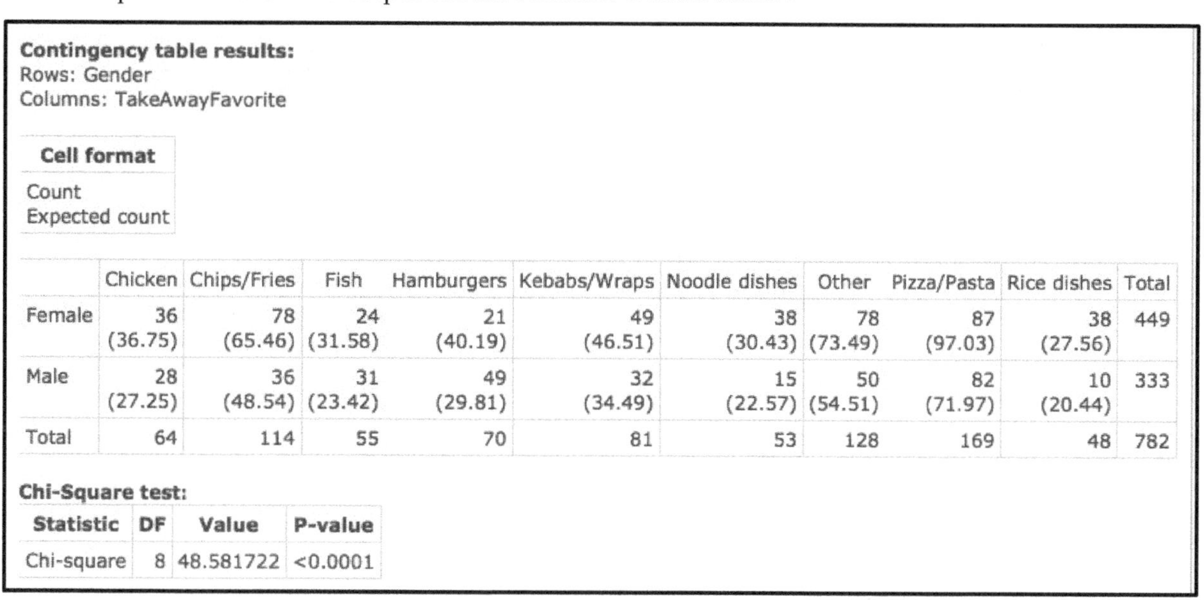

Contingency table results:
Rows: Gender
Columns: TakeAwayFavorite

Cell format
Count Expected count

	Chicken	Chips/Fries	Fish	Hamburgers	Kebabs/Wraps	Noodle dishes	Other	Pizza/Pasta	Rice dishes	Total
Female	36 (36.75)	78 (65.46)	24 (31.58)	21 (40.19)	49 (46.51)	38 (30.43)	78 (73.49)	87 (97.03)	38 (27.56)	449
Male	28 (27.25)	36 (48.54)	31 (23.42)	49 (29.81)	32 (34.49)	15 (22.57)	50 (54.51)	82 (71.97)	10 (20.44)	333
Total	64	114	55	70	81	53	128	169	48	782

Chi-Square test:

Statistic	DF	Value	P-value
Chi-square	8	48.581722	<0.0001

a. This analysis is a hypothesis test; write the null and alternate hypotheses in the context of the variables being used.

b. Check the conditions for the hypothesis test. There are two conditions to consider; one of the conditions is clearly met, while the other one may raise questions. Discuss.

c. The expected value for the response "Pizza/Pasta" for female students in about 97. Show how this number was calculated.

d. The expected value for the response "Pizza/Pasta" for female students in about 97. Explain what this expected value means in the context of the Chi-Square test.

e. Starting from the "Female/Chicken" cell and moving right-wards, show the first two or three terms in the calculation of the test statistic $\sum \frac{(O-E)^2}{E}$, and the last term ("Male/Rice dishes").

f. Either by using a Chi-Square Distribution Chart or by using software, find the critical value for the degrees of freedom given if we use $\alpha = 0.05$.

g. Compare the value of the test statistic (48.58) to the critical value and explain how the comparison explains the size of the *p*-value. (A sketch of a χ^2 sampling distribution may help.)

h. Is the test statistically significant? Give a reason for the answer.

i. So, do the data afford evidence that amongst Australian secondary level students, there is a gender difference in "favourite take away" food? In other words, interpret the outcome of the Chi-Square

B. Our second statistical question is:

Do the preferences for "take away" food amongst high school students differ in different parts of Australia?

- The variable *ResidState* records the state in which the student resides. Get a two-way table and Chi-Square analysis for the variables *ResidState* and *TakeAwayFavorite*. Use software with the file ***CASAustralia2011Combined***

j. The output shows that we should not trust the analysis because one of the conditions is clearly not met. (The software output may even bring that to the reader's attention.) What is the condition and how is the condition violated?

http://commons.wikimedia.org/wiki/Catego 1

- One way around the difficulty seen in part j is to group some of the categories. The variable *State2* groups the states and territories of Australia so as to avoid the problem. The other solution is to simply collect more data. Get a two-way table and Chi-Square analysis for the variables *State2* and *TakeAwayFavorite* using the file ***CASAustralia2011Combined.***

k. Confirm that df = 24 is correct for the table shown in the output you get.

- Use software to show the ***Chi-Square Distribution*** with df = 24 where the test statistic is 31.73 and the p-value is 0.1338 to get a picture like the one shown here.

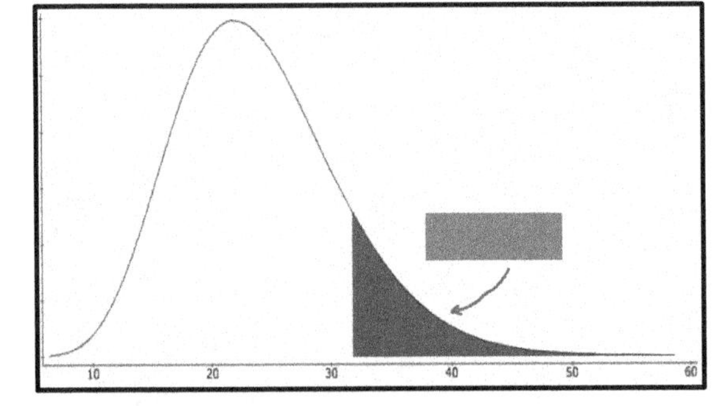

l. On the picture made (or a rough sketch of it) indicate the test statistic and the *p*-value.

m. In general, one can say that the [bigger/smaller] the Chi square test statistic is, the [bigger/smaller] the *p*-value will be. Make at least one sentence by choosing the correct choices for "bigger" and "smaller." [Playing with the software Chi-Square Distribution calculator may be useful.]

n. Is the Chi-Square hypothesis test statistically significant? Give a reason for the answer.

o. So, how does our test answer the statistical question? Do we have evidence that there are regional differences in the preferences of high school students in Australia for take away food?

4. *Which weight-loss program is best? Another weight-loss experiment* In the UK, a large experiment was conducted to evaluate the effectiveness of weight-loss programs as against not following a weight-loss program. [*Helen Truby, et al. Randomised controlled trial of four commercial weight loss programmes in the UK: initial findings from the BBC "diet trials"* BMJ, doi:10.1136/bmj.38833.411204.80 (published 23 May 2006)]

Here is how the study in the UK was conducted:

- Overweight and obese volunteers from across Great Britain were *randomly allocated* to four different weight- loss programs or to no weight-loss program.
- The costs of the weight-loss programs were paid for by the researchers.
- Those who were allocated to the "no-program" treatment were promised that, after the study, the researchers would then pay for a weight-loss program for them.
- Records of weight, as well as other medical variables, were kept over a period of time.

In the Chi-Square Test of Independence analysis we are looking at the participants who lost 10% or more of their weight compared with those who lost less than 10%, and comparing these numbers by the weight loss program that the subjects used. (Rosemary Conley is a UK weight-loss program.)

a. What makes this study an experimental study and not an observational study?

b. What is the factor?

c. What is the response variable?

d. What are the null and alternate hypotheses for this test of independence? Write them in the context of the data.

e. Are the conditions met for this test of independence?

f. What is the meaning (in words) of the expected counts in the context of the experiment?

g. Copy this graphic of the chi-square sampling distribution for $df = 3$ and show the test statistic and the p-value on it.

From Summary Statistics — Test for Independence
Column attribute (categorical): unassigned
Row attribute (categorical): unassigned

		Programme: Atkins	Weight Watchers	Slim-Fast	Rosemary Conley	Row Summary
WeightLoss	Lost less than 10%	22 (25.2)	30 (29.6)	33 (26.4)	22 (25.8)	107
	Lost 10% or more	18 (14.8)	17 (17.4)	9 (15.6)	19 (15.2)	63
	Column Summary	40	47	42	41	170

```
Column attribute:         Programme
  Number of categories:   4
Row attribute:            WeightLoss
  Number of categories:   2
Ho: Programme is independent of WeightLoss
Chi-square:   7.011
DF:           3
P-value:      0.072
```

The numbers in parentheses in the table are expected counts.

no data — Function Plot
y = chiSquareDensity(x, 3)

h. If $\alpha = 0.05$, is this hypothesis test statistically significant? Give at least one reason for your answer.

i. If we decided to be less strict and set $\alpha = 0.10$, so that a rare result would be one that would come about by sampling variation 10% instead of 5% of the time, would the test be statistically significant?

j. Consider the p-value and the test statistic and use them to give a good interpretation of the results of this test in the context of the experiment.

k. If someone from the UK asked you: "Does it really make a difference which program I choose?" how would you answer this person?

l. If the p-value had been 0.023, would your answer to the question in k differ or be the same?

m. If the p-value had been 0.023, would you be able to tell from that which program is best or worst? Explain why or why not. (*Hint: Would you need to do more analysis?*)

5. *The Titanic* Who survived on the *Titanic?* Specifically, was there an association between whether a passenger survived and whether the passenger was a first-, second- or third-class passenger?

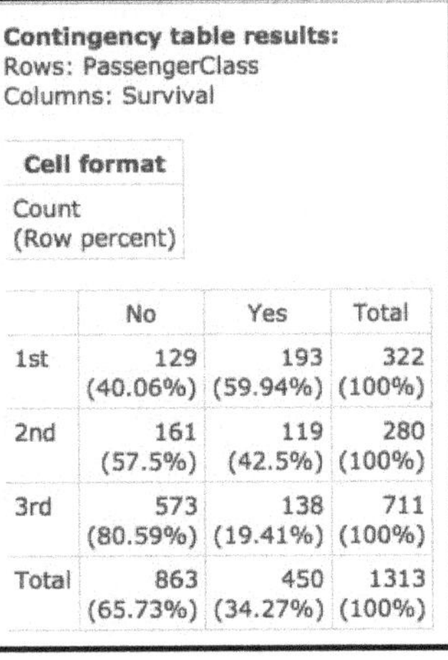

Contingency table results:
Rows: PassengerClass
Columns: Survival

Cell format
Count (Row percent)

	No	Yes	Total
1st	129 (40.06%)	193 (59.94%)	322 (100%)
2nd	161 (57.5%)	119 (42.5%)	280 (100%)
3rd	573 (80.59%)	138 (19.41%)	711 (100%)
Total	863 (65.73%)	450 (34.27%)	1313 (100%)

- Using the file ***TitanicPassengers*** get a two-way table similar to the one shown here with proportions or percentages calculated so that *PassengerClass* is regarded as the explanatory variable, and *Survival* as the response variable.

a. From the percentages in the two-way table, do you think that survival was independent of class, or not independent of passenger class? Give reasons using the percentages.

b. To test whether there is an association (or relationship) between the two categorical variables *PassengerClaas* and *Survival* it is appropriate to use a Chi square Test for Independence. Explain why this Chi Square hypothesis test is appropriate but not a Comparing two proportions test such as was done in §4.3.

c. Formulate your ideas into a statistical question appropriate for a Chi Square Test for Independence. [*Hint:* Look at the statistical questions that were used for Exercise 4.]

d. Set up the null and alternate hypotheses for Chi Square Test for Independence in the context of the variables being used here.

- Use the file ***TitanicPassengers*** and software to get output that shows the expected frequencies and the results of the Chi Square test. The output should be similar to what is shown here.

Contingency table results:
Rows: PassengerClass
Columns: Survival

Cell format
Count Expected count

	No	Yes	Total
1st	129 (211.64)	193 (110.36)	322
2nd	161 (184.04)	119 (95.96)	280
3rd	573 (467.32)	138 (243.68)	711
Total	863	450	1313

Chi-Square test:

Statistic	DF	Value	P-value
Chi-square	2	172.29912	<0.0001

e. Show how the expected counts were calculated for the "Yes-1st Class" cell and the "No-Third Class" cell.

f. Complete this sentence in the context of this hypothesis test: "The expected counts are the counts expected if the____________ __________were true and the variables were _________________."

g. Are the numbers in the cells big enough for the conditions to be met (for the Chi-Square Test for Independence)? Give a reason for your answer. (Will you look at the observed or the expected counts?)

h. Show how the test statistic was calculated by putting the numbers in the formula. You can check your self by doing the calculation, but the important thing is that you show where the numbers go. The software output should show all the numbers necessary.

i. For $df = 2$, consult the ***Chi-Square Distributions Chart*** and find the *Critical Value* for χ^2 for $\alpha=0.05$. Is the χ^2 for this test above that *Critical Value?* Does that make out result "rare" or "reasonably likely"?

j. The p-value given is < 0.0001. Is our test *statistically significant?* Give a reason for your answer.

k. Give a good interpretation of the results of this test in the context of the data—surviving on the *Titanic*. What do the data say about who was more likely to survive on the *Titanic?*

(Note: We have an entire population and not a sample, but the test for independence still applies. The null hypothesis is in effect saying: "If the survivors were chosen at random *without regard* to their class on the ship, what would the distribution of survivors look like?")

- Using the file ***TitanicPassengers*** repeat the analysis done above, but filtering or grouping by *Gender.* (The choice will depend un the software being used.) That is, get the two-way table and the Chi-Square Test for the relationship between *PassengerClass* and *Survival*, but just for the females, and then just for the males. The output should be similar to what is shown here.

l. Does the association between passenger class and survival hold up when broken down by gender? Give a reason for the answer based upon the output.

m. [*Review*] Look over the percentages or proportions as they are presented in the output. The probability of surviving if a passenger was both female and 1st class was 0.937. The probability of surviving if a passenger was both a male and 3rd class was 0.116. Use the notation of §1.2 and the letters and symbols S for survived, and 1st and 3rd for the class and M and F for genders to express these probabilities.

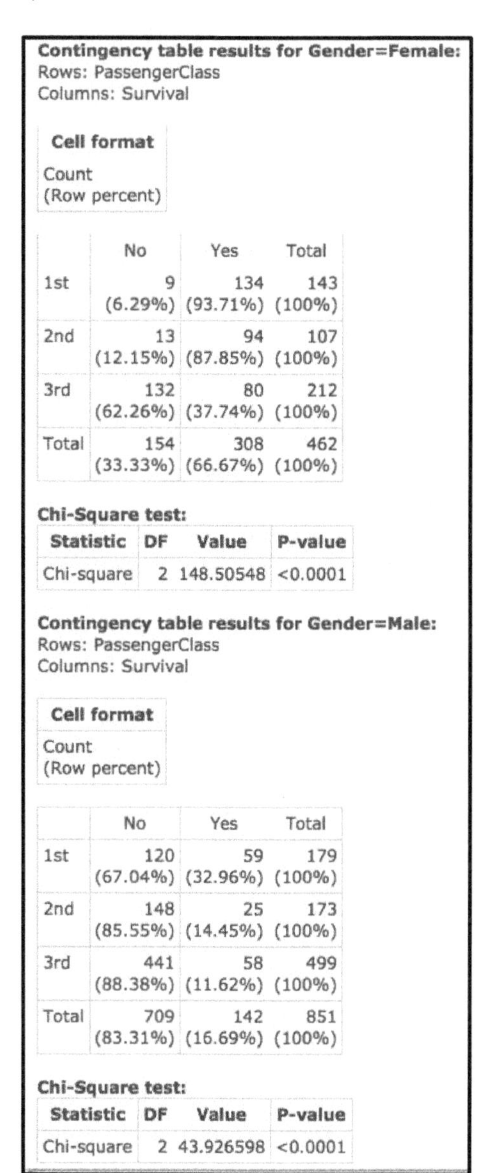

Contingency table results for Gender=Female:
Rows: PassengerClass
Columns: Survival

Cell format
Count (Row percent)

	No	Yes	Total
1st	9 (6.29%)	134 (93.71%)	143 (100%)
2nd	13 (12.15%)	94 (87.85%)	107 (100%)
3rd	132 (62.26%)	80 (37.74%)	212 (100%)
Total	154 (33.33%)	308 (66.67%)	462 (100%)

Chi-Square test:

Statistic	DF	Value	P-value
Chi-square	2	148.50548	<0.0001

Contingency table results for Gender=Male:
Rows: PassengerClass
Columns: Survival

Cell format
Count (Row percent)

	No	Yes	Total
1st	120 (67.04%)	59 (32.96%)	179 (100%)
2nd	148 (85.55%)	25 (14.45%)	173 (100%)
3rd	441 (88.38%)	58 (11.62%)	499 (100%)
Total	709 (83.31%)	142 (16.69%)	851 (100%)

Chi-Square test:

Statistic	DF	Value	P-value
Chi-square	2	43.926598	<0.0001

§4.5 Exercises: Confidence Interval, Hypothesis Test...?

1. ***Who Skips Breakfast?*** In both the Australian and Canadian Census at School questionnaires, there were questions concerning what students had for breakfast. We will be interested in those brave souls among the high school students who reported that they skipped breakfast. We have two statistical questions:
   - *"Who is more likely to skip breakfast: an Australian or a Canadian student? Or are they equally likely to skip breakfast?"* And if there is a difference in the proportions then:
   - *"An estimate of the difference in proportions if there is one."*

- Open the file ***CASComparisonAusCan*** and get a two-way table showing the numbers of students who said "yes" to *SkipsBreakfast* by the Country of the student. The output should resemble the table shown.

Australia Canada Comparison

		SkipsBreakfast		Row Summary
		No	Yes	
Country	Australia	480	91	571
	Canada	475	125	600
Column Summary		955	216	1171

S1 = count( )

   a. Calculate the proportion who skipped breakfast (that is, "yes" to *SkipsBreakfast*) among the Australian students and the proportion who skipped breakfast among the Canadian students. Use the correct *conditional probability* notation (the notation from§1.2) to express these probabilities.

   b. Our statistical question is: "Is there a difference in the proportion who skips breakfast between the population of high school students in Australia and the population of high school students in Canada?" For this question, choose the best procedure and explain your choice.
      - A confidence interval for one proportion? (§4.1)
      - A hypothesis test for one proportion? (§4.2)
      - A confidence interval for the difference of proportions? (§4.3)
      - A hypothesis test to compare proportions? (§4.3)

   c. For your choice, use notation (not the conditional probability notation) that will distinguish between the *population proportions* of students who skip breakfast in Australia and students who skip breakfast in Canada. (*Hint:* Will you have "hats" or will you have "no hats"?)

   d. Express the sample proportions with the correct notation. (*Hint:* "hats" or no "hats"?)

   e. We want a hypothesis test (so that helps you with question b). Set up the null and alternate hypotheses.

   f. Set up the test statistic for your test, showing the symbol for it and the formula. You do not actually have to complete the calculation (although you may) because we will have software do the calculation. But to show the calculation, you will have to calculate a third proportion and give it the correct symbol.

- Get software to do the hypothesis test for comparing proportions p_A and p_C, the proportions skipping breakfast for the Australian students and the Canadian students. Use the file ***CASComparisonAusCan.*** The output should be equivalent to what is shown here.

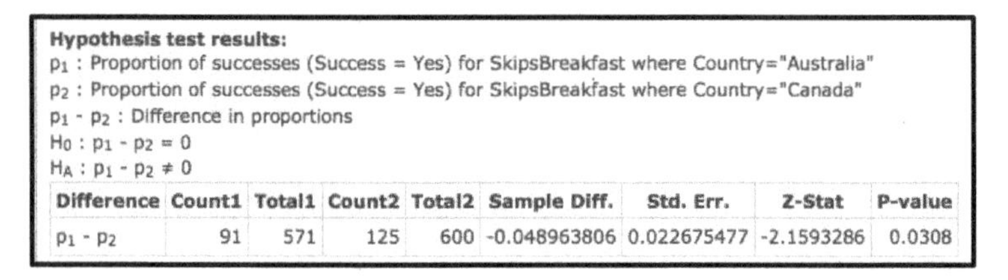

Hypothesis test results:
p_1 : Proportion of successes (Success = Yes) for SkipsBreakfast where Country="Australia"
p_2 : Proportion of successes (Success = Yes) for SkipsBreakfast where Country="Canada"
$p_1 - p_2$: Difference in proportions
$H_0 : p_1 - p_2 = 0$
$H_A : p_1 - p_2 \neq 0$

Difference	Count1	Total1	Count2	Total2	Sample Diff.	Std. Err.	Z-Stat	P-value
$p_1 - p_2$	91	571	125	600	-0.048963806	0.022675477	-2.1593286	0.0308

g. What is the value of the test statistic? Using $\alpha = 0.05$ and the appropriate critical value, is the value of the test statistic "reasonably likely" or "rare"? Give a reason for your answer.

h. What is the *p*-value? Is it bigger or smaller than $\alpha = 0.05$?

i. Does the evidence of parts g and h lead you to declare the test *statistically significant* or not? (The answers to parts g and h should point in the same direction; either both should give you reason to say that the test is significant or both should say that the test is not significant.)

j. Do you have evidence against the null hypothesis? Explain.

k. From the results, can you say that there *is* a difference or that there is *no* difference in the likelihood of skipping breakfast between Canadian students and Australian students?

l. Now we want to have an *estimate* of the difference in the proportions of Australian and Canadian students who skip breakfast. Choose the best procedure and give a reason for your choice.

   - A confidence interval for one proportion? (§4.1)
   - A hypothesis test for one proportion? (§4.2)
   - A confidence interval for the difference of proportions? (§4.3)
   - A hypothesis test to compare proportions? (§4.3)

m. Your estimate will be in the form of an interval. You should be able to predict from the results of your hypothesis test whether the interval will *exclude* or *include* zero. Choose the correct answer (exclude or include) and give a reason for your answer.

- Use the file ***CASComparisonAusCan*** to get a ***Confidence Interval for Comparing two proportions***. The results should be equivalent to what is shown here.

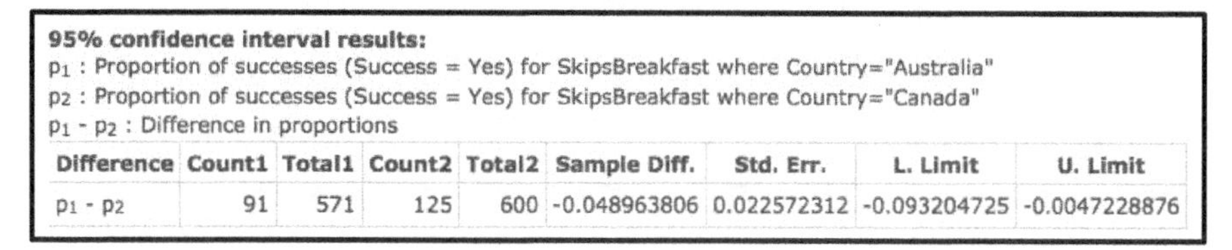

95% confidence interval results:
p_1 : Proportion of successes (Success = Yes) for SkipsBreakfast where Country="Australia"
p_2 : Proportion of successes (Success = Yes) for SkipsBreakfast where Country="Canada"
$p_1 - p_2$: Difference in proportions

Difference	Count1	Total1	Count2	Total2	Sample Diff.	Std. Err.	L. Limit	U. Limit
$p_1 - p_2$	91	571	125	600	-0.048963806	0.022572312	-0.093204725	-0.0047228876

n. Here is an interpretation of the confidence interval that is wrong. "We can be 95% confident that in the populations of all high school students in Australia and Canada, between 0.5% and 9.3% of students skip breakfast." What is wrong? (*Hint:* the proportion of students who skip breakfast is about 15% or 20%, not less than 1% to 9%. What do the numbers 0.5% to 9.3% mean?)

o. Give a correct interpretation of your estimate in the context of the data.

p. If we want an estimate of the *population* proportion of just the *Australian* high school students who skip breakfast, choose the correct procedure. and carry out the procedure by hand

   - A confidence interval for one proportion? (§4.1)
   - A hypothesis test for one proportion? (§4.2)
   - A confidence interval for the difference of proportions? (§4.3)
   - A hypothesis test to compare proportions? (§4.3)

   Carry out the procedure "by hand" (using a calculator) and then check the results using software with the file ***CASComparisonAusCan.***

2. ***Who Skips Breakfast: Males or Females?*** Our statistical question is:

Are male or female students more likely to skip breakfast? Or is there no gender difference?

We intend this statistical question to be an inferential question, using the Census @ School data.

a. Do we want a hypothesis test or a confidence interval for our question? Give a reason for your answer.

b. Are we looking at the difference of proportions or are we looking at a single proportion against some standard? Give a reason for your answer.

We are actually going to do the analysis twice, once for the Canadian sample and once for the Australian sample. So for the Canadian students we will compare the proportions of male and female students who skip breakfast and then, for the Australian students, we will again compare the proportions of male and female students who skip breakfast.

CASComparisonCanadaOnly

		SkipsBreakfast		Row Summary
		No	Yes	
Gender	Female	227 0.793706	59 0.206294	286 1
	Male	248 0.789809	66 0.210191	314 1
Column Summary		475 0.791667	125 0.208333	600 1

S1 = count ( )
S2 = rowProportion

- Use the file ***CASComparisonCanadaOnly*** (with the data for Canada only) to get two-way table similar to the one shown here that shows the relationship between *Gender* and the categories of *SkipsBreakfast*. (It may be possible and convenient to use the file ***CASComparisonAusCan*** with an appropriate filter.)

c. Using appropriate notation, set up the null and alternate hypotheses for a hypothesis test to test whether males or females are more likely to skip breakfast. Use the correct notation.

d. Calculate the sample proportions from your summary table and give them the correct symbols.

e. Also calculate $\hat{p}$ and say what this $\hat{p}$ means in the context of the data.

- Using the ***CASComparisonCanadaOnly*** have software do the hypothesis test whose null and alternate hypotheses have been set up in part c. The results should give the test statistic and *p*-value as shown here.

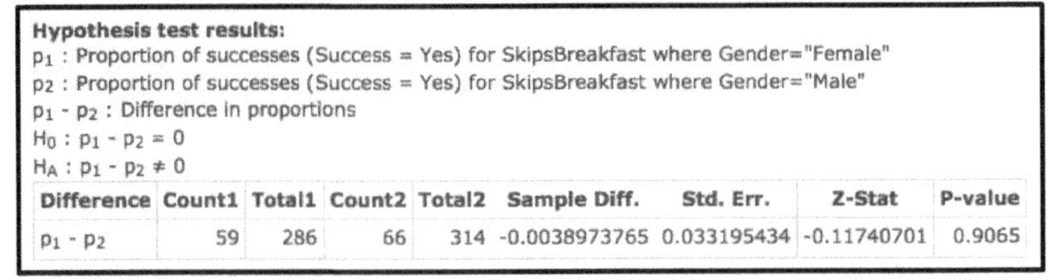

Hypothesis test results:
p_1 : Proportion of successes (Success = Yes) for SkipsBreakfast where Gender="Female"
p_2 : Proportion of successes (Success = Yes) for SkipsBreakfast where Gender="Male"
$p_1 - p_2$: Difference in proportions
$H_0 : p_1 - p_2 = 0$
$H_A : p_1 - p_2 \neq 0$

Difference	Count1	Total1	Count2	Total2	Sample Diff.	Std. Err.	Z-Stat	P-value
$p_1 - p_2$	59	286	66	314	-0.0038973765	0.033195434	-0.11740701	0.9065

f. With the proportions that you have calculated, show how the test statistic was calculated.

g. Is the test *statistically significant?* Give reasons for your answer in terms of the *p*-value and the test statistic shown in the output

h. Interpret the results of the hypothesis test for Canadian students in the context of possible gender differences in the likelihood of skipping breakfast.

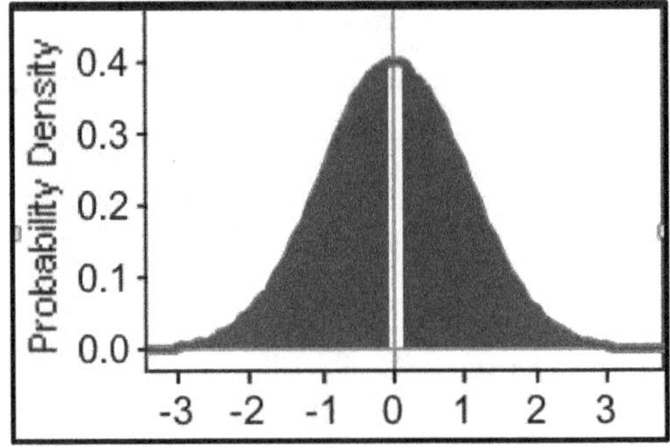

i. Here is a graphic of the sampling distribution for the test. Explain what the shading in the graphic shows, and explain why there is so much shading.

- Using the ***CASComparisonAustraliaOnly*** have software do the hypothesis test to test if there is a gender difference in the likelihood of skipping breakfast amongst high school students in Australia
  - j. Is the test *statistically significant?* Give reasons for your answer in terms of the p-value shown in the output and also the test statistic that was calculated.
  - k. Is there a gender difference in skipping breakfast for Australian high school students? Explain.
  - l. Neither the Australian nor the Canadian Census @ School samples are random samples drawn from the population of high school students in those countries. In Australia, many schools around the country participated in Census @ School and the Australian Bureau of Statistics promoted the project. In Canada, not many schools participated. What are the implications of this information for the analysis in this exercise? Discuss.

3. *San Mateo Real Estate I* Unlike the samples from the Census at School, the sample that you will analyze here is a simple random sample, so we can have confidence in our results. The sample size is $n = 400$. The cases are houses that were sold in San Mateo County in 2007–2008.

- Open the file ***SanMateoRESampleY0708*** and use software to make a two-way table showing the relationship between the variable *Style_2* and *Region*. It should look this.

SanMateoRESample0708

		Style2			Row Summary
		Other	Ranch	Traditional	
Region	Central	72	36	20	128
	Coast	32	6	4	42
	North	52	15	11	78
	South	93	36	23	152
Column Summary		249	93	58	400

S1 = count ( )

We are interested in whether specific style of houses are more probable in some of the "regions" of the county than in other regions. Or do all of the regions have a similar "mix" of styles. So, our statistical question is:

Is there an association between the style of houses sold and the region?

a. Choose which analysis is appropriate for this question, and for each of the other proposed procedures state why each of them would not be correct.
   - Difference of proportions hypothesis test
   - Test of independence
   - Least squares regression line
   - Single proportion hypothesis test

b. Set up the null and alternate hypotheses for the procedure you have chosen.

- Use software with the file ***SanMateoRESampleY0708*** to carry out the procedure that you chose.

c. Are the conditions met for the test you have chosen? Be complete and include both types of sample and whether the numbers in the cells are sufficiently big.

d. the correct procedure involves a "degrees of freedom." Show how the *df* was calculated.

e. Show how the test statistic was calculated; you do not have to complete the calculation but show enough of it so that a reader can see that you know where the test statistic comes from.

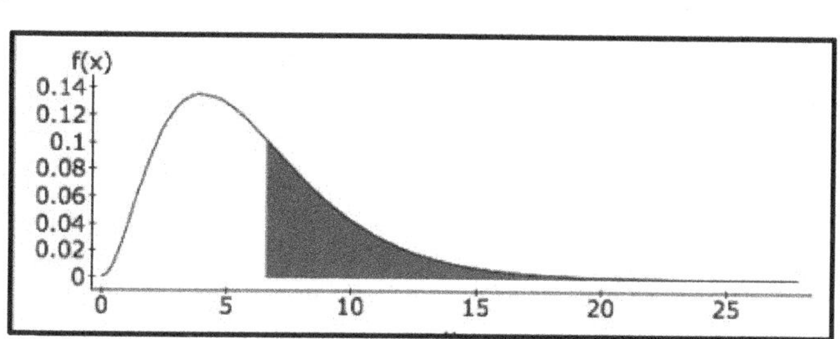

f. Here is the sampling distribution for the test statistic. Sketch a rough copy of the graphic and show the test statistic and the p-value on it.

g. From the evidence that you have, is the test statistically significant? Give a reason for your answer.

h. Give an interpretation of this test in terms of the statistical question stated above.

4. *San Mateo Real Estate II* Here is another exercise with the same random sample of data. The sample size is $n = 400$. The cases are houses that were sold in San Mateo County in 2007–2008. In the data there is a "list price" (the price that the seller was asking) and also a "sale price" (the amount that was actually paid.) Often the sale price is less than the list price, but it also happens that a house may have several potential buyers, and that the sale price ends up to be higher than the list price. And, of course, sometimes the house is sold for exactly the price that was asked. The variable *ListSale* is a categorical variable that records whether the house sold for:

- "over the listed price,"
- "the same as the listed price," or
- "under the listed price."

Our statistical question is:

Are there differences by region in the proportions of houses sold over, the same as, or under the list price?

SanMateoRESample0708

		Style2			Row Summary
		Other	Ranch	Traditional	
Region	Central	72	36	20	128
	Coast	32	6	4	42
	North	52	15	11	78
	South	93	36	23	152
Column Summary		249	93	58	400

S1 = count ( )

- Open the file ***SanMateoRESampleY0708*** and use software to make a two-way table showing the relationship between the variable *ListSale* and *Region*. It should look the graphic above.
  a. the statistical question should be answered with a Chi-Square Test for Independence. Give a reason that this is the appropriate choice.
  b. Set up the null and alternate hypotheses for the Chi Square analysis in the context of the variables measured.
- Use the file ***SanMateoRESampleY0708*** and software to carry out the test. Get the software to show the expected counts.
  c. Explain what the expected counts *mean* in terms of the null hypothesis. Make certain that your answer speaks of the variables specifically.
  d. Are the conditions met for the test you have chosen? Be complete.
  e. In this case you should have noticed that one of the expected counts is less than five. The rule of thumb that we have given in the ***Notes*** that no expected count can be less than five is conservative. Another rule of thumb says that for a table with twelve cells, as many as 20% of the cells can have expected counts smaller than five. We have twelve cells, so we can have $0.20 \times 12 = 2.4$ cells with expected counts smaller than five. If we use this rule of thumb, we can proceed.
  f. Here is the sampling distribution for the test statistic. Copy the plot to show the test statistic and p-value on it.
  g. Give an interpretation of this test in terms of the statistical question stated above.

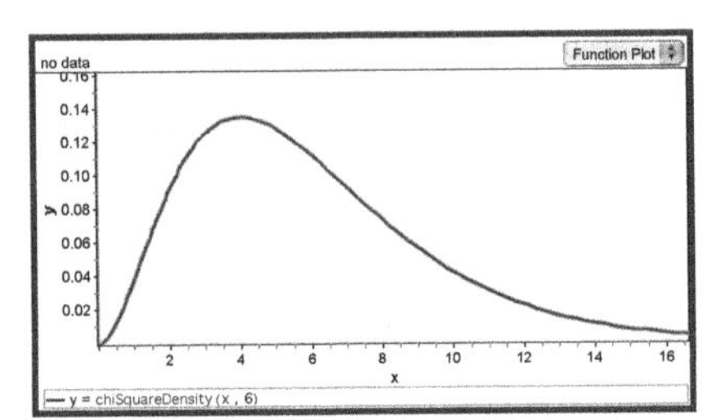

5. *Roller Coasters* Here is a table showing the numbers of steel and wooden roller coasters in two regions. The data are from the roller coaster database. The roller coaster database collects data on as many of the roller coasters in the world that the database builders can have access to.

a. Calculate proportions to indicate whether the proportion of wooden roller coasters differs between the two regions. Use good notation.

b. Is there evidence from your calculations that there may be a difference in the proportion of wooden roller coasters by region? Explain.

Roller Coasters Summer 09

		Construction		Row
		Steel	Wood	Summary
Region	Great Lakes	50	18	68
	West Coast	84	9	93
Column Summary		134	27	161

S1 = count ( )

¬(Region = "West")

c. If the conditions are met, there would be two ways to use formal inference procedures to generalize to the population of all roller coasters in these two regions. What are these two procedures? Explain.

d. Check out the conditions for at least one of the two procedures referred to. Do you have any doubts about whether the conditions are met? What questions would you want answered before proceeding?

e. For both of these procedures (assuming that it is wise to carry them out—which it may not be), the *p*-value comes out to be 0.0048. What does that tell you?

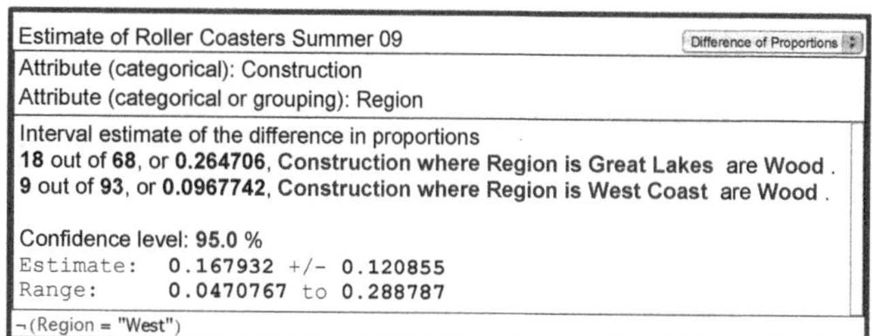
Estimate of Roller Coasters Summer 09 — Difference of Proportions

Attribute (categorical): Construction

Attribute (categorical or grouping): Region

Interval estimate of the difference in proportions

18 out of **68**, or **0.264706**, **Construction where Region is Great Lakes are Wood** .

9 out of **93**, or **0.0967742**, **Construction where Region is West Coast are Wood** .

Confidence level: **95.0** %

```
Estimate:   0.167932 +/- 0.120855
Range:      0.0470767 to 0.288787
```

¬(Region = "West")

f. Here is the output for the confidence interval for the difference of proportions of wooden roller coasters in the two regions. State how this confidence interval agrees with the *p*-value of 0.0048.

g. Suppose you wanted to see if wooden roller coasters in the database were *longer* on average than the steel roller coasters. (We are not worried about the population of all roller coasters; we are only interested in the ones that happen to be in the database in our sample.) Which procedure would you use and why?

- Chi-square test of independence on *Construction* (Wooden and Steel) and *Length* of roller coaster
- Compare means and medians of the variable *Length* by the variable *Construction*
- Get a least squares regression line between the variable *Length* by the variable *Construction*
- Do a difference of proportions hypothesis test on *Length* by the variable *Construction*

h. Suppose you wanted to see if there is a difference in the relationship between the *Speed* and the *Length* of a roller coaster according to its *Construction*. Which procedure would you use and why?

- Chi-square test of independence on *Construction* (wooden and steel) and *Speed* of roller coasters
- Compare means and medians of the variable *Length* and *Speed* by the variable *Construction*
- Get a least squares regression lines relating *Speed to Length* for the two types of *Construction* of roller coasters
- Do a difference of proportions hypothesis test on *Speed and Length* by the variable *Construction*

Special Exercise 6 on Pizza Delivery Data

Pizza Delivery Survey An on-line statistical package program is used by many colleges and universities in teaching statistics. The website hosts surveys from time to time, and the box to the right shows the questions that were asked for a survey on pizza delivery. Here are what the data look like: the sample size is $n = 2119$. The data are in the file ***PizzaDeliverySurvey.***

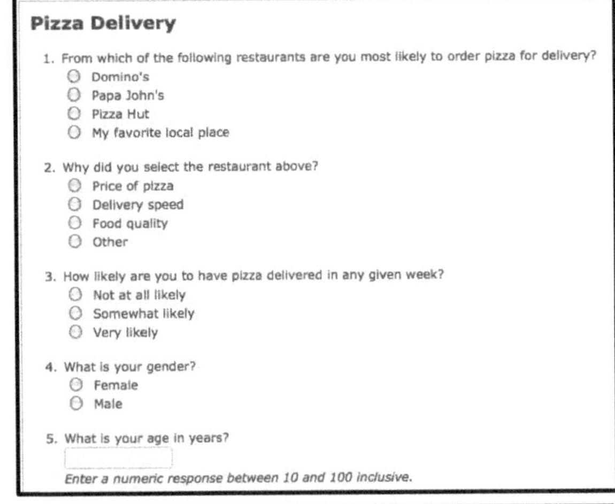

Pizza Delivery

1. From which of the following restaurants are you most likely to order pizza for delivery?
   - Domino's
   - Papa John's
   - Pizza Hut
   - My favorite local place
2. Why did you select the restaurant above?
   - Price of pizza
   - Delivery speed
   - Food quality
   - Other
3. How likely are you to have pizza delivered in any given week?
   - Not at all likely
   - Somewhat likely
   - Very likely
4. What is your gender?
   - Female
   - Male
5. What is your age in years?

   Enter a numeric response between 10 and 100 inclusive.

	Favorite	Reason	Chances	Gender	Age
405	Pizza Hut	Food qu...	Not at all...	Male	53
406	Domino's	Delivery ...	Not at all...	Male	19
407	My favorite local place	Other	Not at all...	Male	21
408	My favorite local place	Food qu...	Not at all...	Male	19
409	My favorite local place	Food qu...	Not at all...	Female	23
410	Papa John's	Food qu...	Not at all...	Male	21
411	Papa John's	Food qu...	Very likely	Female	19
412	My favorite local place	Food qu...	Somewh...	Male	18

The variables measured correspond to the questions shown in the questionnaire:

- *Favorite* (with the categories shown.)
- *Reason* (with the categories shown.)
- *Chances* (with the categories shown.)
- *Gender*
- *Age*

1. ***Revision on sampling*** The questionnaire shown above was put on the website with the invitation to the users of the site to answer the questions. In §3.1 that three types of non-random samples were discussed, as well as three types of random sampling. Here is a review:
   - Voluntary Response samples: where the people or things in the sample are "self-chosen".
   - Convenience samples: where the people or things in the sample are chosen by a researcher, but primarily because the people or things were readily at hand, or convenient.
   - Judgment samples: where the people or things in the sample are chosen by the researcher, but "deliberately."
   - Simple random sampling: where the people or things in the sample are chosen by a chance procedure from a list of the population.
   - Random Cluster sampling: where groups (called clusters) of the people or things are chosen randomly, and then all of the people or things in the groups randomly chosen are used as the sample.
   - Stratified Random Sampling: where people or things are randomly chosen from within groupings (called strata) of the people or things in the population.

   a. Which type of sampling listed above most approximates the way the $n = 2119$ people for this sample were chosen? Give a reason for your answer.

   b. Can we reliably use the inferential techniques (hypothesis tests and confidence intervals) that were covered in Chapter 4 using this sample? Give a reason for your answer.

   c. Can we use the "descriptive" techniques that were covered in the first two chapters (calculating probabilities, means, medians, dot plots, etc.) using this sample? Give a reason for your answer.

2. *Ages of the Respondents and Binomial Distributions*

a. What is the *binwidth* of this histogram?

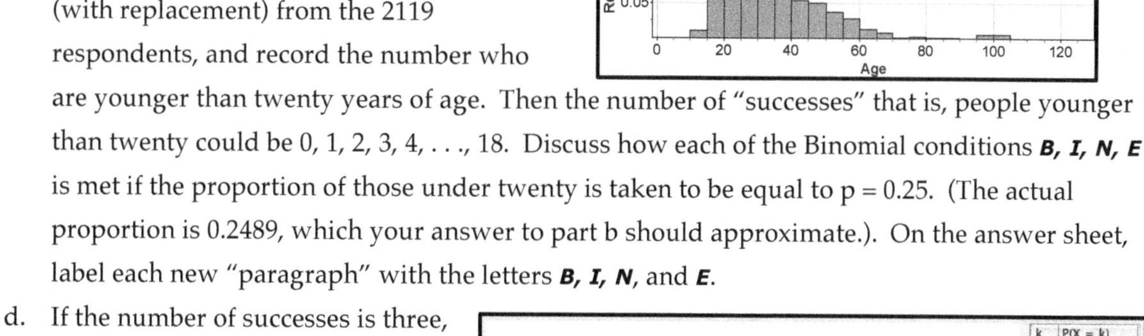

b. From the histogram, estimate the proportion of the respondents who are younger than twenty years of age.

c. Suppose we randomly choose n = 18 people (with replacement) from the 2119 respondents, and record the number who are younger than twenty years of age. Then the number of "successes" that is, people younger than twenty could be 0, 1, 2, 3, 4, . . ., 18. Discuss how each of the Binomial conditions ***B, I, N, E*** is met if the proportion of those under twenty is taken to be equal to p = 0.25. (The actual proportion is 0.2489, which your answer to part b should approximate.). On the answer sheet, label each new "paragraph" with the letters ***B, I, N,*** and ***E***.

d. If the number of successes is three, then the Binomial Model says that the probability is: $P(X=3)=0.1704$. Use the Binomial formula to confirm this probability.

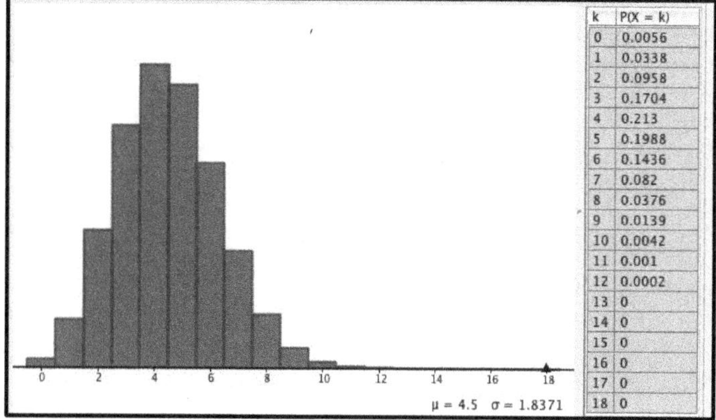

k	P(X = k)
0	0.0056
1	0.0338
2	0.0958
3	0.1704
4	0.213
5	0.1988
6	0.1436
7	0.082
8	0.0376
9	0.0139
10	0.0042
11	0.001
12	0.0002
13	0
14	0
15	0
16	0
17	0
18	0

e. Use the Binomial Distribution shown on the right of the graph to determine the probability that fewer than three of the eighteen are aged less than twenty years. Use the correct notation.

3. *Which Procedure?* The next questions treat the 2119 respondents as the population, and look at a random sample of size $n=120$ to infer to the population of respondents to the survey. (The survey is not that big, and we could certainly analyze it, but this exercise is for practice!)

a. We already know that in the "population" (the respondents to the Pizza Delivery Survey) the proportion aged less than twenty is about 0.25. If we only have the random sample of n = 120, and want to get an estimate the proportion aged less than twenty in the population, which procedures should we use? Give a reason for your choice:

- Hypothesis test to see whether 25% (or some other number) is a good guess for the population proportion.
- Chi-Square Test of Independence for two categorical variables.
- Confidence Interval for one population proportion.
- Confidence interval for the Difference of two proportions.
- Hypothesis test for the Difference of two proportions.

b. All of the choices are inferential procedures. The correct answer to question 1a (Revision on Sampling) is that the data are from a "Voluntary Response Sample", and since this is not a random sample, the answer to question 1b is that we cannot trust the results of inferential procedures (but: we can do descriptive analyses.) Why can we use inferential techniques with our sample of n = 120, but not with the bigger sample of 2119?

c. The correct answer to question 3a is a confidence interval for one proportion, since we want to use our random sample of size $n = 120$ to infer the proportion in what we have taken to be our population (the respondents to the survey). Show that the conditions are met to proceed.

d. Here is the output. Use the correct formula to show how the CI was calculated.

Estimate of Sample of PizzaSurveyData.csv — Estimate Proportion
Attribute (categorical): YoungerTwenty
Interval estimate for population proportion of Younger than 20 in **YoungerTwenty**

```
Count:             26 out of 114, or 0.22807
Confidence level:  95.0 %
Estimate:          0.151 to 0.3051
```

e. Give a good interpretation of the confidence interval in the context of what we have chosen to be the population.

4. ***Which Procedure? Part II*** We shall continue to treat the sample of n = 120 as a sample from the population of the respondents.

a. Our statistical question is: *"Is there a gender difference in the choice of favorite pizza place?"* We will use these data from our sample. Determine which inferential procedure should be used, and give a reason:

Sample of PizzaSurveyData.csv

		Favorite				Row Summary
		Domino's	My favorite local place	Papa John's	Pizza Hut	
Gender	Female	12	22	17	17	68
	Male	8	20	13	10	51
Column Summary		20	42	30	27	119

S1 = count ( )

- Hypothesis test to see whether a given number is a good guess for the population proportion.
- Chi-Square Test of Independence for two categorical variables.
- Confidence Interval for one population proportion.
- Confidence interval for the Difference of two proportions.
- Hypothesis test for the Difference of two proportions.

b. If our statistical question is: "is there a difference between males and females in the proportion who choose 'My favorite local place'?" then determine which inferential procedure should be used and give a reason.

- Hypothesis test to see whether a guess is a good guess for the population proportion.
- Chi-Square Test of Independence for two categorical variables.
- Confidence Interval for one population proportion.
- Confidence interval for the Difference of two proportions.
- Hypothesis test for the Difference of two proportions.

c. If our statistical question is: "How much more (or less) likely are males to choose 'my favorite local place' than females" then determine which inferential procedure should be used, and give a reason.

- Hypothesis test to see whether a given number is a good guess for the population proportion.
- Chi-Square Test of Independence for two categorical variables.
- Confidence Interval for one population proportion.
- Confidence interval for the Difference of two proportions.
- Hypothesis test for the Difference of two proportions

5. ***Theoretical questions:***

a. What is the difference between a "Reasonably Likely Interval" and a "95% Confidence Interval'?

b. True or false, and explain: "It is possible for a 95% Confidence Interval to fail to include the population proportion.

§5.1 Exercises: Inference for Quantitative Variables

1. Confidence Intervals for Duration of Flights to Seattle

A. A Bigger Sample Size: In the **Notes** we calculated the confidence interval for the population mean *ActualDuration* of flights from the San Francisco Bay Area to Seattle using our sample of $n = 20$ flights. (See the **Notes.**) This exercise asks what happens when we increase the sample size.

a. What are the cases for our data: Passengers? Airlines? Flights? Durations of flights? (Remember that we could collect data on more than one variable.)

- Open the file ***OnTimeBayAreaSEA2014***. Get a random sample of $n = 36$ that includes the variable *ActualDuration*. Get summary statistics for this variable that includes the sample mean $\bar{x}$ and sample standard deviation s. The output should resemble but be different from what is shown here, since the samples drawn are random samples.

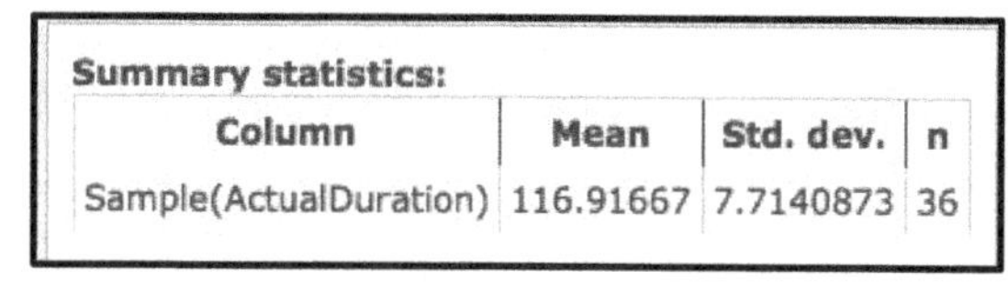

Summary statistics:

Column	Mean	Std. dev.	n
Sample(ActualDuration)	116.91667	7.7140873	36

b. Consult the ***t distribution chart*** or software to get the t^* for the correct *df* and calculate ("by hand" – using a calculator) a 95% confidence interval for the population mean Actual Duration of flights from the Bay Area to Seattle.

- Check the calculations using the file ***OnTimeBayAreaSEA2014*** and software to get a 95% confidence interval for the population mean Actual Duration of flights from the Bay Area to Seattle. Here are software results for the random sample shown above.

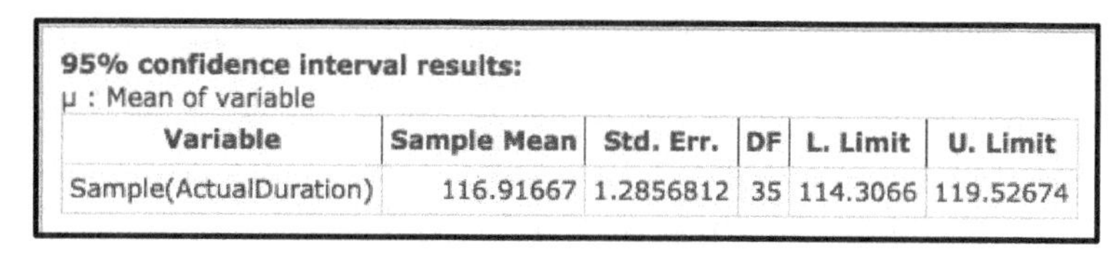

95% confidence interval results:
μ : Mean of variable

Variable	Sample Mean	Std. Err.	DF	L. Limit	U. Limit
Sample(ActualDuration)	116.91667	1.2856812	35	114.3066	119.52674

c. The margin of error is different from the standard error. Report both of them for the calculation of the confidence interval that you have made from your sample.

d. We can express the lower and the upper limits of a confidence interval in this form:

"lower limit" < _____ < ""upper limit"

Which symbol should be put in the blank space between the "< " signs? Should it be μ or should it be $\bar{x}$?

e. The purpose of a confidence interval is to say something about the *population* mean. We do not have to be 95% confident or need an interval for the sample mean; we are 100% certain about it; we calculated it. So the correct answer to part d is μ and not $\bar{x}$ (that is x-bar). The temptation to answer $\bar{x}$ probably comes from the fact that it is easily seen that that x-bar is between the lower and upper limits. Explain from the formula for the confidence interval why it is that x-bar will *always* be between the lower and upper limits. [*Hint:* What does the formula do?]

f. Give a good interpretation of your confidence interval in the context of flights to Seattle. The interpretation should include information on the population ("Flights from the Bay Area to Seattle") and that the interval is for the mean of the variable being measured ("the duration of flights to Seattle from the San Francisco Bay Area") and the units of measurement. ("minutes").

- ***Increasing the sample size***. With the file ***OnTimeBayAreaSEA2014***, draw a random sample of $n = 240$ that includes the variable *ActualDuration*.
Get summary statistics for this variable that includes the sample mean $\bar{x}$ and sample standard deviation s. The output should be different from what is shown here.

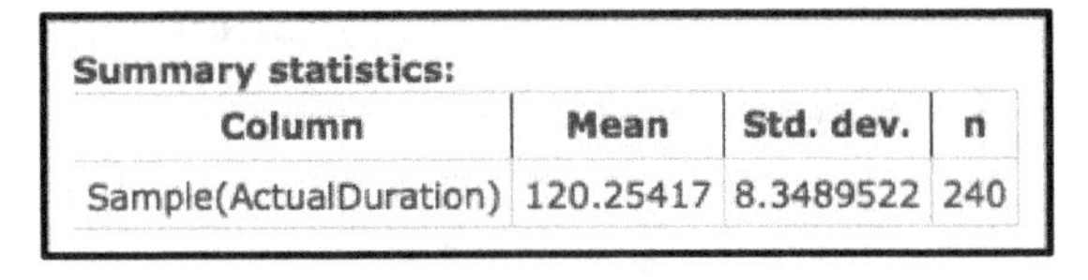

Summary statistics:

Column	Mean	Std. dev.	n
Sample(ActualDuration)	120.25417	8.3489522	240

g. Calculate ("by hand"), using the correct t^*, the 95% confidence interval for the population mean Actual Duration of flights from the Bay Area to Seattle.

- Check the calculations to part g using the file ***OnTimeBayAreaSEA2014.*** (Note: if for hand calculation you depend on the t^* on the t-Distribution Chart then the interval you calculate may be slightly different than what software gives, since it uses a more accurate t^*.)

h. Compare the margin of error for the confidence interval calculated from this new sample of $n = 240$ compared with the margin of error for the sample of $n = 36$. Is it bigger or smaller?

i. You should have a smaller margin of error with a bigger sample size. What does a smaller margin of error imply for the width of the confidence interval? Will the interval be bigger or smaller? Confirm your answer by comparing the widths of the confidence intervals for the two samples (the $n = 36$ and the $n = 240$) you have drawn.

B. More or Less Confident and how wide the confidence interval is.

- Using the $n = 36$ sample go back to software, and change the level of confidence to 99%, and get the confidence interval. Either print that out or copy the information, or leave it on the computer screen. Then go back again and change the confidence level to 90%. The output should resemble what is here.

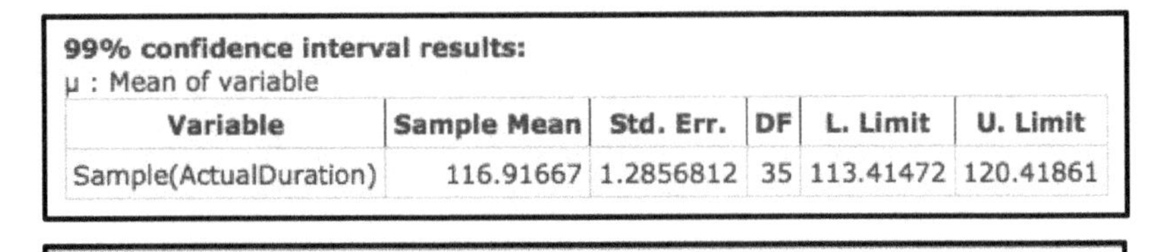

99% confidence interval results:
μ : Mean of variable

Variable	Sample Mean	Std. Err.	DF	L. Limit	U. Limit
Sample(ActualDuration)	116.91667	1.2856812	35	113.41472	120.41861

90% confidence interval results:
μ : Mean of variable

Variable	Sample Mean	Std. Err.	DF	L. Limit	U. Limit
Sample(ActualDuration)	116.91667	1.2856812	35	114.74442	119.08892

j. What in the output for the 99% and the 90% confidence intervals has *not* changed when the confidence level is changed?

k. Does increasing the level of confidence from 95% to 99% *increase* or *decrease* the width of the confidence interval? And changing to 90% Does the interval get wider or narrower?

l. By getting the t^* for 90% and for 99% confirm the software output by hand calculation. What in the calculation causes the 99% confidence interval to be wider and the 90% confidence interval to be narrower?

m. That a 99% confidence interval should be *wider* and a 90% confidence interval should be *narrower* may seem backwards to some people: "Should not a 99% confidence interval be more accurate?" "And a 90% interval be less accurate?" — someone may say. Work out an answer to this person (or to yourself) to counter this thinking. You can ask for help.

2. *Duration of Flights to Seattle: What a confidence interval means.*

Here is the output for a confidence interval based upon one random sample of size $n = 240$.

95% confidence interval results:
μ : Mean of variable

Variable	Sample Mean	Std. Err.	DF	L. Limit	U. Limit
Sample(ActualDuration)	119.95833	0.56182581	239	118.85157	121.0651

a. Give the "standard interpretation" for this confidence interval in the context of the average time a flight takes from the bay Area to Seattle.

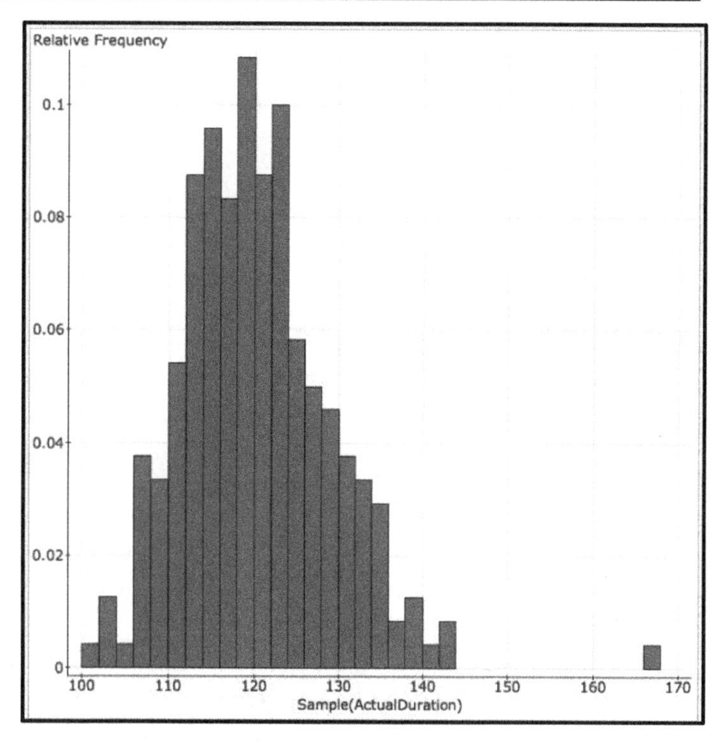

b. Here is a common *wrong* interpretation: "The confidence interval means that 95% of the actual durations of flights are between 118.85 minutes and 121.06 minutes." This plot will help to see that this interpretation is wrong. It shows the actual durations for flights in the sample of $n = 240$. The bars for 119 minutes and 120 minutes are highlighted. Estimate the proportion of flights in this sample that are in the interval: $119 \le ActualDuration < 121$, Is the proportion anywhere near to 95%?

c. The distribution looks roughly Normal. You should be able to make a very rough estimate of the lower and upper limits of 95% of all the Actual Durations. It does not have to be accurate; just a rough estimate. Is this interval bigger than $118.85 < \mu < 121.07$?

d. The phrase "We are 95% confident" means that if the sampling were done repeatedly (over and over again) then 95% of all of the confidence intervals calculated will *include* the actual population mean μ. And we expect 5% will miss the population mean μ. In Exercise 1, all of the flights from the Bay Area to Seattle were considered to be the population, and we drew samples of $n = 36$ and then $n = 240$. So if we did this one hundred times, we would expect 95 of the 100 to include the actual population mean μ. On the next page is software output showing 100 confidence intervals calculated from 100 different samples of size $n = 240$. Each of the horizontal lines is a different confidence interval. For that exercise, how many of the confidence intervals "hit" and how many "missed"?

- Use a ***Confidence Interval Applet*** in software (if available) to make your own version of the plot shown on the next page. Use the variable *ActualDuration*, use a sample size of $n = 240$, and to begin with, get 100 intervals. The output should resemble what is on the next page.

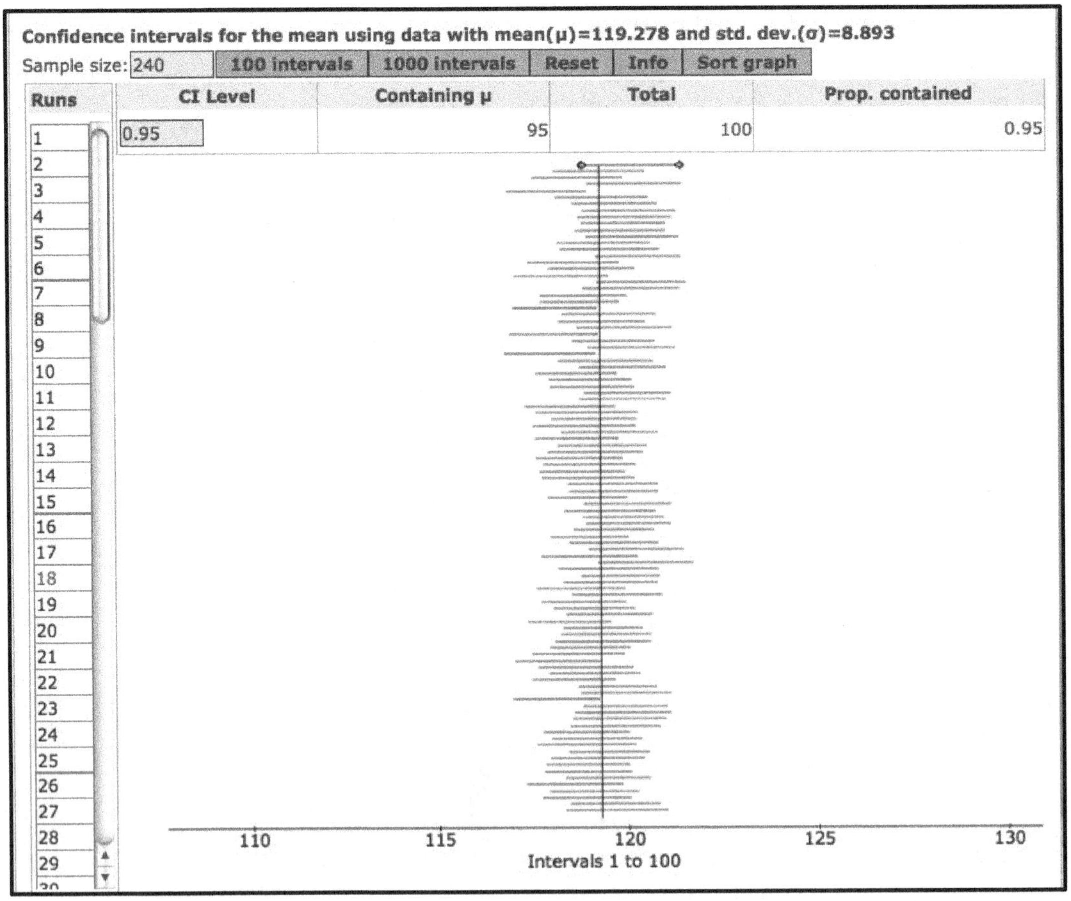

Confidence intervals for the mean using data with mean(μ)=119.278 and std. dev.(σ)=8.893

Sample size: 240 | 100 intervals | 1000 intervals | Reset | Info | Sort graph

Runs	CI Level	Containing μ	Total	Prop. contained
1	0.95	95	100	0.95

Runs: 1, 2, 3, 4, 5, 6, 7, 8, 9, 10, 11, 12, 13, 14, 15, 16, 17, 18, 19, 20, 21, 22, 23, 24, 25, 26, 27, 28, 29

110 115 120 125 130

Intervals 1 to 100

e. Record the number and percentage that included and excluded the actual population mean μ for your run of the ***Confidence Interval Applet***. You may wish to run the applet several times to convince yourself that the thing is working. (You may not get exactly five, sometimes four or six will appear. Randomness rules.)

f. Choose the correct pairing: "A confidence interval that [excludes/includes] the actual population mean μ corresponds to a sample that gets a [rare/reasonably likely] sample mean $\bar{x}$. The choices have been put in alphabetical order: the answer should read: "   " should be paired with "   ".

- Change the sample size from $n = 240$ to $n = 36$ and rerun the ***Confidence Interval Applet.*** You may wish to "Reset" the applet.

g. When the sample size changed to $n = 36$ what do you notice about the length of the confidence intervals? Explain why what you have noticed makes sense from the formula to calculate a confidence interval. [*Hint:* What happens to the margin of error with a smaller sample size.]

h. Record the proportion of the confidence intervals that include the population mean μ.

i. Has using the smaller sample size increased, decreased or not changed your confidence that your interval calculated from your sample includes the actual population mean μ? Give a reason for your answer.

j. The correct answer to part I should be "not changed"; so then, what is the advantage of having a bigger sample size?

3. ***Area of Houses for Sale in San Mateo County (SqFt Again)*** We will look at a *simple random sample* of houses in San Mateo County that were sold in the years 2007–2008. The variable *SqFt* measures the area inside a house. Our goal here is to estimate the population mean area (measured in square feet) of the all of the houses that were sold in the Central region of San Mateo County in 2007–2008. In the sample

- Open the file ***SanMateoRESampleY0708.*** Recall that these data are randomly sampled from all of the houses that were sold in 2007-2008. Using software, get:
  - A dotplot or a histogram of the variable *SqFt*
  - Summary statistics including the mean and standard deviation for the variable *SqFt.*

a. What are the cases for these data?

b. Using the correct notation, write down the sample mean and sample standard deviation.

c. Is the sample distribution left-skewed, symmetric, or right-skewed?

d. According to the information on ***Conditions,*** determine whether the two conditions for use of the *t* distributions are met. Refer to the way the sample was drawn and the shape of the sample distribution.

e. Can we safely proceed with the calculations? Why or why not? (*Hint:* Consider the ***15/40 Rule.***)

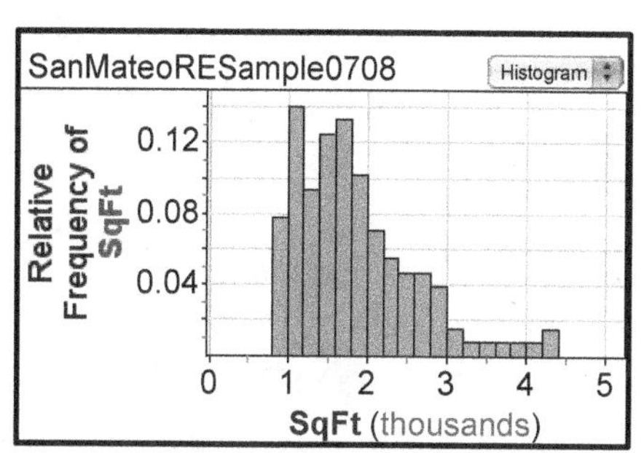

Summary statistics:
Where: Region="Central"

Column	Mean	Std. dev.	n
SqFt	1832.5469	746.4463	128

f. Consult the ***t distribution chart*** or software to get the correct t^* for a 95% confidence interval.

g. Explain why in using the *t* distributions the t^* is not 1.96.

h. Calculate, by hand using a calculator, a 95% confidence interval for the population mean *SqFt* for the Central Region for the population of all houses in San Mateo County sold in 2007–2008.

- Using the file ***SanMateoRESampleY0708*** check the calculations by using software to get the 95% confidence interval for the population mean of the variable *SqFt* for houses in the Central Region of San Mateo County. In using software it will be necessary to filter just for Region = "Central".

i. What is the margin of error for your confidence interval?

j. What is the standard error for your confidence interval?

k. Confused Conrad's answers to questions i and j are the same number. Should his answers be the same or is Confused Conrad confused? Explain.

l. Give a good interpretation of your confidence interval in the context of the data.

- Go to the software estimate and change the confidence level from 95% to 99%.

m. What has happened to the margin of error? Explain (in terms of the ***t distribution chart***) why the change that you have seen is in the direction that you have seen.

n. Does a bigger confidence level lead to a wider or a narrower confidence interval? Explain (mostly to convince yourself that the relationship makes sense).

4. *Obesity in the American population, by age and gender.* On the Centers for Disease Control website (http://www.cdc.gov/nchs/nhanes.htm) the NHANES study is described:

> The National Health and Nutrition Examination Survey (NHANES) is a program of studies designed to assess the health and nutritional status of adults and children in the United States. The survey is unique in that it combines interviews and physical examinations.

One of the most common measures of obesity is $BMI = \dfrac{\text{Weight}}{(\text{Height})^2}$ where weight is measured in kilograms and height is measured in meters. The NHANES data constitute a random sample of the population of people living in the USA.

- Choose the age file ***NHANES2010Age21to30***. Use software to get separately for males and for Females (so use either "filtering" or "grouping" depending on the software used.) All of the questions depend on the output.
  - Dot plots showing the distributions of BMI.
  - Summary statistics including mean and standard deviation for BMI.
  - Confidence intervals for the means of BMI.

Here is an example of what the output may be.

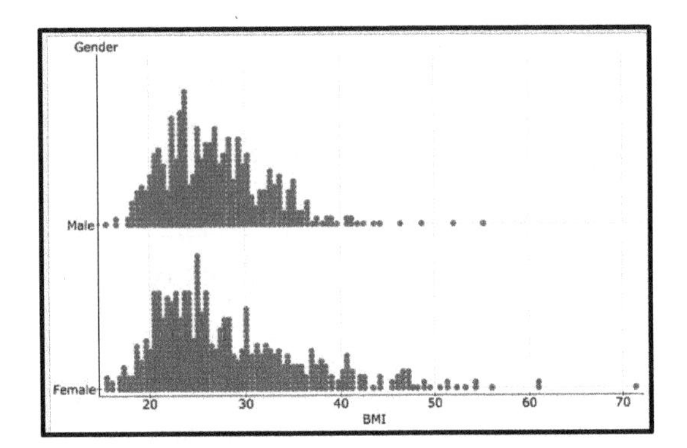

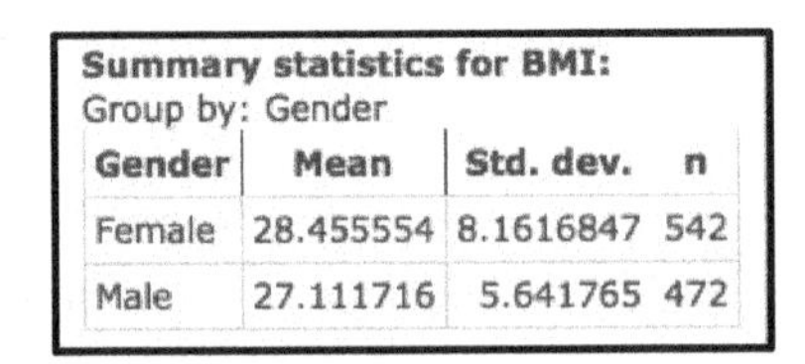

Summary statistics for BMI:
Group by: Gender

Gender	Mean	Std. dev.	n
Female	28.455554	8.1616847	542
Male	27.111716	5.641765	472

95% confidence interval results:
Group by: Gender
μ : Mean of BMI

Gender	Sample Mean	Std. Err.	DF	L. Limit	U. Limit
Female	28.455554	0.35057438	541	27.7669	29.144207
Male	27.111716	0.25968328	471	26.601435	27.621997

a. What are the cases for these data?

b Describe the shapes of the sample distributions of BMI.

c. Determine whether the two conditions for use of the *t* distributions are met. Explain with reference to the way the sample was drawn and the shape of the sample distribution, and the rules of thumb for using the *t* distributions.

d. The output should give both the standard deviation and the standard error. Show for wither the males or for the females how the standard error is derived from the standard deviation.

e. Consult the ***t distribution chart*** or software to get the correct t^* to calculate a 95% confidence interval. It should not

f. Show how the calculations for the 95% confidence interval for estimating the mean BMI were carried out. Do this either for the CI for the females or for the males.

g. Find the margin of error of the confidence interval whose calculation was shown in part f.

i. Give a good interpretation of the confidence intervals in the context of the data. What do they say about the average BMI for males and females in the age group that was chosen.

- An alternate measure of obesity is simply waist circumference. In the data the variable is *WaistCir*, measured in centimeters. (There are 2.54 cm/in, so 85 cm is about 33.5 in). For the same age group, and for males and females, use software to repeat the analysis for *WaistCir* that was done for *BMI*.

  j. Are the conditions for using the t procedures still valid for *WaistCir* as they were for ***BMI***? Give a reason for the answer.

  k. The meaning of confidence intervals: Can it be that the mean waist circumference in the population of USA residents (of which we have random samples) for the age group chosen is the same for males and females? Give a reason for the answer based upon the confidence intervals that have been calculated.

  l. In the analyses that have been done, the number of males and the number of females have been different. What about the calculations that have been done makes this difference irrelevant.

5. *Roller Coaster G-Force* The amount of acceleration felt in a roller coaster ride is sometimes measured by concept of g-force, which compares acceleration to the force of gravity. It is one of the variables for which the Roller Coaster Database (www.rcdb.com/) has some (but not much) data. Our sample from the database for roller coasters in America is found in the file ***RollerCoastersAmericaSample.*** In that sample there are just $n = 31$ roller coasters where the information about *GForce* was provided. The data in the database cannot be thought of as a random sample of all the roller coasters in America, and the data in the file ***RollerCoastersAmericaSample*** is also not a random sample of the database.

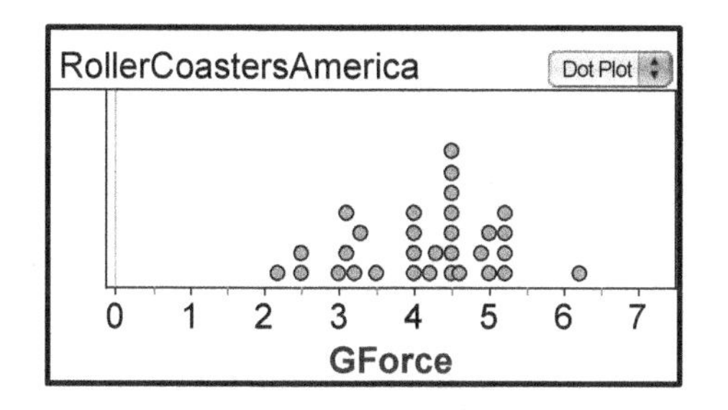

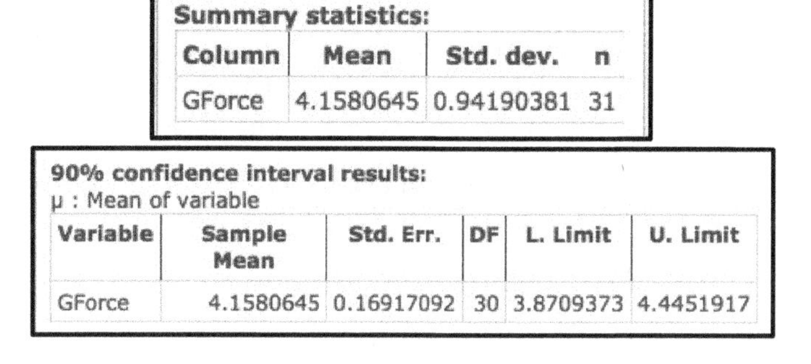

Summary statistics:

Column	Mean	Std. dev.	n
GForce	4.1580645	0.94190381	31

90% confidence interval results:
μ : Mean of variable

Variable	Sample Mean	Std. Err.	DF	L. Limit	U. Limit
GForce	4.1580645	0.16917092	30	3.8709373	4.4451917

a. What are the cases for these data?

b. Is *GForce* a quantitative or categorical variable?

c. Use the correct notation for the sample mean and sample standard deviation.

d. Show exactly how software calculated the 90% confidence interval, showing the correct t^* and where the numbers go.

e. Confused Conrad expresses the confidence interval as $3.387094 \leq \bar{x} \leq 4.44519$, seeing that $\bar{x} = 4.15805$ is in that interval. Even so, CC is wrong. What is his mistake? Correct his mistake.

f. Will a 95% confidence interval be wider or narrower? Give a reason for your answer.

g. Judging by the dot plot, and the ***Rule of Thumb***, is the Normality Condition for the t – procedures problematical? [If your software has "bootstrapping" facility, it is worthwhile running it on these data to see the approximate sampling distribution generated.]

h. The data are not a random sample from the population of roller coasters in America. Can we trust our calculations even though we can do the calculations?

6. ***Age of Mother at First Birth and Length of Pregnancy.*** The data are a simple random sample from the population of all births registered in 2006.

AgeMother	The age of the mother of the child in years.
FirstBirth	Two categories: "First Birth", "Later Birth" indicating whether the birth was the mother's first birth or a later birth.

- Open the file ***MilleBirthsSample***. Use software to get:
  - A dotplot of the variable *AgeMother* filtered for just the first births.
  - Summary statistics, including mean and standard deviation, for just the first births.
  - A confidence interval estimate of the mean age at first birth for *all* births in 2006.

  The output should resemble this:

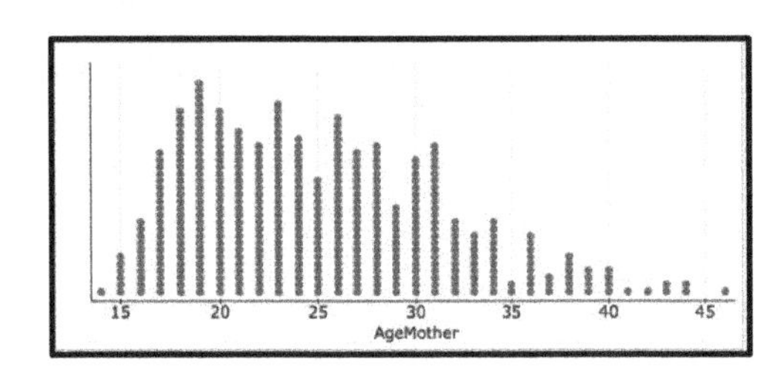

Summary statistics:
Where: FirstBirth="First Birth"

Column	Mean	Std. dev.	n
AgeMother	25.126168	6.3778721	428

95% confidence interval results:
Where: FirstBirth="First Birth"
μ : Mean of variable

Variable	Sample Mean	Std. Err.	DF	L. Limit	U. Limit
AgeMother	25.126168	0.30828608	427	24.520221	25.732115

a. One of the conditions for doing a confidence interval is that the population distribution be Normal, which we judge by looking at the sample distribution. Is this sample distribution Normal? If not, can we proceed to get a trustworthy confidence interval? Give a reason.

b. The other condition is that the sample be a random sample. Is that condition met?

c. If you write the confidence interval as 24.52 <   < 25.73, what symbol should be placed between the "less than" signs? Should it be μ or $\bar{x}$? Give a reason for the choice.

d. Find the margin of error for this confidence interval. (It is not the same as the "standard error.")

e. Will the width of a 99% confidence interval be wider or narrower? Give a reason based upon the numbers that go into the formula.

f. Give a good interpretation of the confidence interval in the context of the data. A good interpretation should mention mother's age at birth for first births and the population.

- Go to: www.cdc.gov/nchs/data/databriefs/db21.pdf .

g. Do the data here agree with what this download is saying? Give a reason for your answer.

- Change the filter (or grouping) so as to repeat the analyses for "Later Births".

h. For the later births, are the conditions for the calculations still met? Give reasons based on the way the sample was collected and the sample size or the appearance of the dot plot..

i. Interpret the confidence interval for the mean age of mother's for "later births".

j. Compare the interval for the later births

§5.2 Exercises on Hypothesis Testing for One Mean

1. ***Flying to LAX*** If you start a search for a flight from Oakland to Los Angeles on Delta something like this may appear on your screen. The flight was scheduled to leave at 6:19 AM (but actually left five minutes early) and was scheduled to take 89 minutes. The actual duration of the flight was actually 74 minutes, and the arrival time was 22 minutes early. Nice.

The advertised times for the duration of fights from the Bay Area vary somewhat, but we will practice the ideas of hypothesis testing by hypothesizing (making an educated guess) of an elapsed time (or duration of flight) of μ = **80** minutes. We have data for the first quarter of 2014 from all three major Bay Area airports to LAX, so we should be able to test the idea that the duration average is eighty minutes. Our statistical question is:

For flights from the Bay Area to LAX, is the mean flight duration equal to or different from eighty minutes?

- Open the file ***OntimeBayAreaLAX2014*** and use the ***Sample*** function of software to get a random sample of flights with sample size of n= 36 that includes the variable *ActualDuration*.
- Using software and the sample chosen from ***OntimeBayAreaLAX2014*** get a graphic showing the distribution of *ActualDuration* of the sample chosen. This can be a dot plot or a histogram or a box plot. The idea is to assess the shape of the distribution.
- Get summary statistics, including mean and standard deviation. Here is an example of possible output; the output from a your sample of n = 36 will be different.

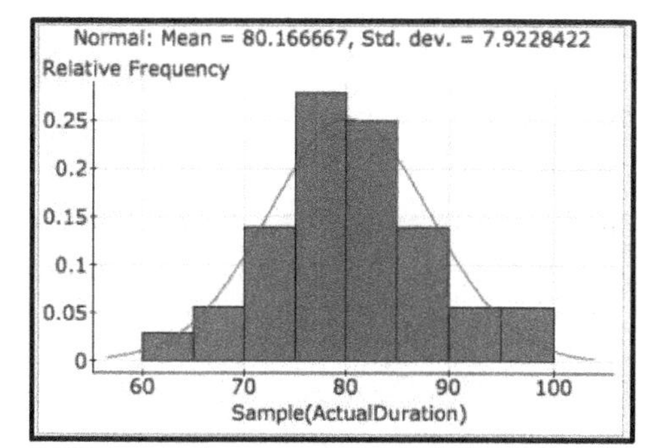

a. What are the cases for these data? (Airlines, airports, flights, travelers, tickets?)

b. Record the sample mean and sample standard deviation for the variable *ActualDuration* for your sample, using the correct notation.

c. Inspect the graphic for whether there is evidence of extreme skewness or outliers. State why it is important to check for skewness and extreme outliers.

Inspecting the sample distribution puts you in the position of almost all researchers who cannot see the shape of the population distribution. For this exercise, the sample suggests that while there is some skewness in the variable *ActualDuration*, it is not extreme. So, because of the ***robustness*** of the t procedures, we can proceed with a hypothesis test.

d. Which symbol should be used for the 80 min in the context of a hypothesis test: $\overline{x}$, $\hat{p}$, μ_0, σ , μ , s?

e. Confused Conrad can't decide between $\overline{x}$ and $\hat{p}$ to answer question d. You say: "Neither of those!" Convince CC why these two symbols cannot be the answer to part d.

f. ***Step 1:*** Set up the null and alternate hypotheses for the test announced just before part d.

g. ***Step 2:*** Conditions. How are the *two* conditions for using the t conditions met for this test?

h. ***Step 3:*** Calculate the test statistic for the test. See the example in the ***Notes.***

i. ***Step 4:*** Find an approximation to the p value using the ***t distribution Chart*** and the "little box" method. See the example in the ***Notes.***

- Get software to do the hypothesis test. The output for the example above is given here, but your own will be different.

Hypothesis test results:
μ : Mean of variable
H_0 : μ = 80
H_A : μ ≠ 80

Variable	Sample Mean	Std. Err.	DF	T-Stat	P-value
Sample(ActualDuration)	80.166667	1.3204737	35	0.12621733	0.9003

j. Check your calculation of the test statistic with what the output says.

k. Does the p-value given agree with the approximation you determined for part i? Give a reason.

l. Make a sketch of the sampling distribution and show the p value and test statistic. For the example shown above, the sketch looks like this. (The huge amount of shading starting at $t = 0.126$ shows that the p-value is 0.90. Your result may well be different.)

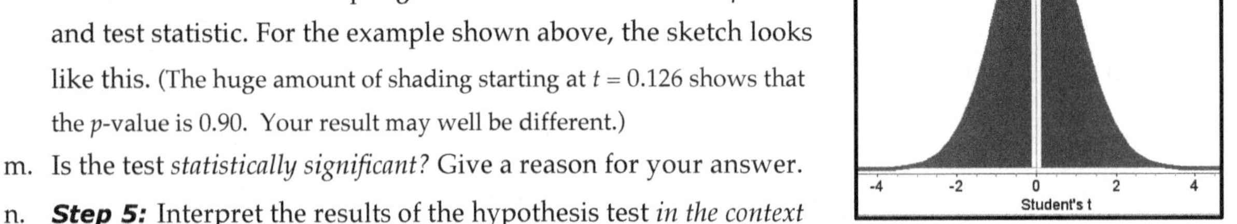

m. Is the test *statistically significant?* Give a reason for your answer.

n. ***Step 5:*** Interpret the results of the hypothesis test *in the context of the data* on duration of flights to Los Angeles from the SF Bay Area.

o. According to the section ***How we can read an hypothesis test from a confidence interval,*** we may be able to predict the outcome of a confidence interval from an hypothesis test. For this problem, will a 95% confidence interval *include* or *exclude* the $\mu_0 = 80$ minutes? Explain why.

p. Using the sample mean $\overline{x}$ and sample standard deviation s from your sample of $n = 36$, calculate ("by hand" using a calculator, not software) a 95% confidence interval for the mean duration of flight μ from the SF bay Area to LAX. Show your work.

- Use software to get a 95% confidence interval for your sample to check the calculations in part p.

s. Check your confidence interval calculation with the output. Is $\mu_0 = 80$ minutes included in or excluded from your confidence interval? Does what happened agree with your answer to part o?

t. Give a good interpretation of your confidence interval in the context of the data.

u. Suppose we did a one-sided hypothesis test so that $H_a : \mu < 80$ instead of the two-sided test. Which of the following would be the same and which would be different. If different, how would it be different?. – the test statistic? – the p-value?

v. With the alternative hypothesis for the one-sided test, $H_a : \mu < 80$, state in the *context of the hypothesis test* (which means that something has to be said about durations of flights) what a ***Type I error*** for the one-sided test would be.

2. ***Pulse Rate for Children*** According to the "Health Calculator" shown here, children between the ages of eight and fourteen should have an average pulse rate of 84 beats/minute. Other sources (e.g. http://pediatrics.about.com/od/pediatricadvice/a/Normal-Pulse-Rates-For-Kids.htm) give a slightly higher figure. We have a data set from the Centers for Disease Control that is part of a national random sample of health indicators. The cases in the data set are children ages four to fourteen, but pulse rate was measured only for the children aged eight through fourteen. A different measure was used for the younger children. The file ***NHANESChildren4-14.*** Is a random sample of children from an NHANES study that only includes children aged four through fourteen. For the children 8 – 14, the sample size was $n = 210$. Our statistical question is:

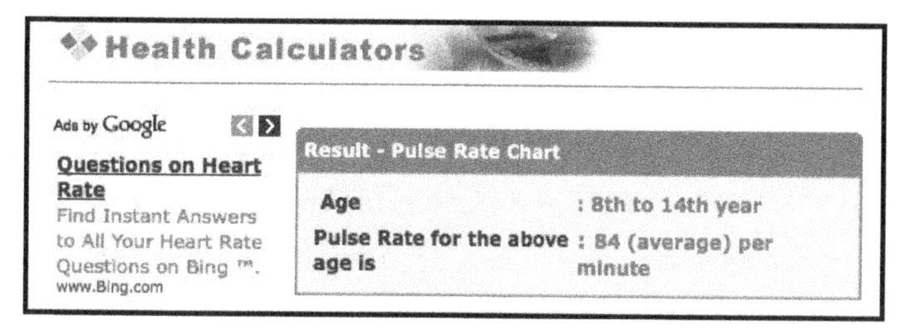

For children eight to fourteen, is the mean pulse rate equal to or different from eighty-four beats per minute?

a. Set up the null and alternate hypotheses for the hypothesis test implied by the statistical question written in italics above.

- Open the file ***NHANESChildren4-14*** and get software to give a graphic showing the distribution of the variable *Pulse.* Also get the sample mean and standard deviation for *Pulse.*
- Using software, get the results for the hypothesis test implied by the statistical question.

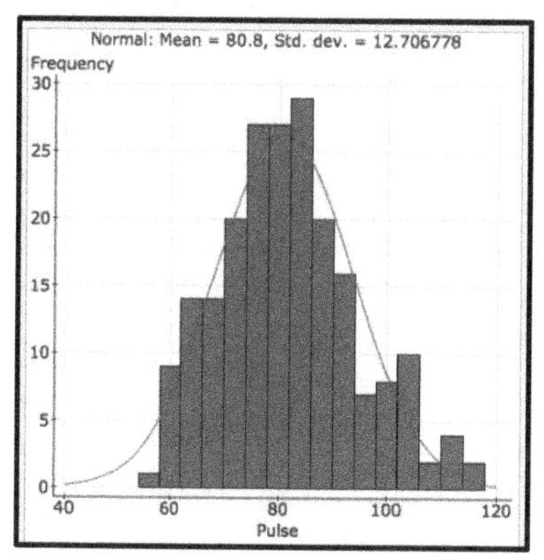

b. The *two* conditions for our hypothesis test are met with the NHANES sample that we have. Explain why this is. Use the graph and the information in the introduction.

c. Show how the test statistic was calculated.

d. Stat Sam locates the test statistic on the ***t-distribution chart*** and sees that the test statistic is "off the chart." SS says, "Ah, yes, significant!" She is right. Explain how she knows the test is significant just from the test statistic being "off the chart." (You may wish to draw a picture.)

e. Interpret the p-value in the context of the hypothesis test.

f. What should you do to get an *estimate* of the population mean pulse rate for children aged eight to fourteen based upon our data? After explaining what you should do, either do the calculation or get software to do the calculation.

g. Interpret the results of your calculation in part f in the context of the data.

h. "Since the sample size is large, the ***power*** of this hypothesis test is big." Explain the meaning of the word ***power*** in this sentence and in the context of the data.

3. *Flying to Las Vegas* Here is another take on the duration of flights from the SF Bay Area. This time we will look at flights to Las Vegas. We will start by drawing a sample of $n = 126$ from the record of flights from the Bureau of Transportation Statistics (BTS).

- Open the file ***OnTimeBayAreaLasVegas2014*** and use the ***Sample*** function of software to get a random sample of flights with sample size of $n = 126$ that includes the variable *ActualDuration*.

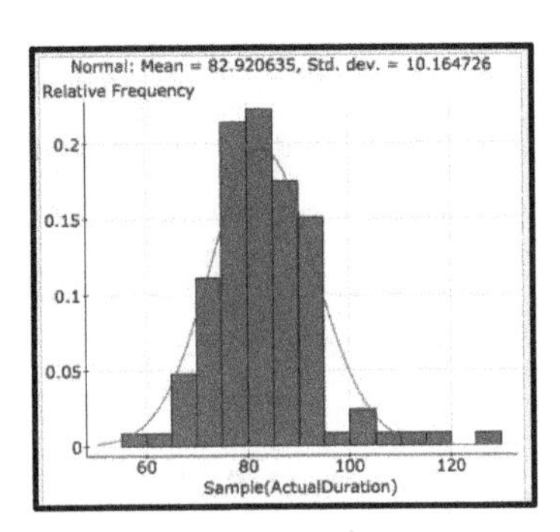

- Using software and the sample chosen from ***OnTimeBayAreaLasVegas2014*** get a graphic showing the distribution of *ActualDuration* of the sample chosen. This can be a dot plot or a histogram or a box plot. The idea is to assess the shape of the distribution.
- Get summary statistics, including the mean and standard deviation.

Here is an example of possible output; the output from a different sample of $n = 126$ will be different.

a. We will get an *estimate* of the mean population duration of flights for the flights from the SF Bay Area to Las Vegas. Does "get an estimate" imply that we will:

   – (I) carry out a hypothesis test OR – (II) find a confidence interval?

   Choose one and give a reason for your choice.

b. With the information that you have (the graphic and what you are told about the sample), determine whether it is possible to have trustworthy results using the t procedures. (*Hint:* Consider ***conditions*** and the ***Rules of Thumb.)***

• c. Do the calculations for what you decided in part a. Check the calculations by opening the file ***OnTimeBayAreaLasVegas2014*** and getting software to calculate a confidence interval for the sample you have drawn. The confidence interval will resemble the one shown here, but will be a different interval.

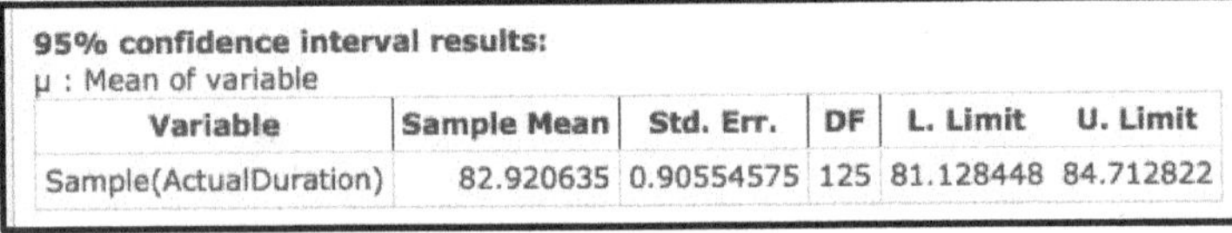

95% confidence interval results:
μ : Mean of variable

Variable	Sample Mean	Std. Err.	DF	L. Limit	U. Limit
Sample(ActualDuration)	82.920635	0.90554575	125	81.128448	84.712822

d. Give a valid interpretation of the confidence interval.

The BTS actually records the *scheduled* duration of each flight, and the *scheduled* mean duration for all of the flights in our "population" of flights is about eighty-seven minutes (rounded to the nearest minute). We will use this eighty-seven minutes as our hypothesized mean for a hypothesis test.

Is the mean of ActualDuration of flights less than eighty-seven minutes?

e. Set up the null and alternate hypothesis test implied by the italicized statistical question above. Will this be a one sided or a two sided test?

• f. Get software to do the hypothesis test using the sample you have drawn from ***OnTimeBayAreaLasVegas2014***, but show (on paper) how the test statistic was calculated.

g. Interpret the p-value shown in the software output in the context of the data.

h. Does the outcome of the hypothesis test agree with what the confidence interval appears to say? Explain very briefly.

i. Explain what a ***Type II error*** would be in the context of this hypothesis test. You need to say something about the mean duration of flights!

Special Exercise 4: Real Estate Data Work on Means

The scenario: Your cousin from Seattle says that there has been a trend in the last twenty years for houses to be bigger in square feet. Her idea is that the mean square feet for houses built in the last twenty years is greater than 2,200 square feet.

- Open the file ***SanMateoRESampleY0506.*** This is a simple random sample of $n = 300$ drawn from all of the houses sold in the year 2005-2006.
- Get a graph and summary statistics of the variable *Sq_Ft* ffor the houses whose age is under 21; that is, filter for *Age* < 21.

Your output may resemble what is here.

Your cousin from Seattle says that there has been a trend in the last twenty years for houses to be bigger in square feet. Her idea is that the mean square feet for houses built in the last twenty years is greater than 2,200 square feet.

1. What is the shape of the distribution of the Sq_ft of houses younger than 21 years?

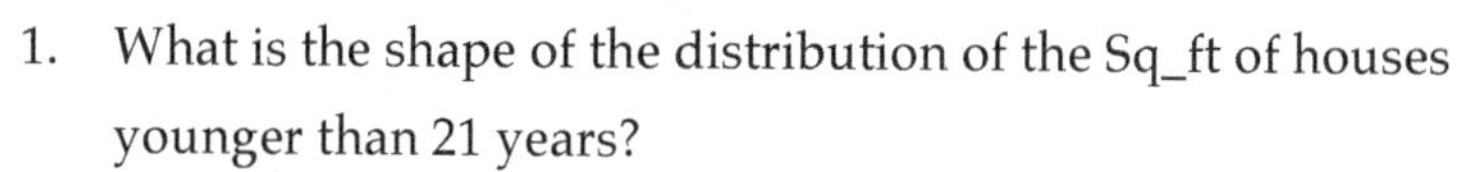

2. Record the mean and standard deviation and the count (the n) of this subsample. Does this distribution have a mean higher than 2,200 square feet?

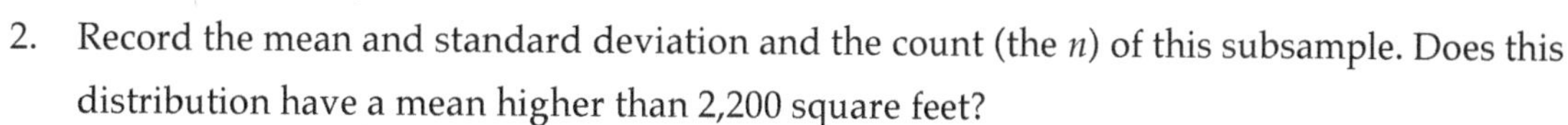

3. a. Are the analyses you have done to answer parts questions 1 and 2 *descriptive* or *inferential* analyses? Give a reason for your answer.

   b. To answer the question: "Is the mean area (in square feet) of houses built in the last twenty years in all of San Mateo County greater than 2,200 ft^2 or not?" should we use:
   - (I) descriptive analyses only?
   - (II) A hypothesis test for one population proportion?
   - (III) a chi-square test of independence?
   - (IV) a hypothesis test for one mean?

   Give a reason which "excludes" the alternate choices.

4. Your cousin wants you to do a hypothesis test. Set up the null and alternate hypotheses for your cousin's question. (*Hint:* Consider what your cousin says to be a challenge to the accepted idea that houses have 2,200 square feet or less.) Will the test be a one-sided or two-sided test?
5. Check whether the conditions are met. Relate the appearance of the sample distribution to the condition that the population distribution must be Normal and to the ***Rule of Thumb.*** (Note: the sample is a SRS from the population, and filtering does not destroy the randomness.)
6. Using the information recoded in answer to question 2, you should be able to calculate the test statistic. Show your work.
7. Use the t table to get an estimate for the p value for this test, using the "little box" procedure. Show your work.

•8 With software, check the calculations using the file ***SanMateoRESampleY0506***. Is the test statistic 3.128? (If the test statistic is negative and it may be that the filtering for *Age* < 21 was not done.)

9. Write a conclusion to the test for your cousin. Interpret the p-value in the context of her ideas. If you have any doubts about the validity of the test, express those as well and state why you have these doubts.

10. a. In the context of the scenario, explain what a ***Type I error*** would be.

    b. In the context of the scenario, explain what a ***Type II error*** would be.

11. Your cousin wants an *estimate* for the population mean size of houses built in the last twenty years.

    a. Which procedure should be used?

       - (I) a hypothesis test for one mean?
       - (II) a confidence interval for one proportion?
       - (III) a confidence interval for one mean?

    b. Carry out the procedure you have selected in part a, showing your calculations.

- With software, check the calculations of the confidence interval using the file ***SanMateoRESampleY0506***. Remember to filter for *Age* < 21.

    c. Record the t^* used in your hand calculation. The t^* used in the hand calculation should not be 3.128. (It is a common error to confuse t^* and the t that is the test statistic in a hypothesis test.)

    d. What is the margin of error for your estimate?

    e. What is the standard error for your estimate?

12. Interpret the confidence interval in a way that makes sense to your cousin in Seattle (your answer must say something about the size of houses).

13. You think of another way to analyze your cousin's idea that the newer houses are larger. You recognize that both variables *Age* and *Sq_Ft* are quantitative. To analyze the relationship between two quantitative variables, which techniques will you use?

    - (I) compare means, medians, standard deviations between two or more groups?
    - (II) chi-square test of independence?
    - (III) a scatterplot and least squares regression line and r^2?
    - (IV) a confidence interval for the variable *Age*?

[14. ***Extra Credit Adventure.*** Do a <u>complete</u> job of analyzing the *relationship* between the variables *Age* and *Sq_Ft*. Write a paragraph explaining to your cousin what you have found. (*Hint:* From your analysis, can you say that newer houses are larger? Can you say that there must be other variables besides when the house was built that affect the size of the house? What statistic informs you? What do the graphics tell you?)]

5. *Flights to Hawaii: Confidence intervals, Hypothesis Tests and Power*

This exercise is designed especially to show the differences between one- and two-sided hypothesis tests. It again looks at the variable *ActualDuration* for a random sample of flights. This time the random sample of flights is from the bay Area to Hawaii.

- Open the file ***OntimeBatAreaHawaii2014*** and use ***Sample*** in the software to get a random sample of $n = 61$ flights from the SF Bay Area to Hawaii that includes the variable *ActualDuration.*
- Have software make a graphic of the sample distribution of the variable *ActualDuration,* and also get the sample mean and sample standard deviation. The result will resemble what is shown here, but will of course not be the same.

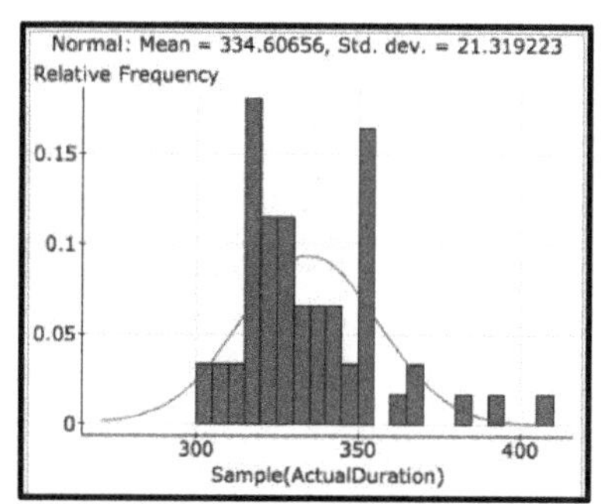

a. We will use t procedures for confidence intervals and hypothesis tests. Judging from what is said in the introduction, and considering the ***Rule of Thumb,*** can we proceed? Explain exactly why you know we can proceed.

b. By hand, calculate a 95% confidence interval for the population mean *ActualDuration* using the information you have so far. Since $n = 61$, the t distribution chart should be able to give you a number to use for the t^*.

- Have software get a 95% confidence interval for the population mean duration of flights from the Bay Area to Hawaii derived from the sample that you have chosen. Here is the result from the sample chosen above. You should have something similar, but not exactly the same.

c. Give a good interpretation of the confidence interval in the context of the duration of flights from the SF Bay Area to Hawaii.

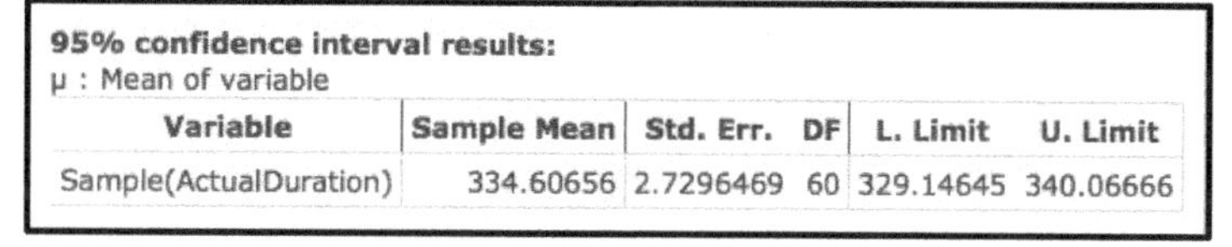

95% confidence interval results:
μ : Mean of variable

Variable	Sample Mean	Std. Err.	DF	L. Limit	U. Limit
Sample(ActualDuration)	334.60656	2.7296469	60	329.14645	340.06666

d. There is a principle that the result of a *two-sided* hypothesis test can be read from whether the confidence interval *excludes* or *includes* the hypothesized μ_0; determine whether the hypothesis test whose hypotheses are: $H_0: \mu = 330$, $H_a: \mu \neq 330$

would be *statistically significant* or *not statistically significant given what you see in your confidence interval.* Explain your reasoning in the context of the test and the confidence interval.

- Have software check your reasoning by actually doing the hypothesis test. For the example we have been using here, we get these results; but your results will not be the same, though you may well come to the same conclusion.

Hypothesis test results:
μ : Mean of variable
$H_0 : \mu = 330$
$H_A : \mu \neq 330$

Variable	Sample Mean	Std. Err.	DF	T-Stat	P-value
Sample(ActualDuration)	334.60656	2.7296469	60	1.687602	0.0967

e. Show how the test statistic was calculated in the hypothesis test that you had software do.

f. Make a sketch of a test statistic distribution similar (but of course not the same) as the one for the example used here. Base the start of the shading on the results of the test that you have done. What does the shading represent? Is it:
   - (i) "*rare*" test statistics or
   - (ii) the p-value for this test?

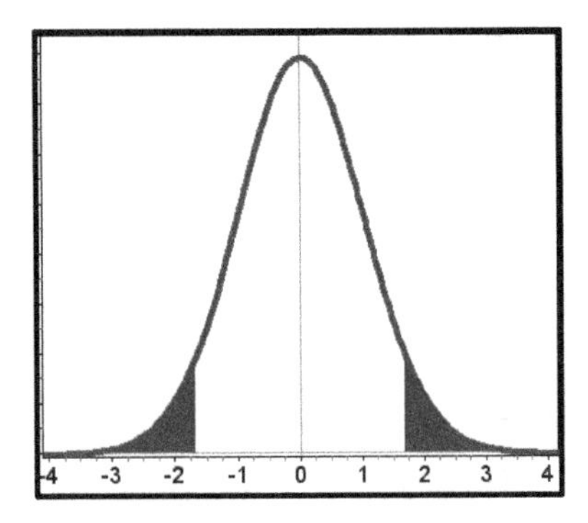

g. Someone has the idea that the hypothesis test should be one-sided so that the null and alternate hypotheses are

   $H_0: \mu = 330$
   $H_a: \mu \neq 330$. For this hypothesis test, will the test statistics be the same or different than the one that you calculated? Give a calculation or a reason for your answer.

h. For the one-sided hypothesis test of part g, what should the p-value be? (You should be able to determine the answer by using the p-value that the software calculated for the two-sided test.)

- Have software do the one-sided test. Here is the result for the example that we have been using. The result you get should be similar but not the same. Notice that our one sided-test was "significant" (just!) whereas the two sided test was not.

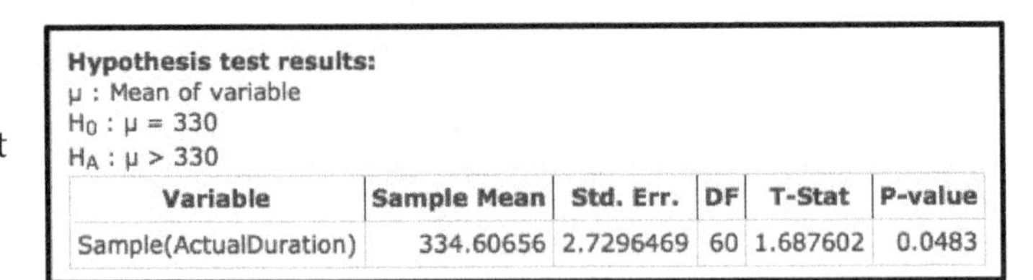

Hypothesis test results:
μ : Mean of variable
$H_0: \mu = 330$
$H_A: \mu > 330$

Variable	Sample Mean	Std. Err.	DF	T-Stat	P-value
Sample(ActualDuration)	334.60656	2.7296469	60	1.687602	0.0483

i. Does the output agree with your calculations in parts g and h? Is the one-sided test statistically significant? Was the two-sided test statistically significant?

j. For the one-sided test (the one that was not statistically significant), explain in the *context* of the data what a ***Type II error*** would be.

- Use the ***Hypothesis Test Power Calculator*** (if available) to determine the *power* of the hypothesis test with the sample size $n = 61$, if *actually* the true mean is $\mu = 335$ minutes so that the difference between the true mean and the hypothesized mean $\mu_0 = 330$ is $\mu - \mu_0 = 335 - 330 = 5$. Consult the ***Software Supplement.***

k. [Optional: only if software has a ***Power Calculator***.] Report the power as calculated by the ***Hypothesis Test Power Calculator*** and interpret the number in the context of the data. That is: what does the power mean in the context of *this* hypothesis test?

- l. Use the ***Hypothesis Test Power Calculator*** to find the sample size that would be necessary to have power at the 85% level to detect a difference of three between the true mean and the hypothesized mean: that is $\mu - \mu_0 = 3$. Is the sample size required bigger than the sample size of $n = 61$ that was used for this hypothesis test?

§5.3 Exercises on Paired Comparisons

1. ***More Blood Pressure Data and Other Analyses*** This exercise will investigate the second component of blood pressure, ***diastolic blood pressure,*** but it also reviews what we have done in the past weeks and even before that. Expect to reach back into what we have done before. Since in the ***Notes*** we only looked at the data for females, this exercise will also look only at females. The sample is a random sample of the NHANES data, which itself can be regarded as a random sample of the American population. The variables we will examine are

   Diastolic1 — *First blood pressure measurement*

   Diastolic2 — *Second blood pressure measurement*

   DiffDist2Dist1 — *The difference between the two measures, where: DiffDist2Dist1=Diastolic2 - Diastolic1*

   a. According to the description just above, what are the cases for this collection?

- Open the file ***NHANESBloodPressure*** and get a graph of the three variables just for the females (so, filter by *Sex = "Female"*) on the same plot. It should resemble the plot shown here.
- Have software get means and standard deviations) for the three variables as well filtered for just the females.

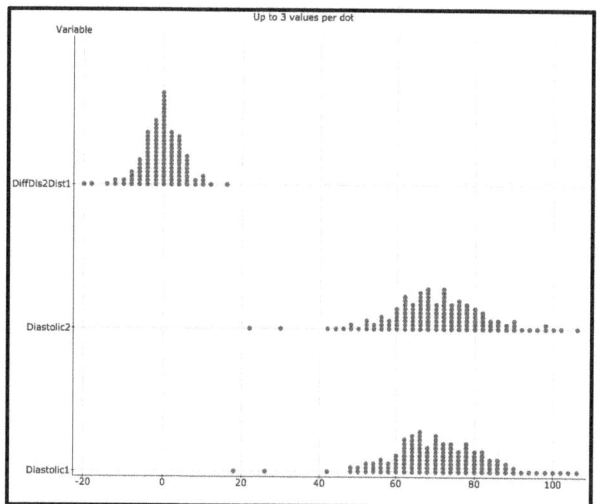

   b. In the context of the data, what does *DiffDist2Dist1 = - 6* mean?

   c. Read the definition of ***paired comparison data*** in the ***Notes.*** Explain how the variables *Diastolic2* and *Diastolic1* fit into that definition of paired comparison data.

   d. Using the correct notation helps to focus on what is being done: what symbols should be used for the numbers –0.3665 and 4.987? [Notice that the *n* is different: the reason for that is that there were only 322 women who has both measures taken.]

Summary statistics:
Where: Sex="Female"

Column	Mean	Std. dev.	n
Diastolic1	70.956989	11.764457	372
Diastolic2	70.435754	11.625822	358
DiffDis2Dist1	-0.36645963	4.9874477	322

   e. We want to have an *estimate* of how *much* the second measure of diastolic blood pressure differs from the first for the population of people from which our sample is drawn. What inferential procedure should we use? Given what we want, should be get a confidence interval, or should we get a hypothesis test? Give a reason.

   f. Are the conditions met for the inferential procedure you have chosen? Or do you have doubts about one or more of the conditions? Be complete. [For the Normality condition, which plot should we look at? For whether the sample is random, see the introduction to the Exercise.]

g. Confused Conrad, when he answers the question about conditions, writes: "Since $0.36646 * 322 = 118 > 10$, we know that the sample size is large enough and the conditions are met." CC has made at least two errors. First, CC is confused about what procedure he is using, and, secondly, CC has also forgotten to consider one of the conditions that need to be considered. Explain what his two mistakes are. (The two mistakes are common ones.)

- Have software do the calculations for a confidence interval for the variable *DiffDist2Dist1* but filtered for the women. (*Sex = "Female"*). See the ***Software Supplement*** under ***Inference for Paired Comparisons.***

h. Show how software calculated the confidence interval using information for the means and standard deviations.

i. What is the margin of error for the estimate?

j. What is the standard error for the estimate? The answer to this question should not be the same as the answer to part i.

k. What does the null hypothesis $H_0 : \mu_{Diff} = 0$ mean in the context of the measuring blood pressure?

l. Does your confidence interval include zero? What does this say about the outcome of the hypothesis test where $H_0 : \mu_{Diff} = 0$ and $H_a : \mu_{Diff} \neq 0$? Specifically, will the hypothesis test be statistically significant or not? Give a reason for your answer.

- Have software carry out the hypothesis test for the variable where $H_0 : \mu_{Diff} = 0$ and $H_a : \mu_{Diff} \neq 0$ using the variable *DiffDist2Dist1* but filtered *Sex = "Female"*. See the ***Software Supplement*** under ***Inference for Paired Comparisons.***

m. Show how the test statistic was calculated using the correct formula using information for the means and standard deviations.

n. Check the degrees of freedom given. What formula was used for this?

o. Is the test statistically significant? Do we reject the null hypothesis? [The answer here should agree with the answer given to part l above.]

p. Interpret the outcome of your hypothesis test and the information from the confidence interval in the context of measuring blood pressure. (Your answer to part l should help you.)

The website http://www.new-fitness.com/Blood_Pressure/numbers.html says that the optimal blood pressure should be 120/80 or less, and "normal blood pressure" is 130/85, where the convention is that the "fraction" refers to ***systolic/diastolic.*** We will use these as the standard and ask the question:

Do American women (or men) have optimal (or normal) blood pressure?

In our data we have the average of all the systolic measurements in the variable *Systolic* and the average of all the diastolic measurements in the variable *Diastolic.* You have two choices to make: whether to look at systolic or diastolic and whether to look at "Optimal" or "Normal" blood pressure. (If you choose to analyze the data on males, change the filtering from *Sex = "Female" to Sex = "Male".*)

q. Decide whether you will use a confidence interval or a hypothesis test to answer the statistical question. Give a reason for your answer. Your answer should look forward to the possible results that you can get. (It is possible to answer our statistical question with either a confidence interval or a hypothesis test.) ***PTO***

r. If you decide on an hypothesis test, clearly state your null and alternate hypotheses, do some descriptive work to look at the distribution of *Systolic* or *Diastolic*, get the sample mean and sample standard deviation, and calculate the test statistic (using the correct notation). You may use software to do the actual calculations, including carrying our the test. **OR**

r′. If you decide on a confidence interval, do some descriptive work to look at the distribution of ***Systolic*** or ***Diastolic***, get the sample mean and sample standard deviation, and calculate the confidence interval (using the correct notation). You may use software to do the actual calculations.

s. Clearly, in a short paragraph, state how your analysis answers the statistical question.

2. ***City and Highway Miles per Gallon (MPG)*** This exercise uses a random sample of the Federal Government's Guide to Fuel Economy for 2014 cars. The data look like this.

FuelEconomyGuide2014Sample

	Manufac...	Division	Carline	EngineDi...	NumCyl	CityMPG	HwyMPG	CombinedMPG	ClassVehicle	TransmissionType
33	Chrysler ...	Dodge	Charger ...	6.4	8	14	23	17	(c) Midsize to Large	(a) Automatic
34	Ford Mot...	Ford	F150 PIC...	5	8	15	21	17	(f) Truck or Van	(a) Automatic
35	General ...	Buick	ENCORE	1.4	4	25	33	28	(e) SUV	(c) Semi-Automatic
36	BMW	BMW	535i	3	6	20	30	23	(c) Midsize to Large	(b) Manual
37	Porsche	Porsche	911 Carr...	3.4	6	19	26	21	(a) Subcompact	(b) Manual
38	Toyota	TOYOTA	SEQUOI...	5.7	8	13	17	15	(e) SUV	(c) Semi-Automatic
39	Honda	Acura	RLX	3.5	6	28	32	30	(c) Midsize to Large	(c) Semi-Automatic
40	General ...	Chevrolet	CRUZE	1.4	4	26	38	30	(c) Midsize to Large	(c) Semi-Automatic
41	Hyundai	HYUND...	SONATA	2.4	4	24	35	28	(c) Midsize to Large	(a) Automatic

Some of the variables measured for each model of car (e.g. Dodge Grand Caravan) are:

Manufacturer	*Name of the company making the car.*
Division	*Make of the car.*
Carline	*Model of the car.*
EngineDisplacement	*Engine displacement, measured in liters (indicates the "size" of the engine)*
NumCyl	*Number of cylinders in the engine.*
CityMPG	*Miles per gallon calculated for urban driving ("stop and go").*
HwyMPG	*Miles per gallon calculated for highway driving.*
CombinedMPG	*Miles per gallon for a combination of city and highway driving.*
ClassVehicle	*Whether the car is a subcompact, compact, midsize etc*
TransmissionType	*type of transmission (automatic, manual, etc.)*

For most cars (but not all), highway MPG is greater than city MPG (hence more economical) but how much bigger? In this exercise we will answer the statistical question:

How much bigger on average is highway MPG than city MPG?

- Open the file ***FuelEconomyGuide2014Sample*** and have software get graphics showing the distributions of the variables *CityMPG* and *HwyMPG*. The plots should be something like the ones shown here.

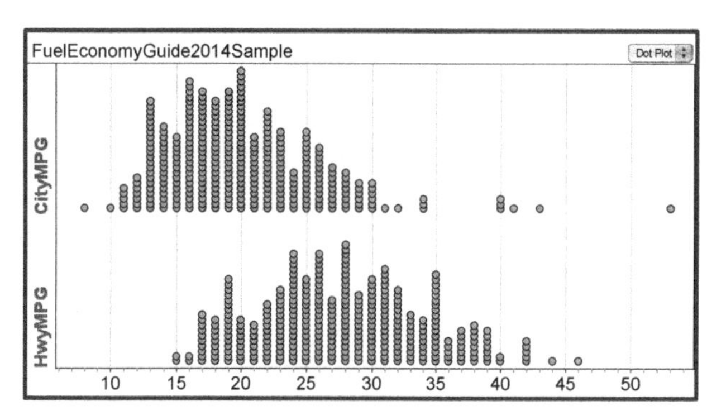

a. What are the cases for these data?

b. Read the definition of ***paired comparison data*** in the **Notes**. Explain how the variables *CityMPG* and *HwyMPG* fit into that definition.

- Use software to create the variable *DiffHwy = HwyMPG - CityMPG.* Consult the ***Software Supplement*** on the topic: ***Creating New Variables.***
- Have software get the mean, standard deviation and the count for the new variable *DiffHwy = HwyMPG - CityMPG*

c. Choose a good inferential procedure to answer the question in italics in the introductory paragraph to this exercise. Give a reason for your choice.

d. Are the conditions met for the inferential procedure you have chosen? Or do you have doubts about one or more of the conditions? Be complete about your judgment.

- Use software to do the calculations for the inferential procedure chosen in part c.

e. Show how the upper limit of the confidence interval that should have been chosen as the desired procedure was calculated.

f. Interpret the results in the context of the data. (You need to say something about fuel economy in the city and on the highway.)

[Parts g and h are review of using confidence intervals and hypothesis tests.]

g. Suppose the government decreed that the average *HwyMPG* for trucks and vans should be 22 MPG. That is the standard. You suspect that the average is less than that, and you have these sample data. Is it true that pickup trucks average 22 MPG or is it less, as you think? Carry out an appropriate inferential procedure, showing all steps. If software is used to confirm the calculations, filter by *ClassVehicle ="(f) Truck or Van".*

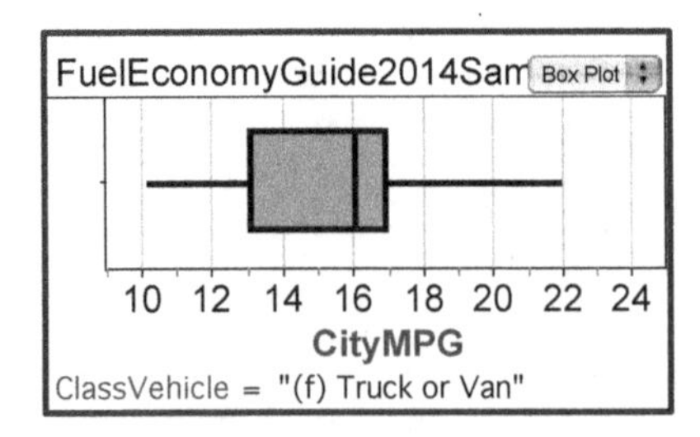

h. Is what you have just done a "paired comparison" problem? Give a reason for your answer.

i. With a hypothesis test, state what a Type II error would be in the context of the hypotheses used in part g.

- j. [Optional] If the software being used has Hypothesis Test Power Calculator, use it to find the power of the test if the difference between the true mean and $\mu_0 = 22$ is one mile per gallon? What is the power if the difference between the true mean and $\mu_0 = 22$ is two miles per gallon?

3. ***Cost of Fuel*** This exercise uses data collected on prices of gasoline and diesel fuel by the Energy Information Administration. The data are collected each week from gas stations, and the data we have start from 1994 and run to 2008, although some of the data collection started later in 1994. We can regard the data as randomly sampled from the sellers of fuel. Generally, there are three grades of gasoline: regular, mid-grade, and premium, and we expect that each higher grade will cost more than the grade below. The relationship between the price of diesel fuel and gasoline is more difficult to predict. At times it appears to be higher, at times lower. (See: The Energy Information Administration: http://www.eia.gov/dnav/pet/pet_pri_gnd_dcus_nus_w.htm.)

- Open the file ***FuelPrices.*** The data look like this. However, the first 35 rows do not have data for the midgrade or the premium prices; so scroll down.)

Each row in the case table gives prices for a *week* and what is shown are the prices for the different types of fuel for that week. So the rows are "weeks" and they start from March 28, 1994, and end at November 24, 2008.

The first four variables give the price for that week of regular, mid-grade, premium, and diesel. The prices are in cents. The last four variables represent the *differences* in cost on a specific date between different types of fuel for a specific week: "diesel and premium," etc. Here are the definitions:

FuelPrices

	Date	Regular	MidGrade	Premium	Diesel	DieselP...	DieselReg	PremReg	MidReg	<r
756	09/15/2008	386.7	397.3	411.5	402.3	-9.2	15.6	24.8	10.6	
757	09/22/2008	373.2	384.7	398.2	395.8	-2.4	22.6	25	11.5	
758	09/29/2008	364.4	375.6	389.5	395.9	6.4	31.5	25.1	11.2	
759	10/06/2008	348.5	360.9	375.2	387.5	12.3	39	26.7	12.4	
760	10/13/2008	310.9	323.5	337.8	365.9	28.1	55	26.9	12.6	

DieselPrem = Diesel − Premium	*DieselReg = Diesel − Regular*
PremReg = Premium − Regular	*MidReg = MidGrade − Regular*

a. What are the cases for these data? (Try to be exact in the description; think of what the numbers in each row give you.)

b. Give a reason that the variables *DieselPrem = Diesel − Premium, DieselReg = Diesel − Regular , PremReg = Premium − Regular, MidReg = MidGrade − Regular* make these data able to analyzed by "Paired Comparison" methods.

c. From your experience (and before looking at the data), what is your *guess* about the average difference between mid-grade gasoline and regular gasoline: ten cents, nine cents, eight cents, twelve cents? If we do a hypothesis test to test your idea with these data, what symbol should be used for your idea?

d. Set up the null and alternate hypotheses to test your idea.

- e. Use software to do a descriptive analysis of the data to determine whether the conditions are met for the hypothesis test. The conditions are met, but you need to say exactly *why* they are met.

- f. Use software to get the sample mean and sample standard deviation for the relevant variable and assign to these the correct symbols.

- g. Calculate the test statistic by hand and use it to approximate a *p*-value. Then use software to check your calculations.

h. Interpret the hypothesis test in light of your hypotheses. Do the data shoot down your idea about the difference between regular and mid-grade, or are the data consistent with your idea?

i. Explain what a ***Type II error*** would be in the context of your guess for the population mean difference.

j. We want to get an estimate of the mean difference between diesel and regular. Which inferential procedure (CI or HT) will we use, and which variable will we use?

k. Check the dot plot of the distribution of the variable you chose in part j. This is always a good practice. However, why can we proceed with the analysis whatever shape the dot plot has?

l. Either by hand or using software, carry out the procedure you chose in part j and interpret your results in the context of the data.

4. ***Grades for Calculus 2 in Costa Rica*** Grades were recorded for several years for students completing Calculus 2 courses at the Instituto Tecnológico de Costa Rica. There is one case per student. The exams appear to be marked out of 100, and the Course Grade is also given out of 100. The first exam has a reputation for being much harder than the others. The topics for each exam are the same each semester. The data are found in the file ***GradesCalculus2***.

	Semester	Sex	Exam1	Exam2	Exam3	Course_grade
98	1999-2	F	79	81.9	62.5	76.4
99	1999-2	F	78.3	68.6	71.4	68.9
100	1999-2	M	70.3	83.3	77.5	75.7
101	1999-2	M	58.3	72.2	75.4	55.8
102	1999-2	M	84.7	91.4	78.6	80.1
103	1999-2	M	74	73.6	76.8	78.4
104	1999-2	M	91.3	92.8	98.2	84.2
105	1999-2	M	86.3	95.8	90.0	86.4

a. What are the cases for the data? Are the cases "Grades," or are they "students"? Give a reason for your answer.

b. Consider the structure of the data, with the grades for Exam1, Exam 2, Exam3, and the Course Grade for each student. What about this structure fits the definition of "Paired Comparison?"

• c. See the comments above about the first exam compared with the others. Use software with the file ***GradesCalculus2*** to get *descriptive* (and not inferential) statistics and graphics to confirm that the data show that "the first exam has the reputation of being much harder than the others."

d. So, given that the first exam is harder than the others, we can ask: *How much better do we expect a students taking Calculus 2 at the ITCR to do on average?* For this question, do we want: (i) A hypothesis test, or (ii) A confidence interval? Give a reason for your answer.

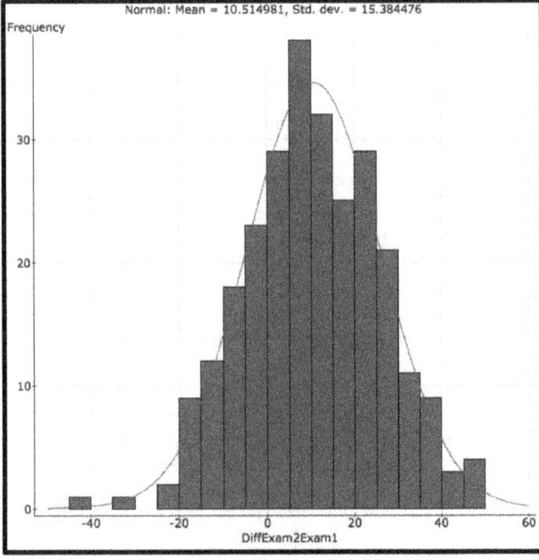

• e. Use software to create the variable *DiffExam2Exam1* = *Exam2 - Exam 1* and get summary statistics and a graphic for this variable. The graphic should resemble what is here, although the graphic could be a dot plot or box plot. On the graphic showing the distribution, what is the meaning of the negative values in the context of the data?

f. The correct answer to what we want (the question in part d) is a "Confidence Interval" since we want an estimate of "how much better a student does on average." Are the conditions met for calculating a confidence interval in this instance? Recall that there is a condition that has to do with the way the data were collected from what we think is the population, and another that is related to the sample size. Discuss both in your answer.

• g. Have software calculate the 95% confidence interval for the difference between the Exam 2 and Exam 1 scores. Using the summary statistics for the variable *DiffExam2Exam1* show how the lower limit of the confidence interval was calculated. (If you are using a chart with a t distribution, you may get a slightly different answer because the t^* you use will not be as accurate as what software uses.)

h. What is the margin of error for the confidence interval that has been calculated? How is it different from the "standard error"?

i. Taking into account that you may not be convinced that the sample that you have is a random sample of the population of all Calculus 2 students at the ITCR (for all years!), give a reasonable interpretation of the confidence interval that has been calculated.

j. Repeat the analysis but for the difference *DiffExam3Exam2 = Exam3 - Exam2.* What is your conclusion about what is expected for students' scores on Exam 3 compared with Exam 2? Give reasons for your answer based upon the analysis.

5. *Flights from SFO: Distance, Schedule Duration, Actual Duration* The data we will examine are a random sample of flights from the SFO (the San Francisco airport) for the first three months of 2014. The random sample has $n = 540$. The data are in the file ***OnTimeSFO2014Sample***.

OnTimeSFO2014Sample

	DEST_C...	DES...	CRSD...	Departu...	TAXI_IN	ArrivalD...	CRSDur...	ActualD...	DiffCRS...	AirTime	DISTANCE
363	Los Ang...	CA	1442	25	12	21	90	86	4	53	337
364	New Yo...	NY	1510	-9	5	-41	330	298	32	277	2586
365	New Orl...	LA	1223	48	2	35	242	229	13	205	1911
366	San Die...	CA	844	-9	6	-13	86	82	4	64	447
367	Chicago...	IL	1251	27	19	19	248	240	8	198	1846
368	Newark,...	NJ	2120	-2	11	-32	320	290	30	270	2565
369	Los Ang...	CA	1052	-2	14	10	91	103	-12	61	337
370	Medford...	OR	1950	19	5	22	91	94	-3	73	329
371	Washin...	VA	600	-8	8	-15	300	293	7	268	2419

Unlike previous samples, there is not a specific destination; rather, the destinations include all of the places in the USA to which it is possible to fly from SFO. International flights are not included.

CRSDuration	*Scheduled duration of the flight in minutes; i.e. the "advertised" duration.*
ActualDuration	*Actual Duration of the flight in minutes, as recorded.*
DiffCRSActual	*DiffCRSActual = CRSDutaion - ActualDuration The difference between scheduled duration of flight and actual duration of flight*
DISTANCE	*The distance in miles for the flight as scheduled.*

a. As a reminder: what are the cases for these data? Airports, Airlines, Flight times, Flight Distances, Flight Durations, Flights, Airplanes?

b. Give a reason that the variables *CRSDuration* and *ActualDuration* can be regarded as "Paired Data" in the context of the data we have here.

c. Give a reason that the variables *DISTANCE* and *ActualDuration* can *not* be regarded as "Paired Data" in the context of the data we have here.

- d. Some exploration of the data: Using the file ***OnTimeSFO2014Sample*** have software make a dot plot of the distribution of the variable *ActualDuration.* What is a good description of the shape of this distribution?
- Have software make a scatterplot of the relationship of *Distance* and *ActualDuration* with *Distance* as the response variable. The result should resemble this graphic.

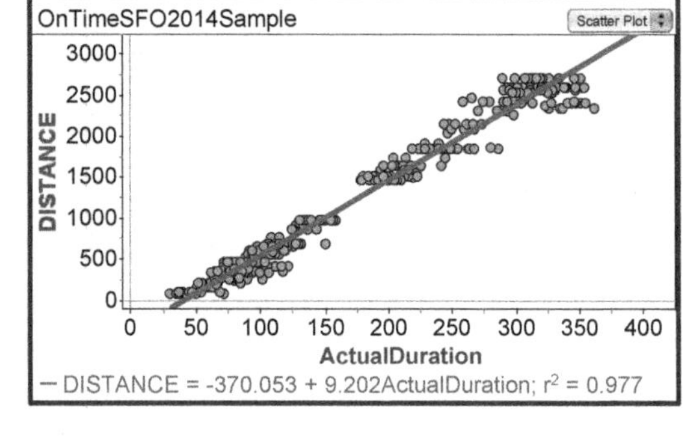

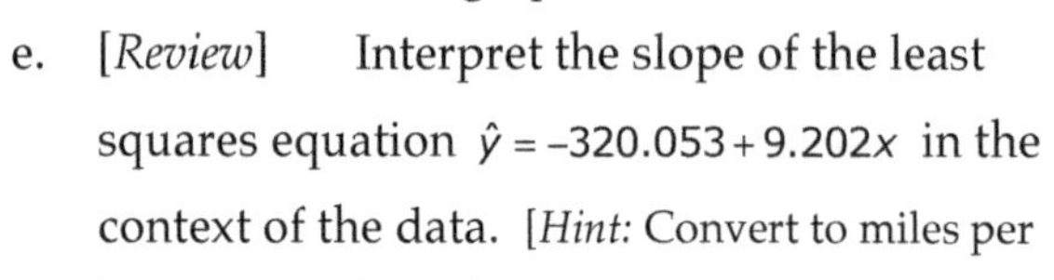

e. [*Review*] Interpret the slope of the least squares equation $\hat{y} = -320.053 + 9.202x$ in the context of the data. [*Hint:* Convert to miles per hour. Also: why is there a gap in the distance distribution?]

We can use the variable *DiffCRSActual* to answer this statistical question:

"Do we expect flights from SFO to take a shorter time, the same time or a longer time than scheduled?

- Have software get graphics showing the distributions of the three variables *CRSDuration, ActualDuration* and *DiffCRSActual.* The plots should resemble these.

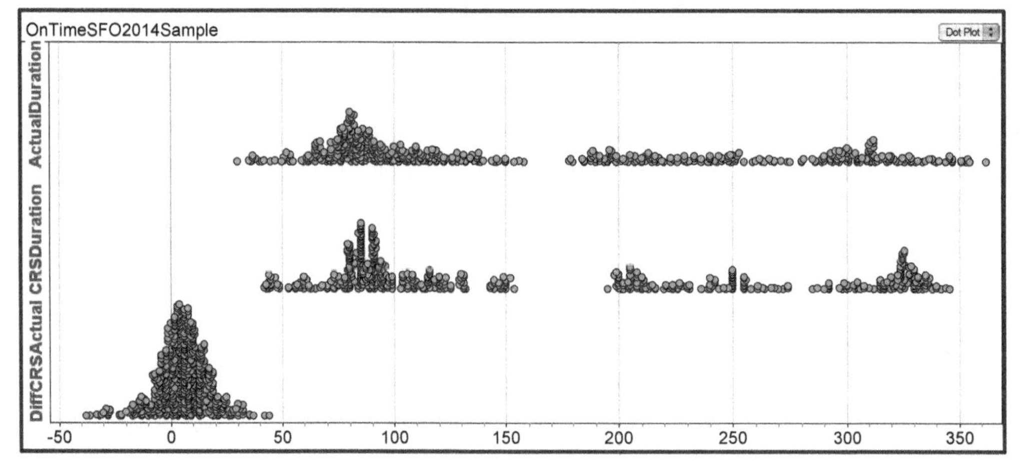

f On the graphic for the variable *DiffCRSActual* notice that there are negative numbers. What does a negative value for this variable mean in the context of scheduled duration and actual duration?

g. Are the conditions for using the *t* procedures for paired comparisons met in this situation? Is the sample a random sample and does the relevant distribution meet the Normality Condition? What does the ***Rule of Thumb*** say for this situation? Explain.

h. Here again is the statistical question: *"Do we expect flights from SFO to take a shorter time, the same time or a longer time than scheduled?* Which will be more informative: a confidence interval or a hypothesis test? Explain your reasoning.

- Have software make a confidence interval or do a hypothesis test using the variable *DiffCRSActual.*

i. Interpret the results of the analysis in the light of the statistical question in the context of the expectation of actual duration of flights being shorter longer or the same as the scheduled duration.

§5.4 Exercises: Comparing One Measure for Two Collections

1. **Blood Pressure Data Again** In the **Notes** for this section we found that there was a significant difference between males and females in mean ***systolic*** blood pressure. Will we find the same pattern for the second component of blood pressure, that is, ***diastolic*** blood pressure? Our statistical question is:

   Is there a difference between men and women in average (or mean) diastolic blood pressure?

   We will use the same data set as we did for the study of Systolic blood pressure. These data are a random sample of data collected as part of the NHANES study, and that study is a random sample of the population of the USA residents.

- Open the file ***NHANESBloodPressure*** and have software get a graphics to compare the distributions of the variable *Diastolic* grouped by *Sex* and filtered for the age group twenty to thirty.
- Also, have software get summary statistics for the distributions of variable *Diastolic* grouped by *Sex* and filtered for the age group twenty to thirty. (To filter, use: $Age < 31$)

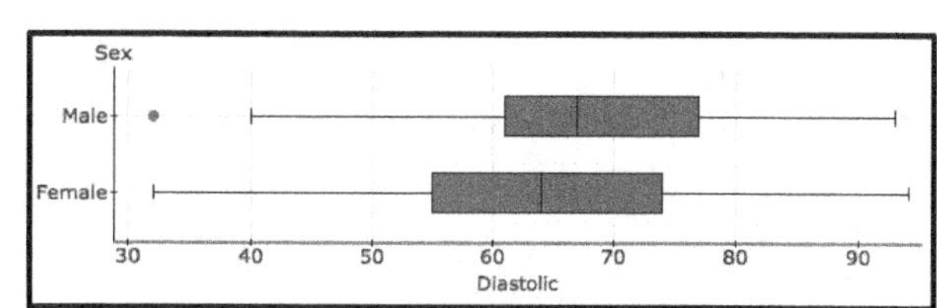

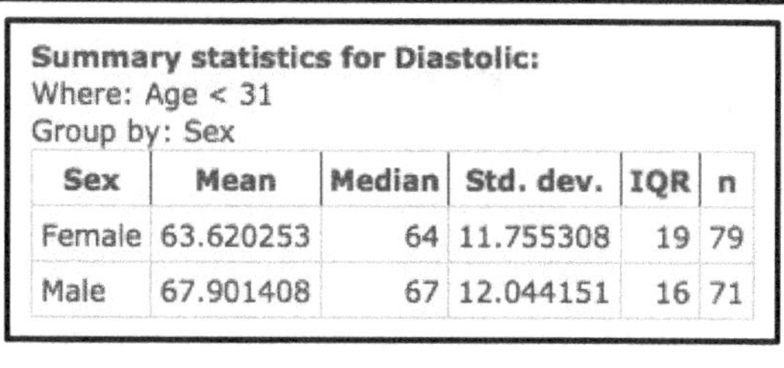

Summary statistics for Diastolic:
Where: Age < 31
Group by: Sex

Sex	Mean	Median	Std. dev.	IQR	n
Female	63.620253	64	11.755308	19	79
Male	67.901408	67	12.044151	16	71

a. How will you characterize the shapes of the distributions?

b. From the information presented, does it appear that, on average, men have higher or lower *Diastolic* blood pressure than females? Cite a reason for the answer.

c. Use appropriate symbols for the means and standard deviations given in the summary statistics. Should the symbols be μ_f, μ_M etc. , or should they be $\bar{x}_F, \bar{x}_M$ etc.

d. Another way of stating our statistic question (above in italics) is: We want to know *whether* (or *if*) the difference between males and females that we see in sample mean *Diastolic* blood pressure can be trusted to infer to the population. Does this statistical question suggest that we carry out a hypothesis test or calculate a confidence interval? Give a reason for the answer.

e. Are the conditions met for the inferential procedure you have chosen? Be complete.

f. The intended answer to part d is that you should do a hypothesis test. Set up the null and alternate hypotheses for the hypothesis test, using good notation.

g. For your alternate hypothesis, do you want a two-sided test or a one-sided test? Choosing either one is defensible, but you must choose one, and you must give a reason for the one you chose. However, the reason cannot be that the sample mean for females is lower than the sample mean for males.

h. Both this exercise and the exercise in the last section were on blood pressure. Write how you know that you should use the methods of this section rather than the methods of paired comparison to do the hypothesis test.

- Using the file ***NHANESBloodPressure*** have software carry out the calculations for the hypothesis test. Consult the ***Software Supplement*** on ***Inference for Two Independent Means***.
  i. Using the summary statistics, show how to calculate the test statistic for the hypothesis test.
  j. Explain how the degrees of freedom shown in the software output was calculated. It is sufficient to cite the formula that was used (though you may want to check the calculations!)
  k. Use the ***t Distribution Chart*** to get an approximation of the p-value using the test statistic and the degrees of freedom that software has calculated. The p-value calculated by software should agree with the approximation.
  l. Make a sketch of the t sampling distribution like the one on the right; on it, show your test statistic and, *by* shading, show the p-value.

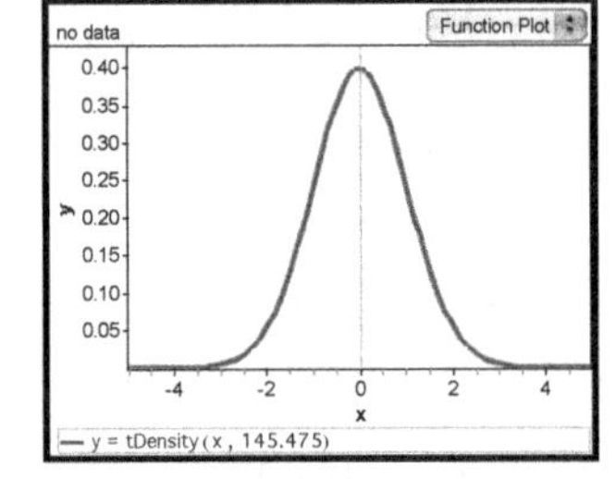

  m. The next plot is Confused Conrad's answer to question l, showing his p value of 0.03 with a vertical line just next to the mean of the sampling distribution. What is CC's confusion? How should he show the p-value?

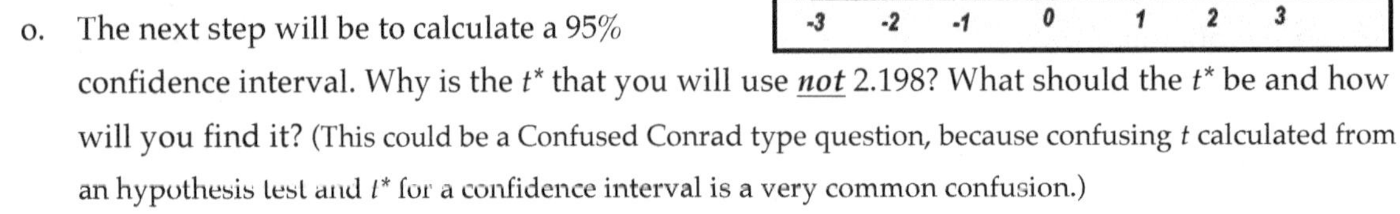

  n. Interpret your hypothesis test in the context of the data.
  o. The next step will be to calculate a 95% confidence interval. Why is the t^* that you will use ***not*** 2.198? What should the t^* be and how will you find it? (This could be a Confused Conrad type question, because confusing t calculated from an hypothesis test and t^* for a confidence interval is a very common confusion.)
- Using the file ***NHANESBloodPressure*** have software carry out the calculations for the confidence interval. Consult the ***Software Supplement*** on ***Inference for Two Independent Means***.
  p. Show how the upper limit of the confidence was calculated.
  q. Give a good interpretation of the confidence interval in the context of the data.
- Remove the age filter (*Age* < 31) from the software hypothesis test calculation, and thus get the results for all ages, and not just those younger than 31. The results got should agree with the output shown here.
  r. There should be a hypothesis test for the difference between men and women on the mean level of diastolic blood pressure, but now it is for all ages. Interpret this hypothesis test in the context of the data.
  s. Describe a Type II error in the context of the hypothesis test in part q. (Context means that the answer must say something about diastolic blood pressure and gender.)

Test of NHANESBloodPressure — Compare Means

First attribute (numeric): Diastolic

Second attribute (numeric or categorical): Sex

Ho: Population mean of **Diastolic** for **Female** equals that for **Male**

Ha: Population mean of **Diastolic** for **Female** is not equal to that for **Male**

	Female	**Male**
Count:	427	380
Mean:	69.9555	72.0184
Std dev:	11.7709	13.0823
Std error:	0.569634	0.671109

Using unpooled variances

Student's t:	-2.344
DF:	767.434
P-value:	0.019

2. *More Work on the Real Estate Data*

- This work will also be done using ***SanMateoRESampleY0506*** that is a random sample of the Real Estate Data for the population of houses in San Mateo County that were sold between June 2005 and June 2006.
- Here is the scenario: You have a cousin who is a Realtor in Seattle, who has some questions about the real estate market in San Mateo County, and she has learned that you have a random sample of data on recent house sales in the county She is asking you to use this random sample to give her some idea of the *population* of recent house sales in San Mateo County.

  Background: In Seattle, your cousin specializes in selling "big, old" houses. Her idea is to see if she can concentrate her business on big, old houses here in San Mateo County. So here are your Realtor cousin's questions:

A. First Question: I want to know whether the average (that is the mean) sale price of big, old houses is significantly greater than the houses for sale that are not big and old. By "big, old," I mean houses whose area is more than 1,800 square feet in the house and also whose age is more than fifty years old.

B. Second Question: If you find that the big, old houses are selling for more on average, I would like to have an estimate of how much more they sell for, on average.

For this analysis, the variable *BigOld* has been created that has two categories:

- *"Big Old"* — *Houses that are more than 50 years old and have Sq_Ft greater than 1800 ft2.*
- *"Not Big Old* — *Houses that are either 50 years old or younger or 1800 ft2 or less.*

Also, we shall use variables for the list prices and sale prices that express the prices in thousands of dollars, so as to avoid huge numbers: Hence $ListThou = \frac{List\_Price}{1000}$ and $SaleThou = \frac{Sale\_Price}{1000}$

a. To answer her first question, does your cousin want a significance test or a confidence interval? Give a reason for your answer based on what she says.

b. To answer her *second* question, does your cousin want a significance test or a confidence interval? Give a reason for your answer based on what she says.

c. Will we use the formulas for means or for proportions? Give a reason for your answer.

d. Will we use the formulas for *one* sample/population or the formulas for *two* samples/populations? Give a reason for your answer.

- Using the file ***SanMateoRESampleY0506*** have software get:
  - A graphic or graphics showing the distributions of the variable *SaleThou* grouped by the categories of the variable *BigOld.*
  - Summary statistics, including the means and standard deviations of the variable *SaleThou* grouped by the categories of the variable *BigOld.*

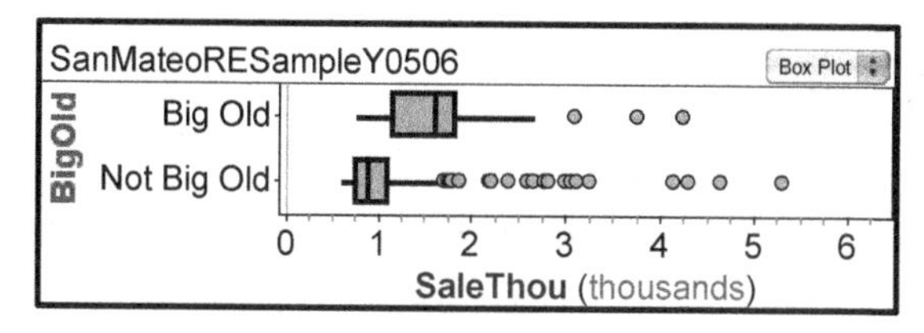

Summary statistics for SaleThou:
Group by: BigOld

BigOld	Mean	Median	Std. dev.	IQR	n
Big Old	1636.9091	1590	711.92934	720	47
Not Big Old	1048.7063	860	640.40681	392.5	253

e. Record the mean and standard deviation and the count (the n) of the "Big Old" and "Not Big Old" houses, using the correct notation. [Subscripts "B" and "N" are useful.]

f. What shapes do the distributions have? Can we safely proceed with our analysis? In other words, are the conditions met for using the t procedures? Consult the ***Notes*** about conditions and apply the ***Rule of Thumb.***

A. Answering the first question

g. To answer the first of your cousin's questions, you will use a hypothesis test. State why a hypothesis test makes sense, given the nature of the question.

h. Set up the null and alternate hypotheses for the hypothesis test using good notation.

- Use software with the file ***SanMateoRESampleY0506*** to calculate the test statistic and the p-value for the hypothesis test for a difference of means in the sale prices of houses between the "Big Old and "Not Big Old" houses. (Use the variables *SaleThou* and *BigOld*.)

i. Using the summary statistics, show how the test statistic was calculated in the software output.

j. In the software output, there is a number for DF. Show the formula that calculated this number.

k Is the test statistically significant? Do we have sufficient evidence to reject the null hypothesis? Give a reason for your answer.

l. Write a conclusion for your cousin stating what you have found from your hypothesis test. Make certain that it makes sense to a Realtor

B. Answering the second question

m. The second question was to get an estimate of *how much more* the "Big Old" houses sell for (on average) than the "Not Big Old" houses. Explain why a confidence interval makes sense to answer this question.

- Use software with the file ***SanMateoRESampleY0506*** to calculate the confidence interval for the population difference of means in the sale prices of houses between the "Big Old and "Not Big Old" houses. (Use the variables *SaleThou* and *BigOld*.) The output should be equivalent to what is shown here, although if "pooled Variance was chosen, the number will be slightly different.

n. Identify the number in the output shown here that is the margin of error, and show how it was calculated. Make certain that the correct value for the t^* is used. (The correct number is not 5.281 and not 1.96.)

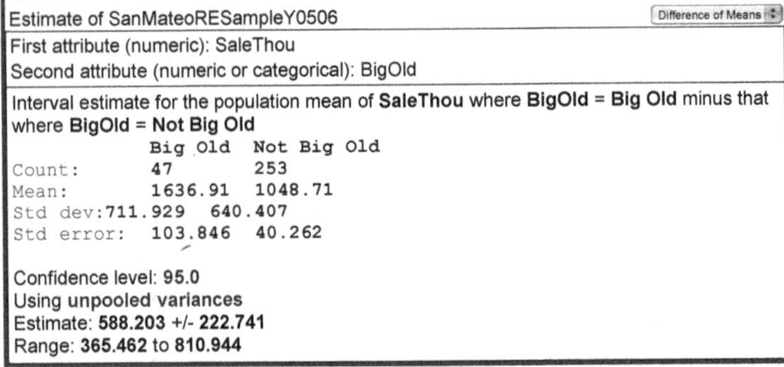

Estimate of SanMateoRESampleY0506 — Difference of Means

First attribute (numeric): SaleThou

Second attribute (numeric or categorical): BigOld

Interval estimate for the population mean of **SaleThou** where **BigOld** = **Big Old** minus that where **BigOld** = **Not Big Old**

```
             Big Old   Not Big Old
Count:       47        253
Mean:        1636.91   1048.71
Std dev:711.929   640.407
Std error:   103.846   40.262
```

Confidence level: **95.0**

Using unpooled variances

Estimate: **588.203** +/- **222.741**

Range: **365.462** to **810.944**

o. You should see Range: 365.462 to 810.944. Write this in the form: 365.462 <   < 810.944. What should be in the blank space?

p. Write a conclusion for the confidence interval output for your cousin, stating what you have found from your confidence interval calculations. Make certain that it makes sense to a Realtor.

[q. Extra: there is a variable called *OnMarket* that records the number of days on the market. Was there a significant difference between BigOld and NotBigOld houses in the time on the market in 2005–2006? Answer with the appropriate procedure.]

3. *Miles per Gallon (MPG) in cars and what affects it.* This exercise uses a random sample of the Federal Government's Guide to Fuel Economy for 2014 cars. The data can be found in the file ***FuelEconomyGuide2014Sample*** and the spreadsheet looks like this.

FuelEconomyGuide2014Sample

	Manufac...	Division	Carline	EngineDi...	NumCyl	CityMPG	HwyMPG	CombinedMPG	ClassVehicle	TransmissionType
33	Chrysler ...	Dodge	Charger ...	6.4	8	14	23	17	(c) Midsize to Large	(a) Automatic
34	Ford Mot...	Ford	F150 PIC...	5	8	15	21	17	(f) Truck or Van	(a) Automatic
35	General ...	Buick	ENCORE	1.4	4	25	33	28	(e) SUV	(c) Semi-Automatic
36	BMW	BMW	535i	3	6	20	30	23	(c) Midsize to Large	(b) Manual
37	Porsche	Porsche	911 Carr...	3.4	6	19	26	21	(a) Subcompact	(b) Manual
38	Toyota	TOYOTA	SEQUOI...	5.7	8	13	17	15	(e) SUV	(c) Semi-Automatic
39	Honda	Acura	RLX	3.5	6	28	32	30	(c) Midsize to Large	(c) Semi-Automatic
40	General ...	Chevrolet	CRUZE	1.4	4	26	38	30	(c) Midsize to Large	(c) Semi-Automatic
41	Hyundai	HYUND...	SONATA	2.4	4	24	35	28	(c) Midsize to Large	(a) Automatic

Some of the variables are explained in §5.3 Exercise 2, which compared Highway MPG to City MPG. But what features of cars are related to high or low fuel economy, in general? Here are four variables about cars that may be related to fuel economy:

EngineDisplacement	*Engine displacement, measured in liters (indicates the "size" of the engine)*
NumCyl	*Number of cylinders in the engine.*
ClassVehicle	*Whether the car is a subcompact, compact, midsize etc*
TransmissionType	*type of transmission (automatic, manual, etc.)*

- Open the file ***FuelEconomyGuide2014Sample*** and have software get summary statistics for the variable *CombinedMPG* grouped by the values of the variable *NumCyl*. The result should resemble what is here.

Summary statistics for CombinedMPG:
Group by: NumCyl

NumCyl	Mean	Median	Std. dev.	IQR	n
3	36.333333	36	0.57735027	1	3
4	27.903226	27	4.642363	5	124
5	24	24	1.4142136	2	2
6	21.831776	21	2.9889292	3	107
8	16.567568	16.5	2.0876747	3	74
10	15	15	1	2	3
12	14.333333	14.5	0.81649658	1	6
16	10	10	NaN	0	1

a. [*Review*] Use the measures of center to describe the relationship between the numbers of cylinders a car has and fuel economy as measured by miles per gallon. Cite the relevant numbers.

b. [*Review*] Use the measures of spread (or variability) to describe the variation of MPG within the cars having the same number of cylinders. Also, in the explanation state why it is that the software reports "NaN" ("Not calculable") for the one 16-cylinder car, which is the Bugatti Veyron shown here.

c. [*Review*] Notice that the number of cases (n) differs for the six-cylinder cars, and the eight-cylinder cars. Explain what it is about our formulas for mean and standard deviation that makes it possible to compare categories that have different numbers of cases. [But also notice that there only two five-cylinder cars, only three ten-cylinder cars, only three three-cylinder cars.]

- To analyze these data on number of cylinders and the relationship with fuel economy, it will be better to compare just two categories of cars: those with eight or more cylinders, and those with fewer than eight. Have software make a new variable in the file ***FuelEconomyGuide2014Sample***, called *ManyCylinders* that has just two categories: "8 or More" and "Less than 8." Consult the ***Software Supplement*** on ***Creating new variables***.

- With the file ***FuelEconomyGuide2014Sample*** have software get a graphic showing the distributions by *CombinedMPG* grouped by the categories of the variable *ManyCylinders*. Also, get summary statistics for *CombinedMPG* grouped by the categories of *ManyCylinders*.

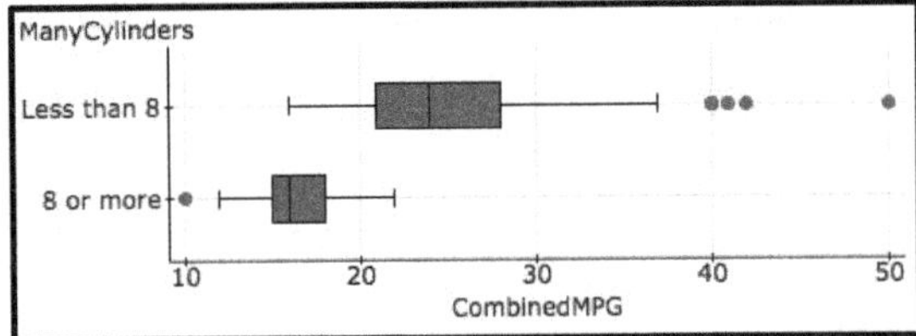

Summary statistics for CombinedMPG:
Group by: ManyCylinders

ManyCylinders	Mean	Median	Std. dev.	IQR	n
8 or more	16.27381	16	2.1862175	3	84
Less than 8	25.224576	24	5.0927974	7	236

d. From the evidence of the graphics and the means, does it appear that a hypothesis test for the difference of independent means would be statistically significant? Give a reason for your answer. [Of course, you can get software to confirm your idea.]

e. It appears that it would be more informative to get an estimate of the difference in the population of cars of the difference in the *CombinedMPG* between the tow categories of cars. Discuss either how the conditions are met for getting a confidence interval for the difference of means, or why the conditions are not met.

- Use software with the file ***FuelEconomyGuide2014Sample*** to get a confidence interval for the difference in the mean *CombinedMPG* between the cars with "Less than 8" cylinders compared with cars with "8 or More" cylinders.

f. Give a good interpretation of the confidence interval in the context of the data.

g. People who know cars will say that one reason the cars with eight or more cylinders have worse fuel economy is because those cars have bigger engines. *EngineDisplacement* measures the size of the engine in a car, where the measure is given in liters. If we want an estimate of how much bigger the engines in cars with eight or more cylinders are on average compared with cars with fewer than eight cylinders, will we get a confidence interval or a hypothesis test? Give a reason for your answer.

- Use software with the file ***FuelEconomyGuide2014Sample*** to do the calculations for the inferential procedure you decided upon in answer to part g.

h. Give a good interpretation of the results of the analysis in the context of the data. That is, what does the analysis say about the difference in average (mean) displacement in liters for the eight or more cylinder cars compared with the cars that have fewer than eight cylinders.

i. The best choice for inferential procedure for the question posed in part g is a confidence interval. This is because a confidence interval gives you an "estimate of how much bigger" the engines of the eight or more cylinder cars are compared with the fewer than eight cylinder cars. However, from the results of the confidence interval you should be able to say whether the associated hypothesis test (that is, testing *whether* there is a difference in means) will be statistically significant or not. Explain how, and state in this instance whether the test was statistically significant.

j. [Review] Which statistical techniques should be used to investigate the relationship between the variables *EngineDisplacement* and *CombinedMPG*. Notice that both of these variables are quantitative. Are the techniques of §5.4 releevant? Why or why not?

4. *Grades for Calculus 2 in Costa Rica* Grades were recorded for several years for students completing Calculus 2 courses at the Instituto Tecnológico de Costa Rica. There is one case per student. The data are found in the file ***GradesCalculus2***. These data were investigated in §5.3 looking at paired comparison data.

	YearSem...	Sex	Exam1	Exam2	Exam3	CourseGrade
87	1998-1	M	76.6	82.9	91.3	87.3
88	1998-1	M	35.3	60.7	16.7	41.5
89	1998-1	F	83.8	85.7	86.7	84.5
90	1998-1	F	48.4	62.1	37.7	50.2
91	1998-1	M	32.2	48.2	28.3	39
92	1998-1	M	84.7	74.3	95.0	85.1

Now we will look at some of the exam and course grades as they are related to gender and the semester the class was held. Two other variables will be considered here:

Semester — *The semester the course was taught: "first" or "second"*
Passing — *Whether a student got 70% or more or not: "Pass" and "Not Pass"*

Statistical question: *At the ITCR, is there a gender difference in mean Exam 1 scores in Calculus 2?*

a. From the way the statistical question is stated, what appears to be the intended population?
   - Students in Costa Rica
   - Calculus 2 students at the ITCR
   - Students taking Calculus 2 (wherever they study).
   - Engineering students

   Give a reason for your answer.

b. From the way the statistical question is stated, should we:
   - Carry out a paired comparison hypothesis test
   - Carry out an independent sample hypothesis test
   - Calculate a paired comparison confidence interval
   - Calculate an independent sample confidence interval

   Choose the best answer and give reasons.

- Use software with the file ***GradesCalculus2*** to get *descriptive* (not inferential) statistics and graphics to investigate the difference in Exam 1 scores by gender. The summary statistics should look something like what is presented here.

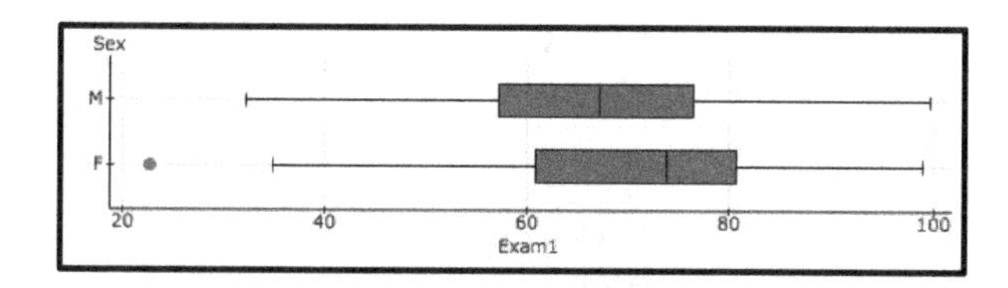

Summary statistics for Exam1:
Group by: Sex

Sex	Mean	Median	Std. dev.	IQR	n
F	70.419355	73.85	15.589675	19.8	62
M	66.36439	67.2	14.969647	19.4	205

c. Discuss whether the conditions are met for using the *t* procedures using these data.

d. The way the statistical question is posed implies a hypothesis test. These data are not "paired comparison (or "paired difference") data. What about the data indicates that they are independent samples and not "paired comparison"?

- Have software do the calculations for an independent sample hypothesis test.

e. Show how the test statistic that software gives you was calculated.

f. Indicate how the "df" was calculated.

g. Is the test statistically significant or not? Give a reason for the answer.

h. Answer the statistical question with the results of the hypothesis test. (The *p*-value should be mentioned as part of the explanation.)

- Have software calculate a 95% confidence interval for the gender difference in mean Exam 1 scores. The result should agree with the output shown here.

Estimate of GradesCalculus2 | Difference of Means
First attribute (numeric): Exam1
Second attribute (numeric or categorical): Sex

Interval estimate for the population mean of **Exam1** where **Sex** = **F** minus that where **Sex** = **M**

	F	M
Count:	62	205
Mean:	70.4194	66.3644
Std dev:	15.5897	14.9696
Std error:	1.97989	1.04553

Confidence level: 95.0
Using unpooled variances
Estimate: **4.05496** +/- **4.4435**
Range: **-0.388531** to **8.49846**

i. First, note whether the 95% confidence interval include "zero difference" between Exam 1 scores for male and female students. Then, explain how the confidence interval agrees with the outcome of the hypothesis test.

Statistical question: *At the ITCR, is there a gender difference in the probability of passing Calculus 2?*

j. Another way to answer the question about the gender difference would be to look at the probability of passing Calculus 2 for males and females. Here are the numbers. Which of the following inferential techniques can be used to answer the statistical question with the numbers given.

GradesCalculus2

		Passing		Row
		Not Pass	Pass	Summary
Sex	F	24	38	62
	M	86	119	205
Column Summary		110	157	267

S1 = count ( )

- Hypothesis test for two independent means (one measure, two collections)
- Hypothesis test to compare two independent proportions
- Hypothesis test for paired comparison (two measures, one collection)
- Chi-Square Test for Independence

Give a reason for the choice.

- Using the file ***GradesCalculus2*** have software carry out the analysis chosen in part j.

k. Answer the statistical question about whether there is a gender difference in the probability of passing Calculus 2 at ITCR using the results of the analysis.

Statistical question: *At the ITCR, is there evidence of a difference in mean Exam 1 scores by semester?*

l. The variable Semester indicates whether the course was taken in the first semester or the second semester. Should the two independent sample procedures (one measurement, two collections) or the paired comparison procedures (two measures, one collection) be used? Give a reason for your answer.

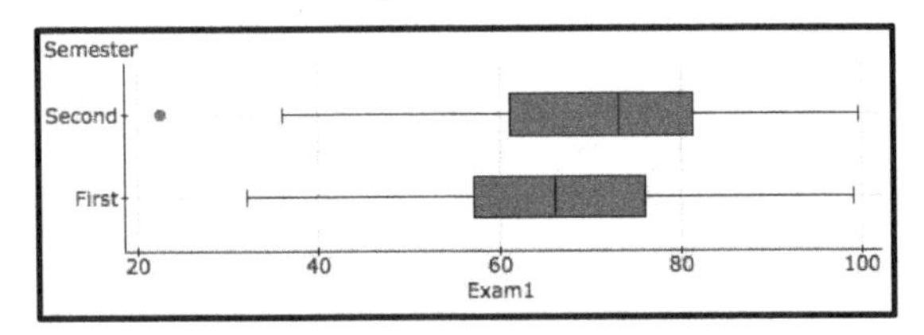

Summary statistics for Exam1:
Group by: Semester

Semester	Mean	Median	Std. dev.	IQR	n
First	66.07	66.05	14.859171	18.9	200
Second	70.995522	73.3	15.650873	20.1	67

m. The variable Semester indicates whether the course was taken in the first semester or the second semester. Which of the following inferential techniques can be used to answer the statistical question about whether there is a "semester difference" in mean Exam 1 scores.

- Hypothesis test for two independent means (one measure, two collections)
- Hypothesis test to compare two independent proportions
- Hypothesis test for paired comparison (two measures, one collection)
- Chi-Square Test for Independence

Give a reason for the choice

- Using the file ***GradesCalculus2*** have software carry out the analysis chosen in part m.

n. Answer the statistical question about whether there is a semester difference in the mean Exam1 scores in Calculus 2 at ITCR using the results of the analysis.

o. State, in the context of students taking a calculus 2 in Costa Rica, what a Type 2 would be.

p. Here is software output for the confidence interval for the difference of the means for students who took Exam 1 in the Second semester compared with students who took Exam 1 in the First semester. First, note whether the confidence interval includes "zero difference" and then explain how the results agree with the conclusion reached by the hypothesis test analysis.

95% confidence interval results:
μ_1 : Mean of Exam1 where Semester="Second"
μ_2 : Mean of Exam1 where Semester="First"
$\mu_1 - \mu_2$: Difference between two means
(without pooled variances)

Difference	Sample Diff.	Std. Err.	DF	L. Limit	U. Limit
$\mu_1 - \mu_2$	4.9255224	2.1817292	108.59344	0.60122427	9.2498205

q. State (or calculate) the margin of error for this confidence interval.

The following part is optional.

- The hypothesis test analysis above showed a p value of 0.026, indicating that we can be confident that getting a sample difference of 4.9256 (either way) or bigger between the semesters would be quite rare if actually there were no difference. What does this mean? It means that if we replicated the test over and over again, we would see the kinds of differences only 2.5% of the time, if actually there were no difference. Some software packages have the capability of simulating this process. Usually it is an Applet called ***Resampling***. Consult the ***Software Supplement*** for Resampling, and specifically for a ***Randomization Test for Two Means.*** Run the applet with these data for the variable *Exam1* comparing the categories "First" and "Second" in the variable *Semester*. The p- value should be similar to the proportion of times that the randomization test gives a value bigger than the difference 4.9256 (or smaller than – 4.9256). Do the test 10,000 times. Here is one run of the Randomization test 10,000 times.

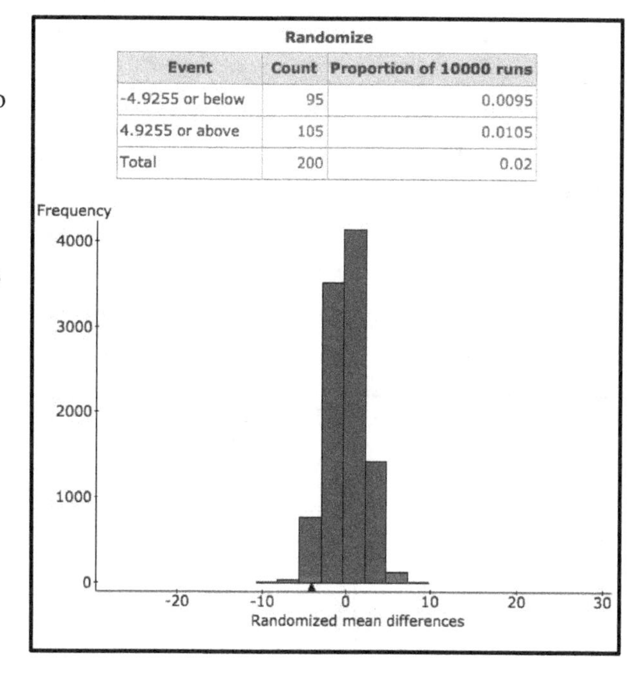
Randomize

Event	Count	Proportion of 10000 runs
-4.9255 or below	95	0.0095
4.9255 or above	105	0.0105
Total	200	0.02

r. Does your run of 10,000 give a proportion "extreme" similar to the p-value? Does the run shown here give a proportion extreme similar to the p-value>

5. Movies Data: Animations and Romance These are data on movies culled from the Internet Movie Database (IMDb) by Hadley Wickham (See: http://had.co.nz/data/movies/description.pdf).

MoviesDataAnimationRomance

	title	year	length	rating	votes	Genre2	LogVotes	FilmPost1990
29	Velocity of Gary, The	1998	100	4.7	570	Romance	2.75587	Film after 1990
30	Betty Boop and the Little King	1936	7	7.3	28	Animation	1.44716	Film 1990 or before
31	Star Trek: Insurrection	1998	103	6.3	13395	Romance	4.12694	Film after 1990
32	Dogfight	1991	89	7.1	1503	Romance	3.17696	Film after 1990
33	Lighthouse Keeping	1946	7	7	29	Animation	1.4624	Film 1990 or before
34	Madeline	1952	8	6.9	12	Animation	1.07918	Film 1990 or before
35	Aggie Appleby Maker of Men	1933	73	5.6	26	Romance	1.41497	Film 1990 or before

The variables we will explore in the file ***MoviesDataAnimationRomance*** are described below. This file is a random sample of the 58,000 in the data set by Wickham, but only includes films in the genre categories Animation and Romance.

length	*length of the film in minutes*
rating	*Average IMDb user rating*
votes	*Number of IMDb users who rated the film*
Genre2	*Whether "Romance" or "Animation"*
LogVotes	*Logarithm (base 10) of the number of votes.*
FilmPost1990	*Whether the film was released after 1990 or not*

a. What are the cases for these data?

b. Since we will be doing inferential analysis, we need to decide what the population is from which the data come. Should we regard the population as:

- all films ever made in the genres Animation and Romance
- all films that get recorded in the IMDb in the genres Animation and Romance
- films in our sample in the genres Animation and Romance
- films in the 58000 in the genres Animation and Romance collected by Wickham

Give reasoning for the answer. (You may well have a class discussion on this matter. It is relevant to consult Wickham's account of how the data were collected and also the IDMb comments on the reliability of their data: See http://www.imdb.com/help/show_leaf?infosource)

General statistical question: *How do films in the genres Animation and Romance differ?*

(Note: the two categories animation and romance are almost mutually exclusive: there are just two films that were in both categories: *Long ji mao* and *Murasaki Shikibu: Genji monogatari*. These are excluded.)

- Open the file ***MoviesDataAnimationRomance*** and using software do descriptive analyses of the data set. Get graphics and numerical summaries for the variables length, rating and votes grouped by the categories of Genre2. Here is an example for the length of films of what you should get.

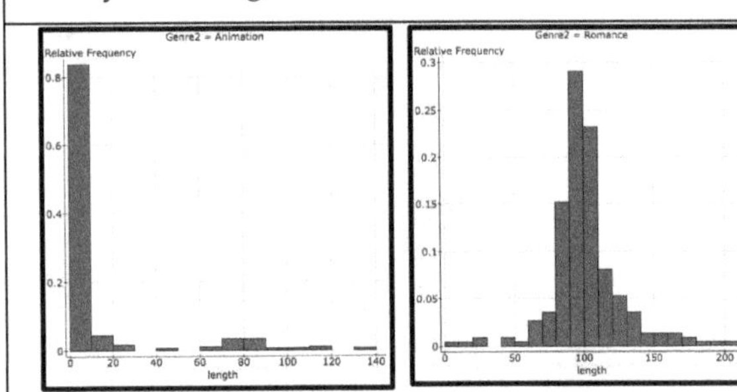

Summary statistics for length:
Group by: Genre2

Genre2	Mean	Median	Std. dev.	IQR	n
Animation	15.333333	7	25.038269	2	189
Romance	100.69643	99	25.576699	18	224

c. [*Review*] What can you say about the distributions of difference in lengths of the films for Animation and Romance films? Discuss shape, center and spread.

d. If we do a hypothesis test for the difference in mean length between the genres Animation and Romance, will the p-value for the null hypothesis of no difference between the means be small or large? Explain. (If uncertain, get software to carry out the hypothesis test; then explain.)

- e. What about average rating? Do Animation films or do Romance films get a higher rating from the IMDb raters? Or can we not say from our sample? Have software (using the file ***MoviesDataAnimationRomance***) carry out descriptive analyses and then either a hypothesis test or a confidence interval for the difference of the two means to answer the question. Here is a graphic of the sample ratings. In the interpretation, cite the p-value as evidence for a difference in the population of films or as evidence that we cannot say that there is a difference.

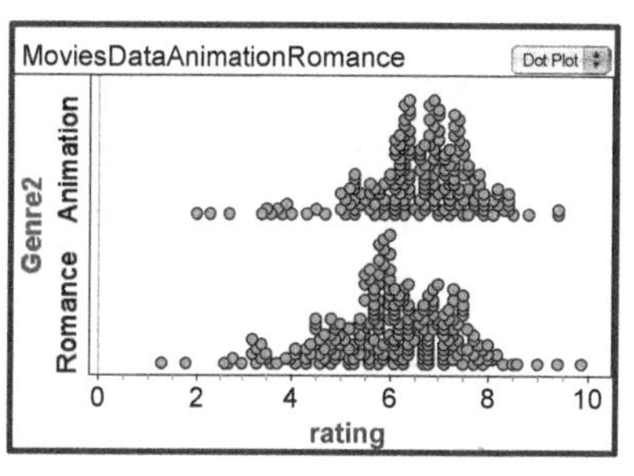

Specific statistical question: *Are films of the animation genre or are films of the romantic genre more likely to have been made after 1990?*

Here are the data on the numbers of films made after 1990 in the two genres.

MoviesDataAnimationRomance

		FilmPost1990a		Row Summary
		Film 1990 or before	Film after 1990	
Genre2	Animation	148	41	189
	Romance	124	100	224
Column Summary		272	141	413

S1 = count ( )

f. What type of *inferential* analysis should be done to answer the statistical question?

- Hypothesis test for two independent means (one measure, two collections)
- Hypothesis test to compare two independent proportions
- Hypothesis test for paired comparison (two measures, one collection)
- Chi-Square Test for Independence

Give a reason for the choice.

- Using software with the file ***MoviesDataAnimationRomance***) carry out either a hypothesis test for two independent proportions or a Chi-Square Test for Independence.

g. Give a good interpretation of the results of the analysis in the context of data. Which of the two genres' films are more likely to be made after 1990, or can we not say from the data?

h. Look at the dot plots (or histograms or box plots) for the distributions of the variable *votes* for the two genres of films. What can you say about the shapes of the distributions?

i. The distribution for *votes* with many films getting fewer than twenty votes, and a few films getting tens of thousands of votes. When this extreme right-skewing happens, statisticians often use the logarithm of the variable. Here, log base ten is used, so a value of one means $10^2 = 100$ votes, and $10^3 = 1000$ votes, and so on. Here is the result of the hypothesis test for the difference of means of *Logvotes* between films in the categories "Animation" and "Romance" in the variable *Genre2*. Is the test statistically significant? Give a reason for your answer.

Test of MoviesDataAnimationRomance (Compare Means)
First attribute (numeric): LogVotes
Second attribute (numeric or categorical): Genre2

Ho: Population mean of **LogVotes** for **Animation** equals that for **Romance**
Ha: Population mean of **LogVotes** for **Animation** is not equal to that for **Romance**

	Animation	Romance
Count:	189	224
Mean:	1.37011	2.1508
Std dev:	0.52078	0.931693
Std error:	0.0378812	0.0622514

Using unpooled variances
Student's t: -10.71
DF: 360.152
P-value: < 0.0001

h. What does this say about the mean numbers of votes for the genres of films?

§5.5 Exercises for Analysis of Variance

1. ***Fuel Economy for Different Categories of Cars.*** In the **Notes**, the first statistical question had to do with the relationship between fuel economy and the type of car, as in this question.

First Statistical Question

To see whether mean MPGHwy differs by the variable CarType: (a) small, (b) large, (c) SUV (d) a Pick-up.

- Open the file ***FuelEconomyGuide2011Sample*** and have software produce a graphic similar to the one shown here, summary statistics, and the ANOVA table, as shown here. Consult the ***Software Supplement*** in the section ***ANOVA***, where this example is shown.

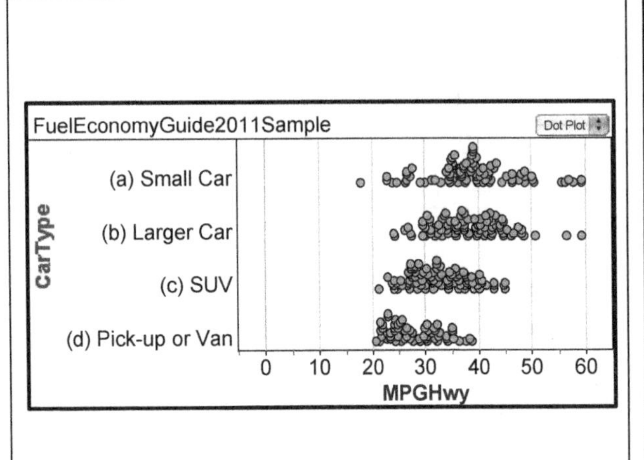

Analysis of Variance results:
Responses in MPGHwy.
Factors in CarType.

Factor means

CarType	n	Mean	Std. Dev.	Std. Error
(a) Small Car	88	39.116814	8.3848052	0.89382324
(b) Larger Car	86	38.308441	6.7467934	0.72752544
(c) SUV	87	32.903215	5.7293088	0.61424637
(d) Pick-up or Van	63	28.121616	4.8991119	0.61723009

ANOVA table

Source	DF	SS	MS	F-Stat	P-value
Treatments	3	5792.9624	1930.9875	43.220901	<0.0001
Error	320	14296.694	44.677168		
Total	323	20089.656			

The first questions (a through d) are designed to become accustomed to the structure of the ***ANOVA*** table. For reference, here are the formulas for *MSG* and *MSE*

$$MSG = \frac{n_1(\bar{x}_1-\bar{x})^2+n_2(\bar{x}_2-\bar{x})^2+\cdots+n_k(\bar{x}_k-\bar{x})^2}{k-1} \qquad MSE = \frac{(n_1-1)s_1^2+(n_2-1)s_2^2+\cdots+(n_k-1)s_k^2}{N-k}$$

a. For the ***Groups*** row in the ***ANOVA*** table, divide the number for the ***Sum of Squares*** (or ***SS***) by the value for the ***Degrees of Freedom*** (or ***DF***). What is the name of that number?

b. For the ***Error*** row, divide the number for the ***Sum of Squares*** (or ***SS***) by the value for the ***Degrees of Freedom*** (or ***DF***). What number did you get? What is the name of that number?

c. Divide the value for ***Mean Square*** for ***Groups*** by the value for the ***Mean Square*** for ***Error***. What is the name of the number that you got?

d. Add the ***Groups*** and ***Error*** values for ***Degrees of Freedom***. Add the ***Groups*** and ***Error*** values for ***Sum of Squares.*** What do you notice about the results?

Second Statistical Question

To see whether in the population of cars sold the mean MPGCity differs by the variable Drive "front-wheel drive," "rear-wheel drive" or "all-wheel drive" for the "Small Cars."

e. On the next page are box plots showing the *MPGCity* by whether a car has "front-wheel drive," "rear-wheel drive," or "all-wheel drive." In the context of an ANOVA test, do the box plots look like the "H_0 *scenario*" or the "H_a *scenario.*" Put into English a reason for your answer.

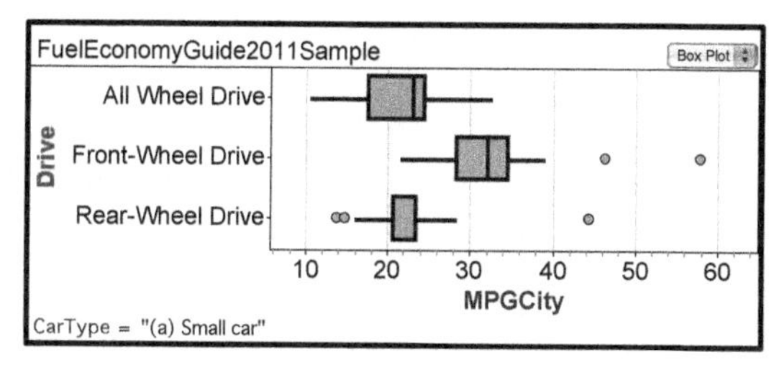

Factor means

Drive	n	Mean	Std. Dev.	Std. Error
All Wheel Drive	21	21.912843	5.642557	1.2313069
Front-Wheel Drive	30	32.40398	7.5054507	1.3703015
Rear-Wheel Drive	37	22.326819	5.0465783	0.82965237

f. Using appropriate notation, write the null and alternate hypotheses for the ANOVA test.

g. For this test $N = 88$, which can be found by adding. What is the value of k?

h. The ANOVA table below deliberately has some of the entries deleted. However, using the relationships between the numbers in an ANOVA table, it should be possible to completely fill in the entries. Copy the table and fill in the missing spaces, using the information given and what you have found about the structure of the ANOVA table.

Source	*Degrees of Freedom*	*Sum of Squares*	*Mean Square*	*F Statistic*	*P Value*
Groups		2070.375			< 0.0000
Error		3187.237			
Total	*87*	5257.612			

To check your answers, follow the bulleted instructions.

- Open the file ***FuelEconomyGuide2011Sample*** and get the ANOVA analysis for the response variable *MPGCity* and the explanatory variable (factor) *Drive.*

i. The p-value is so small that software reports "< 0.0000". Knowing this, interpret the results of the hypothesis test in the context of the data. Does the test confirm what you guessed looking at the graphic?

j. ***Step 2: Checking the conditions***. Because we wanted to look at the ANOVA table, we did not check the conditions. The third condition says that the standard deviations in the population should be equal. However, there is a ***rule of thumb*** about this condition. Read the rule of thumb and how it is applied and apply it to the standard deviations shown. Are the standard deviations here close enough?

k. What are the other two conditions for ANOVA? Are they met for our test?

l. If you do not know much about cars, consult someone who does to answer this question. The question is: is it appropriate to say that a car being a front-wheel drive *causes* the car's fuel economy to be better? Or is reality more complicated?

m. Here is the calculation of the ***MSG***. Using the Summary Table above, give the values that should be placed at A, B, C, D, E, and G. The overall (or grand) mean is: $\bar{x} = 25.663$.

$$MSG = \frac{n_{AW}(\bar{x}_{AW} - \bar{x})^2 + n_{FW}(\bar{x}_{FW} - \bar{x})^2 + n_{RW}(\bar{x}_{RW} - \bar{x})^2}{k-1}$$

$$= \frac{(21)*([A] - 25.663)^2 + ([B])*([C] - 25.663)^2 + ([D])*(22.327 - [E])^2}{[G]-1}$$

n. Use the numbers in the Summary Table and the formula for MSE to calculate the MSE. The ANOVA table gives you the answer.

2. *Full-term births.* It will make sense to many people that the age at first birth should be higher if the mother has more education. We can check our birth data to see whether this idea holds. The random sample we are using is of the 2006 births in the USA, which includes only full-term births.

- have software make a graphic and summary statistics similar to the ones shown. Open the file ***BirthsTwelveHundredSample***. Use the variables *AgeMother* and *EducationMother*. Filter for just the first births using: *BirthOrder = "First Birth"*.

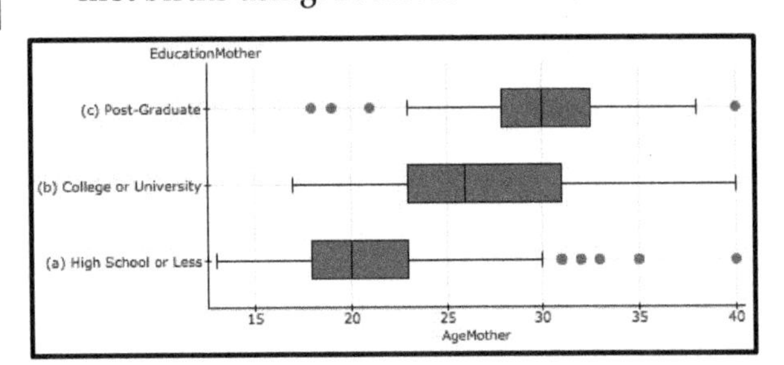

Summary statistics for AgeMother:
Where: BirthOrder="First Birth"
Group by: EducationMother

EducationMother	n	Mean	Std. dev.
(a) High School or Less	158	21.462025	4.8763515
(b) College or University	167	26.814371	5.1197971
(c) Post-Graduate	52	29.807692	4.4941725

First Statistical Question: *Does mean age of the mother for first births differ by educational level?*

a. What are the cases for this collection of data? (Do not confuse the variables with the cases.)

b. We will analyze these data using ANOVA. ***Step 1:*** Set up the null and alternate hypotheses for this hypothesis test using appropriate notation for the four groups.

c. From the numbers and the appearance of the graph, do the data seem to be consistent with the "H_0 *scenario*" or the "H_a *scenario*"? Give a reason for your answer.

d. ***Step 2:*** For the ANOVA test to be trustworthy, three conditions must be met. Check that these conditions are met. For the third condition, use the ***rule of thumb*** given in the ***Notes.***

e. ***Step 3:*** The calculation of the test statistic means calculating the ***Mean Square Groups (MSG),*** the ***Mean Square Error (MSE),*** and then their ratio, the ***F*** test statistic: $F = \frac{MSG}{MSE}$. Calculate the MSG using the formula $MSG = \frac{n_1(\bar{x}_1 - \bar{x})^2 + n_2(\bar{x}_2 - \bar{x})^2 + \cdots + n_k(\bar{x}_k - \bar{x})^2}{k-1}$. To use this formula you need to know that the overall mean is $\bar{x} = 27.359$. What is k? You will be able to check the answers from software output.

f. ***Step 3, Cont.*** Calculate MSE using $MSE = \frac{(n_1 - 1)s_1^2 + (n_2 - 1)s_2^2 + \cdots + (n_k - 1)s_k^2}{N-k}$. (Identify N, and $n_{HSorLess}$, $n_{CollegeUniversity}$ etc. Once again, you will be able to check your answers in this exercise. Keep a record of the sum in the numerator.)

g. ***Step 3, cont.*** Calculate the test statistic $F = \frac{MSG}{MSE}$. From the ***Notes***, you should be able to say whether the value is "big" and consistent with the H_a, or whether the value is "small" and consistent with H_0.

- Using the file ***BirthsTwelveHundredSample*** have software carry out the ANOVA test, using *AgeMother* and *EducationMother*.

h. Do the calculations check?

ANOVA table

Source	DF	SS	MS	F-Stat	P-value
Treatments	2	3729.3099	1864.655	76.512559	<0.0001
Error	374	9114.5946	24.370574		
Total	376	12843.905			

i. ***Step 5: Interpretation***. Do the data support the idea that the mean age at first birth differs by the education of the child's mother? Give a reason.

Second Statistical Question: *Does mean gestation (length of pregnancy) for first births differ by educational level?*

One response to this idea may be to doubt that there should be any relationship between length of pregnancy and length of education. The results may strengthen this doubt. However, a relationship between two variables need not be direct.

- Use software to get the output shown using the file ***BirthsTwelveHundredSample***. Use the variables *Gestation3* and *EducationMother*, again filtering for first births.

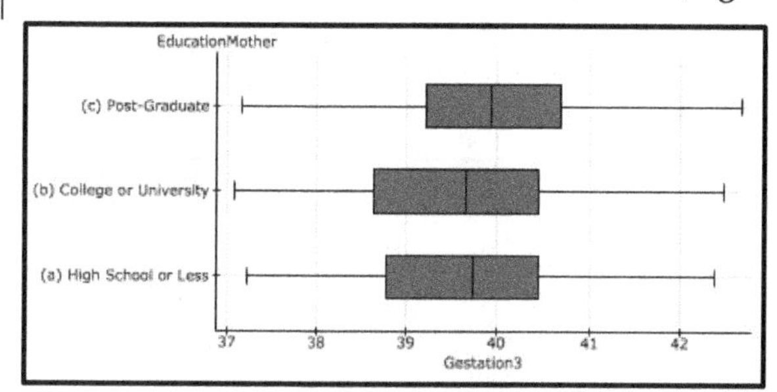

Summary statistics for Gestation3:
Where: BirthOrder="First Birth"
Group by: EducationMother

EducationMother	n	Mean	Std. dev.
(a) High School or Less	158	39.608316	1.1208324
(b) College or University	167	39.621711	1.2379908
(c) Post-Graduate	52	39.858504	1.2094765

j. State why ANOVA is the appropriate inferential technique for analyzing this relationship and not: (I) a t test for comparing means OR (II) a chi-square test for independence.

k. Set up the null and alternate hypotheses for the ANOVA test for the relationship between *Gestation3* and *EducationMother.* From what you see in the graph and Summary Table, do you think that the data are consistent with H_0 or H_a? Give a reason for your answer.

l. Use the information about standard deviations from the ***Summary Table*** and the ***rule of thumb*** to check that the third condition for ANOVA is met.

- Using the file ***BirthsTwelveHundredSample*** have software carry out the ANOVA test. Use the variables *Gestation3* and *EducationMother,* again filtering for first births.

m. From the ANOVA display table (and not by calculating), give the values of MSG, MSE, and F.

n. Use the p-value given in the ANOVA display table to give an interpretation of the ANOVA hypothesis test in the context of the data. You should say something about whether the mean length of pregnancies differs according to the education of the mother.

o. This shows the sampling distribution that we are using for the ANOVA. State what the (blue) *shaded area* in the graphic represents and what the place the shading starts represents.

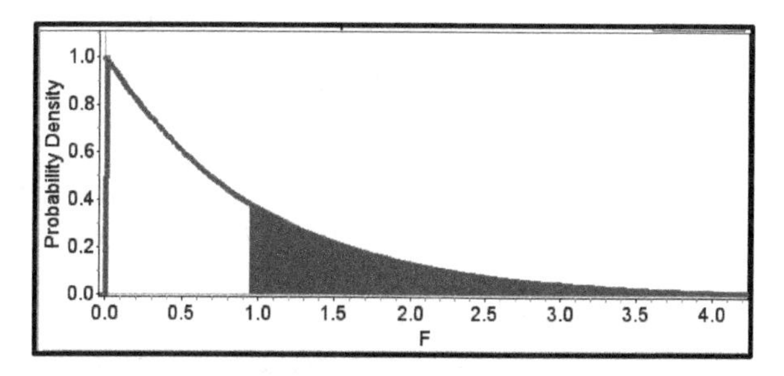

p. ***Choosing the correct procedure.*** If you wish to investigate whether mean *AgeMother* for first births differs between the mothers who are married and the mothers who are not married, which inferential procedure would you use: (I) ANOVA (II) a t test for comparing two means (III) chi-square (IV) some other inferential procedure? Give a reason for your answer. (The variable ParentsMarried has two categories.)

3. *Weight Loss in Israel* Dr. Iris Shai and her colleagues carried out a study in Israel that compared three types of diets in a randomized controlled experiment (www.nejm.org/doi/full/10.1056/NEJMoa0708681). The subjects ($n = 322$; 277 men and 45 women) were randomly assigned to one of three diets: either a "low-fat" diet, a "low-carbohydrate" diet, or a "Mediterranean diet." All of the subjects were employed at the same workplace and ate in a cafeteria at work, so the diets for the main meal of the day (the mid-day meal in Israel) could be closely monitored. They also had relatively high adherence rates for the dieters; they kept to their diets, perhaps because they worked with each other and saw each other every day.

For studies of change, it is important for the research report to include the "baseline" characteristics (the measures at the beginning of the study) of the subjects (or experimental units). Here are the baseline values for the weight-loss study. The numbers are in the form "mean ± sd."

Table 1. **Baseline Characteristics of the Study Population.***

Characteristic	Low-Fat Diet (N=104)	Mediterranean Diet (N=109)	Low-Carbohydrate Diet (N=109)	All (N=322)
Age — yr	51±7	53±6	52±7	52±7
Male sex — no. (%)	89 (86)	89 (82)	99 (91)	277 (86)
Current smoker — no. (%)	19 (18)	16 (15)	16 (15)	51 (16)
Weight — kg	91.3±12.3	91.1±13.6	91.8±14.3	91.4±13.4
BMI	30.6±3.2	31.2±4.1	30.8±3.5	30.9±3.6
Blood pressure — mm Hg†				
Systolic	129.6±13.2	133.1±14.1	130.8±15.1	131.3±14.5
Diastolic	79.1±9.1	80.6±9.2	79.4±9.1	79.7±9.2
Waist circumference — cm‡	105.3±9.2	106.2±9.1	106.3±9.1	105.9±9.1

a. The means of weight (in kg) are almost the same for the three groups at the beginning of the study. Is this a good thing or not a good thing? Give a reason for your answer.

b. If the mean weights at the start are almost the same then should the result of an ANOVA hypothesis test on the *baseline weights* be consistent with the ANOVA null hypothesis or the ANOVA alternate hypothesis? Give a reason for your answer.

c. ***Reading the research report***. On page 235 of Shai, *et. al.* (2008), there is the following sentence. (It may be useful to answer this part and the next part d together.)

All groups had significant decreases in waist circumference and blood pressure, but the differences among the groups were not significant.

Does the part of the sentence that says "All groups had significant decreases in waist circumference..." refer to

- (I) (perhaps multiple) t tests on one mean;
- (II) (perhaps multiple) comparisons of two means; or
- (III) an ANOVA test?

Read carefully, think carefully, and give reasons for your answer. The numbers that the author gave are shown here.

		Weight loss (kg)	
Diet	n	mean	sd.
Low-Carbobydrate	109	-4.6	6.5
Low-Fat	104	-2.9	4.2
Mediterranean	109	-4.4	6.0

d. ***Reading the research report***. Does the part of the sentence quoted in c that says: "...but the differences among the groups were not significant" refer to:

- (I) (perhaps multiple) *t* tests on one mean;
- (II) (perhaps multiple) comparisons of two means; or
- (III) an ANOVA test?

Read carefully, think carefully, and give reasons for your answer.

e. ***Reading the research report***. The sentence following the one above says:

"The waist circumference decreased by a mean [±sd.] of 2.8±4.3 cm. in the low-fat group, 3.5±5.1 cm. in the Mediterranean-diet group, and 3.8±5.2 cm in the low-carbohydrate group (P = 0.33 for the comparison among groups.)"

Is "P = 0.33" a

- (I) *p*-value or
- (II) a test statistic, and is it for:
- (III) (perhaps multiple) *t* tests on one mean;
- (IV) comparison of two means; or
- (V) an ANOVA test?

Based upon your decision, give an interpretation of the "P = 0.33."

f. ***Review.*** In the quote above, the mean and standard deviations are given for changes in waist circumference. Using the sample sizes in the table on the previous page, calculate a 95% confidence interval (or get software to do the calculation) for mean waist circumference for one of the three diets.

g. ***Review.*** To what population should the confidence interval reasonably refer? All people in the world on the diet you have chosen? Israelis? People whose diet program is coordinated with their work? Give an interpretation of the confidence interval.

h. ***Review.*** The correct answer to part c is that the sentence is referring to three different "one mean tests" and is saying that in all (and in each) of the diet groups there was a significant reduction in waist size. Choose one of the groups and carry out the one mean *t* test; find the *test statistic* to confirm that the *t* test is significant. (Use $n_{LF} = 104$, $n_{Med} = 109$, or $n_{LC} = 109$; what should be the logical value of μ_0 for weight loss?).

i. ***Reading the research report***. Compared with most dieting studies, this Israeli study had relatively high retention or "adherence" rates (that is, the proportion who continued with their diets). Here is what the authors say (p. 232):

"...the 24-month adherence rates were 90.4% in the low-fat group, 85.3% in the Mediterranean-diet group and 78% in the low-carbohydrate group (P = 0.04 for the comparison among diet groups.)"

(The actual numbers were 94 out of 104 for *LF*, 93 out of 109 for *Med*, and 85 out or 109 for *LC*.)

What inferential procedure gave the *p*-value of 0.04:

- (I) ANOVA
- (II) Comparison of two means
- (III) Chi-square or
- (IV) Comparing two proportions? Give a reason for your answer.

j. Interpret the *p*-value of 0.04 referred to in part l in the context of the study.

4. ***Arrival Delay to LAX by Bay Area Origin Airport*** This exercise uses a random sample of flights in 2014 from the Bay Area to LAX. The file is ***OnTimeBayAreaLAX2014Sample.***

Statistical question:

Are there differences amongst the three Bay Area airports (OAK, SFO, SJC), in mean ArrivalDelay in LAX?

- With the file ***OnTimeBayAreaLAX2014Sample*** get the graphics and the summary statistics for the variable *ArrivalDelay* grouped by the variable *Origin* similar to what is shown below.

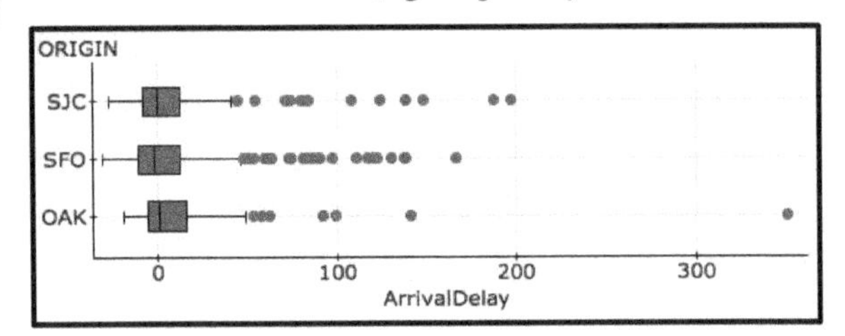

Summary statistics for ArrivalDelay:
Group by: ORIGIN

ORIGIN	n	Mean	Median	Std. dev.	IQR
OAK	105	11.609524	1	41.591303	22
SFO	256	7.7421875	-2	32.100515	23
SJC	145	9.8965517	0	35.928116	20

a. What are the cases for this analysis? Airports? Flights? Airlines? Passengers? Times? (It may help to look at the spreadsheet of the file ***OnTimeBayAreaLAX2014Sample.***)

b. Explain why ANOVA is an appropriate analysis to answer out statistical question, and not:
   - Comparison of two independent means
   - A paired comparison test
   - A Chi-Square Test for Independence

   The answer should specify briefly why each of the others is not appropriate.

c. An ANOVA test has three conditions for its use:
   - A random sample where the groups are independently chosen
   - Normal population distributions,
   - Equal population variances

   From the numbers and information given above which one of these three is problematical for the data that we have here? Give a reason for the answer.

d. [*Review*] Compare the means and the medians for the variable *ArrivalDelay* for each of the three airports. What do the comparisons confirm a bout the shapes of the distributions?

Work-Around 1

The answer to part c is that the distributions are highly right-skewed. Since we have fairly large samples, ANOVA may well be ***robust*** in the face of non-normality. However, it may be that we are just interested in the more common delays, and not the tails of the distribution where the size of the delay is unusual. If so, we can filter for just the arrival delays that are less than 45 minutes.

- With the file ***OnTimeBayAreaLAX2014Sample*** get the graphics and the summary statistics for the variable *ArrivalDelay* grouped by the variable *Origin* but filtered for *ArrivalDelay<45.*

e. Explain what the meaning of the phrase "may well be ***robust*** in the face of non-normality" means in the context of our statistical question.

f. Inspect the graphics and the statistical summaries for the filtered sample. What can you say about the shapes of the distributions?

g. With the filtered sample, is the equal variances condition met sufficiently well to proceed with ANOVA? Give a reason for the answer that may include some rough calculations according to the ***Rule of Thumb*** for equal variances.

h. Set up the null and alternate hypotheses for the ANOVA for the variable *ArrivalDelay* for the categories of the variable *Origin.*

- With the file ***OnTimeBayAreaLAX2014Sample*** have software do the calculations for the ANOVA test for the variable *ArrivalDelay* grouped by the variable *Origin* but filtered for *ArrivalDelay<45*.

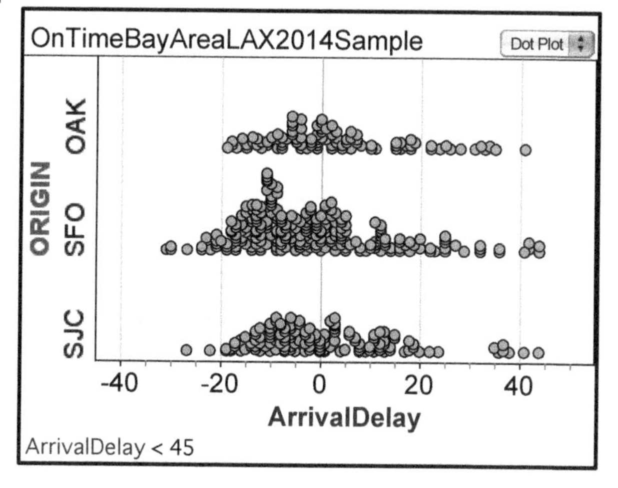

Analysis of Variance results:
Where: ArrivalDelay<45
Responses in ArrivalDelay.
Factors in ORIGIN.

Factor means

ORIGIN	n	Mean	Std. Dev.	Std. Error
OAK	96	2.7708333	13.708606	1.3991287
SFO	228	-1.622807	14.747741	0.97669301
SJC	133	0.67669173	13.752437	1.1924875

ANOVA table

Source	DF	SS	MS	F-Stat	P-value
Treatments	2	1397.9536	698.97682	3.4422041	0.0328
Error	454	92189.617	203.06083		
Total	456	93587.571			

i. What do negative values of *ArrivalDelay* indicate? For flights originating from SFO to LAX, give an interpretation of the mean.

j. Confirm (by a simple calculation) the value of the test statistic from the values for the MSG and MSE as shown in the ANOVA table.

k. Is the test statistically significant? Give a reason for your answer.

l. What does the test say about the null and the alternate hypotheses?

- m. The test *is* statistically significant, since the p value is less than the usual standard of $\alpha = 0.05$. Which airport appears to account for the difference?

n. ***Statistical or practical significance?*** Compare the means for the three airports and inspect the graphic for arrival delays. Is the statistical significance a significant difference for the traveller from the Bay Area to LAX?

Work-Around 2

Another way to analyze the data may be closer to a typical traveller's: What is the probability of being delayed more than fifteen minutes. (The BTS thinks it important enough to record it.) here is a table showing the counts of flights from the three airports in the two categories of the variable *ArrivalDelay15.*

OnTimeBayAreaLAX2014Sample

		ArrivalDelay15		Row
		15 min or less	More than 15 min	Summary
ORIGIN	OAK	74	31	105
	SFO	198	58	256
	SJC	116	29	145
Column Summary		388	118	506

S1 = count ( )

o. Is the variable ArrivalDelay15 a categorical variable or a quantitative variable?

p. Which inferential technique should be used with the variables *ArrivalDelay15* and *Origin?*

- ANOVA
- Comparison of two proportions
- Chi-Square Test for Independence
- Comparison of two independent means

Give reasons for your choice, including reasons for rejecting the "wrong" analyses.

- Using the file ***OnTimeBayAreaLAX2014Sample*** have software do the analysis you have chosen. Do not filter for *ArrivalDelay < 45.*

q. Is there a difference between airports in the probability of being delayed upon arrival in LAX more than 15 minutes? Is it a statistically significant difference? (What is the p-value of the test?). Is the difference a practical one for the Bay Area – LAX traveller?

5. *Playing with Old Cars* The file ***Summer092006Cars*** has data for just the 2006 model year cars that were being sold through the Internet site www.cars.com in the summer of 2009 for these makes and places. We shall be using these two categorical variables:

Place: *Boston, Chicago, Dallas, SF Bay Area* *Make:* *Audi, BMW, Infiniti, Lexus, Mercedes-Benz*

First statistical question:

Are there are differences in the mean miles driven (the variable Miles) for these used cars in among the places: Boston, Chicago, Dallas, and the SF Bay Area (the variable Place)

- Open the file ***Summer092006Cars*** and have software get graphics and the summary statistics for the relationship between the variables *Miles* and *Place*.

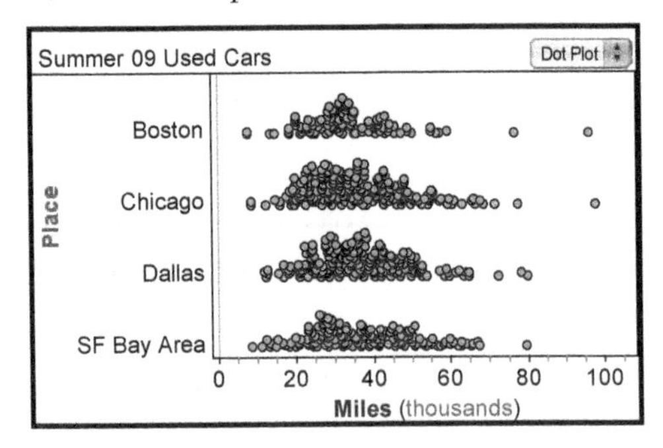

Summary statistics for Miles:
Group by: Place

Place	n	Mean	Std. dev.
Boston	109	33900.697	12354.025
Chicago	213	36336.465	13361.174
Dallas	188	37247.622	12509.428
SF Bay Area	160	37287.8	13661.1

a. Explain why ANOVA is the appropriate inferential procedure and not:
 - (I) a t test to compare two independent means or
 - (II) a Chi-square test for independence.

b. ***Step 1:*** Set up the null and alternative hypotheses for the statistical question, using notation that shows the context (that is, *Place*).

c. ***Step 2:*** There are three conditions for doing ANOVA. Show that two of them are met with no difficulty but that we have reservations about one of the conditions.

d. ***Step 3:*** Without actually doing the calculations (we will get software do that), show where the numbers in the Summary Table should go in the formula for MSG. It will be sufficient to show just two terms. (What is k?)

e. ***Step 3, cont.*** Without actually doing the calculations (we will get software do that), show where the numbers in the Summary Table should go in the formula for MSE. It will be sufficient to show just two terms. (What is N?)

- Have software do the ANOVA calculations using the file ***Summer092006Cars***.

f. From the ANOVA display table, give the value of the MSG, the MSE, and the test Statistic F.

g. ***Step 4*** The ANOVA display gives the p-value. Is the test statistically significant?

h. ***Step 5: Interpretation***. Interpret the p-value in the context of the data. To answer the first statistical question posed above.

Second statistical question:

Are there are differences in the mean miles driven (the variable Miles) for these used cars in among the makes of the cars: Audi, BMW, Infiniti, Lexus, Mercedes-Benz (the variable Make)

- i. Using software and the file ***Summer092006Cars*** do the complete ANOVA analysis to answer the second statistical question. Be clear in the set-up and especially in the conclusions.

§5.6 Exercises on What To Do

1. *Six Statistical Questions about Used Mercedes-Benz'*

The parts of this exercise will present various statistical questions; the task (the challenge, really) is to determine what statistical technique to use. It is then a simple mater to carry out the procedure, either by using software or "by hand". The data are in the file ***Summer09MercedesCClass.*** Here are some of the variables:

Place1	*indicates the place the car was advertised: Boston, Chicago, Dallas, and SF Bay Area.*
Price	*the price listed on the website, in dollars*
Miles	*the reported miles that the car had been driven*
Age	*the age of the used car, in years*
Convertible	*whether the car being sold was a convertible or not*
Body	*indicates the body style of the car: convertible, coupe, sedan, wagon*

Summer 09 Mercedes C-class

	Make1	Place1	Price	Miles	Age	Convert...	Body	Seller	Distance	NoPrice
1	Mercede...	Boston	54987	16507	1.33333	Not Conv...	Coupe	Dealer	6	Price Given
2	Mercede...	Boston	52897	3860	0.416667	Not Conv...	Sedan	Dealer	13	Price Given
3	Mercede...	Boston	50377	31415	3.33333	Not Conv...	Sedan	Dealer	3	Price Given
4	Mercede...	Boston	47000	18500	2.33333	Convertible	Convertible	Private S...	108	Price Given
5	Mercede...	Boston	44899	29029	3.33333	Not Conv...	Sedan	Dealer	105	Price Given

a. List the quantitative variables shown here and the categorical variables shown here.

First statistical question: *Describe and display for the sample at hand the differences in the percentages of convertibles (the variable* Convertible*) being sold in the different places (the variable* Place1*).*

b. Describe the kind of graphic, or table, or measure you will get to do a *descriptive* (not *inferential*) analysis of this question about the proportions of Mercedes convertibles being sold in Boston, etc. What is it that you want to get?

- Use software with the file ***Summer09MercedesCClass*** to do the analysis you chose to do in part b.

c. Express the proportion of convertibles using proper conditional probability notation. Note the place with the highest proportion and the place with the lowest proportion.

Second statistical question: *For the population from which the cars are sampled, can we infer that there is a difference in the proportions of convertibles being sold amongst the places?*

d. For the second statistical question, which inferential technique is appropriate?
   - Hypothesis test for comparing two independent proportions
   - Chi-Square Test for Independence
   - ANOVA
   - Paired Comparison hypothesis test for means

   Give a reason for your choice that indicates why the other choices are not good.

- Use software with the file ***Summer09MercedesCClass*** to do the *inferential* analysis you decided on.

e. Are the conditions for the inferential procedure you have chosen met?

f. What are the null and alternate hypotheses for the inferential technique you have chosen?

g. Interpret the hypothesis test in the context of the question about whether the proportion of convertibles being sold in the four places differs.

Third statistical question: *Describe and display for the sample any differences in the distributions of ages of convertibles being sold compared with cars that are not convertibles. (So compare the distributions of the variable* **Age** *grouped by the variable* **Convertible***)*

h. State what *descriptive* (not *inferential*) techniques you will use to answer this question. Think: do the convertibles tend to be younger, older, more diverse in their ages than the non-convertibles? Describe the kinds of descriptive measures or graphics you will get.

- Use software with the file ***Summer09MercedesCClass*** to do the analysis you chose to do in part h. The output will have both graphics and numerical summaries.

i. Write up a coherent explanation of what the graphics and the numbers tell you in the context of the third statistical question.

Fourth statistical question: *For the population from which the cars are sampled, can we infer that there is a difference in mean age of convertibles as compared with cars that are not convertibles?*

j. What *inferential* technique will you use?
   - Hypothesis test for two independent proportions
   - Chi-Square Test for Independence
   - Hypothesis test for comparing two independent means
   - Paired Comparison hypothesis test

   Give a reason for your choice that indicates why the other choices are not good..

k. Before embarking on the analysis, are the two conditions for the inferential technique that you have chosen met? Explain how they are or, if you think not, why you have reservations.

l. Set up, using the correct notation, the null and alternate hypotheses for your inferential procedure.

m. Confused Conrad has $H_0 : p_{NC} = p_C$ and Forgetful Fiona has $H_0 : \bar{x}_{NC} = \bar{x}_C$ for their answers to part l. Is either correct? What errors are they making?

- Use software with the file ***Summer09MercedesCClass*** to do the analysis you chose.

n. Interpret the results of your *inferential* procedure in the context of the question.

Fifth statistical question: *For the population from which the cars are sampled, estimate how much older the non-convertibles are compared with the convertibles.*

- o. Choose the *inferential* procedure that should be used to answer the fifth statistical question and Use software with the file ***Summer09MercedesCClass*** to do the analysis you chose.

p. Give a good interpretation of the results of your *inferential* procedure.

Sixth statistical question: *For the population from which the cars are sampled, can we infer that there is a difference in mean age of amongst the body styles convertible, coupe and sedans, excluding the wagons) of the cars?*

- q. Choose both *descriptive* and *inferential* techniques to answer the sixth statistical question and Use software with the file ***Summer09MercedesCClass*** to do the analyses you chose. This will involve a filter to exclude the wagons.

r. Give a good interpretation of the results of your analysis in the context of the statistical question.

2. ***Seventh Statistical Question on Year 2006 Mercedes Benzes*** This question uses the same data as question 1 but only looks at the cars for model year 2006. Our question is: *"Do drivers of convertibles drive less than drivers of other body styles of Mercedes Benzes being sold?"*

   Seventh statistical question: *For the population from which the cars are sampled, can we infer that there is a difference in mean miles driven (the variable* ***Miles****) between convertibles and cars being sold that are not convertibles?*

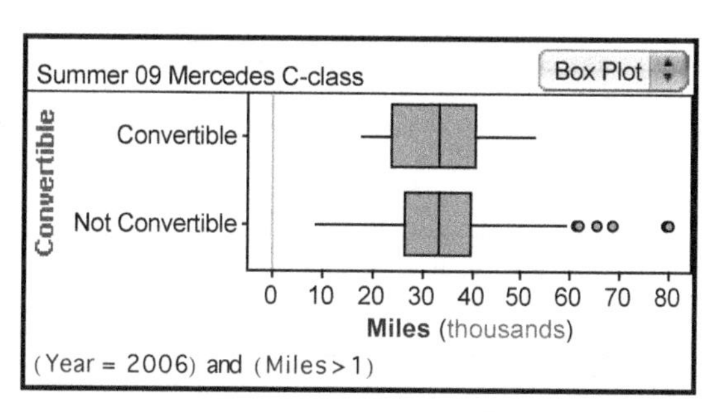

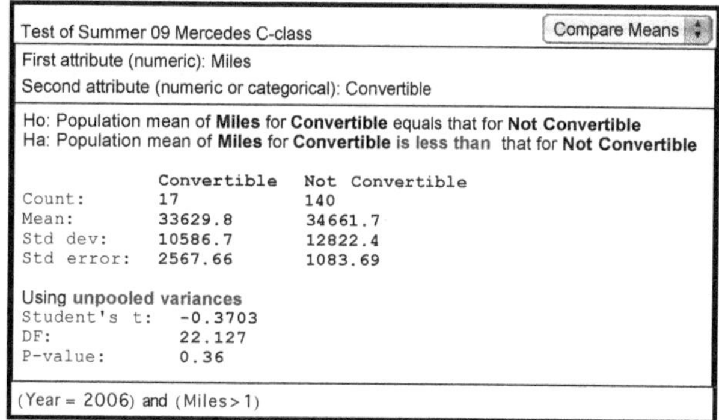

```
Test of Summer 09 Mercedes C-class                          Compare Means
First attribute (numeric): Miles
Second attribute (numeric or categorical): Convertible

Ho: Population mean of Miles for Convertible equals that for Not Convertible
Ha: Population mean of Miles for Convertible is less than that for Not Convertible

               Convertible   Not Convertible
Count:         17            140
Mean:          33629.8       34661.7
Std dev:       10586.7       12822.4
Std error:     2567.66       1083.69

Using unpooled variances
Student's t:   -0.3703
DF:            22.127
P-value:       0.36

(Year = 2006) and (Miles > 1)
```

   a. Is this a one-sided or two-sided test? Explain your answer.
   b. Are the conditions met for the hypothesis test? Explain your answer.
   c. Show how the test statistic was calculated using the numbers given. (Show where the numbers go in the formula.)
   d. Explain how the hypothesis test gives an answer to the statistical question posed above.

3. ***SFO or OAK? Which airport?*** This exercise looks at the differences between Oakland and SFO airports for flights to LAX. The data are in the file ***OnTimeOAKSFOLAX2009Sample.*** The data are a simple random sample of the flight data from the Oakland and San Francisco airports to LAX.

   First statistical question: *For all flights from the Bay Area to LAX is the duration of the flight (the variable ActualDuration in minutes) different from SFO compared with Oakland International (OAK)?*

- Open the file ***OnTimeOAKSFOLAX2009Sample***. Have software produce graphics and summary statistics for the relationship between *ActualDuration* and *Airport*.

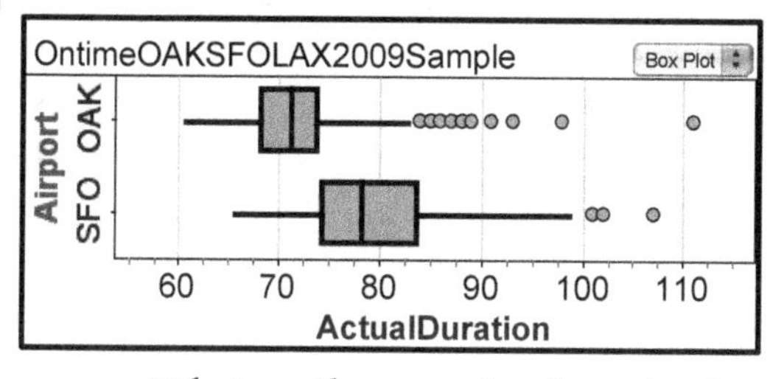

Summary statistics for ActualDuration:
Group by: Airport

Airport	n	Mean	Median	Std. dev.	IQR
OAK	225	71.644444	71	6.790078	6
SFO	200	79.295	78	7.8175151	10

   a. What are the cases for these data?
   b. Interpret the descriptive analysis in the context of the question. Interpret both the graphics and the numbers in the summary table.
   c. What kind of *inferential* analyses are implied by the statistical question?
      - Hypothesis test for two independent means
      - Paired Comparison hypothesis test
      - Either a hypothesis test or a confidence interval for two independent means

      Give a reason for your answer.

d. Use your descriptive analysis and the sample sizes to assess whether the conditions are met for the *inferential* analysis you chose. The data are a random sample of flights.

- Get software to do the *inferential* analysis you chose.

e. You can actually answer this question with either a hypothesis test or a confidence interval. If you chose a hypothesis test, show how the software calculated the test statistic. If you chose a confidence interval, show (approximately) how software calculated that.

f. Interpret the results of your analysis in the context of our statistical question. Use good stat language but also put the answer into a form an "ordinary" airline traveler would understand.

Second statistical question: *For all flights from the Bay Area to LAX does the probability of a departure delay greater than fifteen minutes differ between the two airports?"*

Here is another summary table for the variable that asks whether the departure delay for a flight is over fifteen minutes, or fifteen minutes or less.

Sample of OntimeCombLAX

		Airport		Row Summary
		OAK	SFO	
DelayOver15	Departure Delay 15 min or Less	187	145	332
	Departure Delay Over 15 min	38	55	93
	Column Summary	225	200	425

S1 = count ( )

h. Calculate probabilities, using good notation and in the correct direction, to answer our question in a descriptive sense. Write your conclusion.

i. Which *inferential* analysis will be appropriate for our statistical question? (Your answer should be something like, "A hypothesis test/confidence interval for…" that is, you need to specify the type of procedure and which of the procedures that we have done. There is more than one correct answer, and there is more than one incorrect answer.) Give a reason for your answer.

j. Are the conditions met for your procedure?

- Get software using the file ***OnTimeOAKSFOLAX2009Sample*** to do the *inferential* analysis you chose.

k. Interpret the results of the inferential procedure in the context of the statistical question posed.

4. *Which Airport, continued.* This exercise continues to look at the differences between Oakland and SFO airports for flights to LAX. The data are in the same file: ***OnTimeOAKSFOLAX2009Sample.*** One of the variables that is measured for every flight is the *TaxioutTime*, the time it takes before the plane actually leaves the ground.

First statistical question:

For all flights between the Bay Area and LAX, does mean TaxioutTime differ between the airports OAK and SFO?

a. Besides the graphic shown, what other *descriptive* analyses will be useful to answer our question?

- With the file ***OnTimeOAKSFOLAX2009Sample*** use software to produce the additional descriptive analyses.

b. Interpret the *descriptive* analysis in the context of the question. You should have gotten some numerical summaries to add to the graphic. Comment on what the numbers and the graphic tell you.

c. What kind of *inferential* analyses are implied by the statistical question? Give a reason for your answer.

d. Use your descriptive analysis and the sample sizes to assess whether the conditions are met for the *inferential* analysis you chose.

- With the file ***OnTimeOAKSFOLAX2009Sample*** use software to carry out the inferential analysis.

e. You can actually answer this question with either a hypothesis test or a confidence interval. If you chose a hypothesis test, show how software calculated the test statistic. If you chose a confidence interval, show (approximately) how software calculated that. (The word is "approximately" because the t^* software used may come from a degrees of freedom calculated by the complicated formula shown in §5.4.)

f. Interpret the results of your analysis in the context of our statistical question. Use good statistical language but also put the answer into a form an "ordinary" airline traveler would understand.

Second statistical question: *How is* TaxioutTime *is related to the* ActualDuraton *of the fligh?* You would think that the longer the taxi out time, the longer the flight.

g. What kind of variables are *TaxioutTime* and *ActualDuraton:* categorical or quantitative?

h. What kind of analysis is appropriate to the question asked above about the relationship between *TaxioutTime* and *ActualDuraton?* Give a reason for your answer.

- With the file ***OnTimeOAKSFOLAX2009Sample*** use software to carry out analysis chosen. You may be able to group the analyses by airport, which should be useful.

i. You should have a linear model to answer the question. Interpret the slopes (for the two airports) in the context of the question. (it turns out they are remarkably similar)

j. Interpret the coefficient of determination for the two airports in the context of the question.

5. *Which Airport and which carrier?* This exercise continues to look at the differences between Oakland and SFO airports for flights to LAX. However, the data are random sample in the file: ***OnTimeOAKSFOLAX2014Sample.*** The first statistical question is the same as in Exercise 3.

First statistical question: *For all flights from the Bay Area to LAX is the duration of the flight (the variable* ActualDuration *in minutes) different for SFO when compared with Oakland International (OAK)?*

- Open the file ***OnTimeOAKSFOLAX2014Sample***. Have software produce graphics and summary statistics for the relationship between *ActualDuration* and *Origin*.

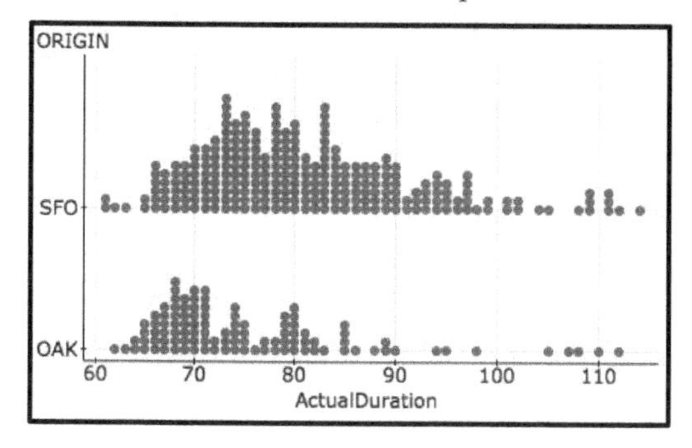

Summary statistics for ActualDuration:
Group by: Airport

Airport	n	Mean	Median	Std. dev.	IQR
OAK	225	71.644444	71	6.790078	6
SFO	200	79.295	78	7.8175151	10

a. The statistical question is the same. Check that the conditions for doing the inferential analysis still hold for this sample.

- With the file ***OnTimeOAKSFOLAX2014Sample*** carry out an inferential analysis that will answer the statistical question.

  b. If a confidence interval was calculated, it should be very close to what is shown here. Explain how the confidence interval results answer the statistical question, and in which way they answer it. Is there a significant difference between SFO and OAK duration of flights to LAX?

95% confidence interval results:
μ_1 : Mean of ActualDuration where ORIGIN="SFO"
μ_2 : Mean of ActualDuration where ORIGIN="OAK"
$\mu_1 - \mu_2$: Difference between two means
(without pooled variances)

Difference	Sample Diff.	Std. Err.	DF	L. Limit	U. Limit
$\mu_1 - \mu_2$	5.4095238	1.2233163	195.97121	2.9969692	7.8220785

  c. Judgment question: is the "significant difference" an important difference? (There can be a variety of answers to this question.)

  d. In the context of the data and the statistical question, explain what a Type I error would be. Explain what a Type II error would be.

Second statistical question: *For all flights from the SFO to LAX is the duration of the flight (the variable ActualDuration in minutes) significantly different for different carriers?*

- With the file ***OnTimeOAKSFOLAX2014Sample*** have software produce graphics and summary statistics that speak to this question. Just below is an example. Use the variables *ActualDuration* and *Carrier*. Notice that in doing this it will be necessary to filter for just the airport SFO.

For the variable *Carrier*, the airline codes are:

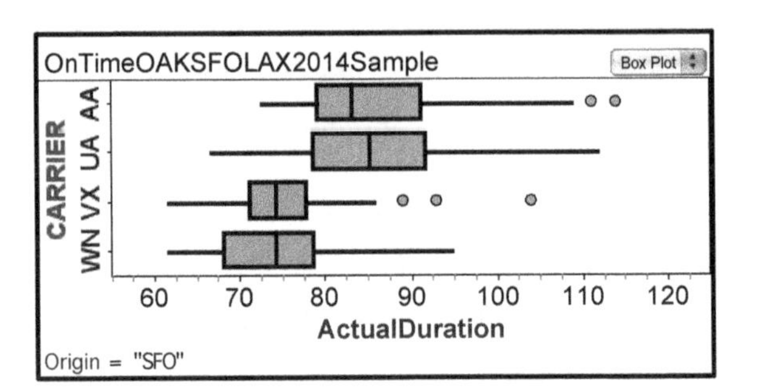

AA	*American Airlines*
UA	*United Airlines*
VX	*Virgin America*
WN	*Southwest Airlines*

  e. Which inferential procedure is the best choice for this statistical question?
     - Chi-Square Test for Independence
     - Paired Comparison hypothesis test for means
     - Independent two-sample hypothesis test for means?
     - Confidence interval for means

     Give a reason for your choice.

  f. Determine whether the conditions are met for the best choice of procedure. To do this, it will be necessary to have the summary statistics of *ActualDuration* grouped by *Carrier*. Explain the conclusions.

  g. Is the ANOVA statistically significant? Interpret the results in the light of the statistical question.

  h. How big are the differences in flight duration? Significant differences are not necessarily meaningful differences.

Standard Normal Probabilities

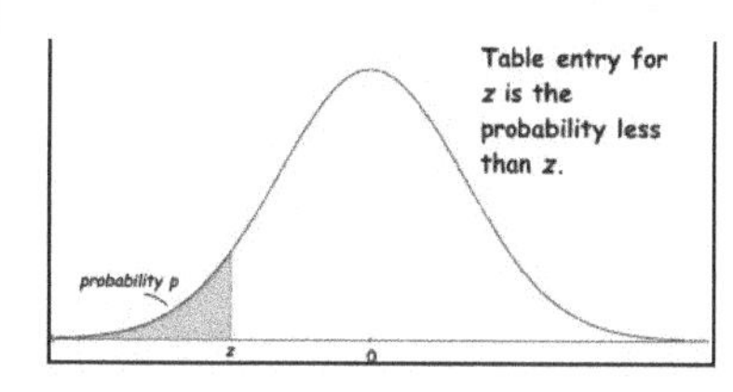
Table entry for
z is the
probability less
than z.
probability p
z
0

z	.00	.01	.02	.03	.04	.05	.06	.07	.08	.09
-3.8	.0001	.0001	.0001	.0001	.0001	.0001	.0001	.0001	.0001	.0001
-3.7	.0001	.0001	.0001	.0001	.0001	.0001	.0001	.0001	.0001	.0001
-3.6	.0002	.0002	.0001	.0001	.0001	.0001	.0001	.0001	.0001	.0001
-3.5	.0002	.0002	.0002	.0002	.0002	.0002	.0002	.0002	.0002	.0002
-3.4	.0003	.0003	.0003	.0003	.0003	.0003	.0003	.0003	.0003	.0002
-3.3	.0005	.0005	.0005	.0004	.0004	.0004	.0004	.0004	.0004	.0003
-3.2	.0007	.0007	.0006	.0006	.0006	.0006	.0006	.0005	.0005	.0005
-3.1	.0010	.0009	.0009	.0009	.0008	.0008	.0008	.0008	.0007	.0007
-3.0	.0013	.0013	.0013	.0012	.0012	.0011	.0011	.0011	.0010	.0010
-2.9	.0019	.0018	.0018	.0017	.0016	.0016	.0015	.0015	.0014	.0014
-2.8	.0026	.0025	.0024	.0023	.0023	.0022	.0021	.0021	.0020	.0019
-2.7	.0035	.0034	.0033	.0032	.0031	.0030	.0029	.0028	.0027	.0026
-2.6	.0047	.0045	.0044	.0043	.0041	.0040	.0039	.0038	.0037	.0036
-2.5	.0062	.0060	.0059	.0057	.0055	.0054	.0052	.0051	.0049	.0048
-2.4	.0082	.0080	.0078	.0075	.0073	.0071	.0069	.0068	.0066	.0064
-2.3	.0107	.0104	.0102	.0099	.0096	.0094	.0091	.0089	.0087	.0084
-2.2	.0139	.0136	.0132	.0129	.0125	.0122	.0119	.0116	.0113	.0110
-2.1	.0179	.0174	.0170	.0166	.0162	.0158	.0154	.0150	.0146	.0143
-2.0	.0228	.0222	.0217	.0212	.0207	.0202	.0197	.0192	.0188	.0183
-1.9	.0287	.0281	.0274	.0268	.0262	.0256	.0250	.0244	.0239	.0233
-1.8	.0359	.0351	.0344	.0336	.0329	.0322	.0314	.0307	.0301	.0294
-1.7	.0446	.0436	.0427	.0418	.0409	.0401	.0392	.0384	.0375	.0367
-1.6	.0548	.0537	.0526	.0516	.0505	.0495	.0485	.0475	.0465	.0455
-1.5	.0668	.0655	.0643	.0630	.0618	.0606	.0594	.0582	.0571	.0559
-1.4	.0808	.0793	.0778	.0764	.0749	.0735	.0721	.0708	.0694	.0681
-1.3	.0968	.0951	.0934	.0918	.0901	.0885	.0869	.0853	.0838	.0823
-1.2	.1151	.1131	.1112	.1093	.1075	.1056	.1038	.1020	.1003	.0985
-1.1	.1357	.1335	.1314	.1292	.1271	.1251	.1230	.1210	.1190	.1170
-1.0	.1587	.1562	.1539	.1515	.1492	.1469	.1446	.1423	.1401	.1379
-0.9	.1841	.1814	.1788	.1762	.1736	.1711	.1685	.1660	.1635	.1611
-0.8	.2119	.2090	.2061	.2033	.2005	.1977	.1949	.1922	.1894	.1867
-0.7	.2420	.2389	.2358	.2327	.2296	.2266	.2236	.2206	.2177	.2148
-0.6	.2743	.2709	.2676	.2643	.2611	.2578	.2546	.2514	.2483	.2451
-0.5	.3085	.3050	.3015	.2981	.2946	.2912	.2877	.2843	.2810	.2776
-0.4	.3446	.3409	.3372	.3336	.3300	.3264	.3228	.3192	.3156	.3121
-0.3	.3821	.3783	.3745	.3707	.3669	.3632	.3594	.3557	.3520	.3483
-0.2	.4207	.4168	.4129	.4090	.4052	.4013	.3974	.3936	.3897	.3859
-0.1	.4602	.4562	.4522	.4483	.4443	.4404	.4364	.4325	.4286	.4247
0.0	.5000	.4960	.4920	.4880	.4840	.4801	.4761	.4721	.4681	.4641

Standard Normal Probabilities

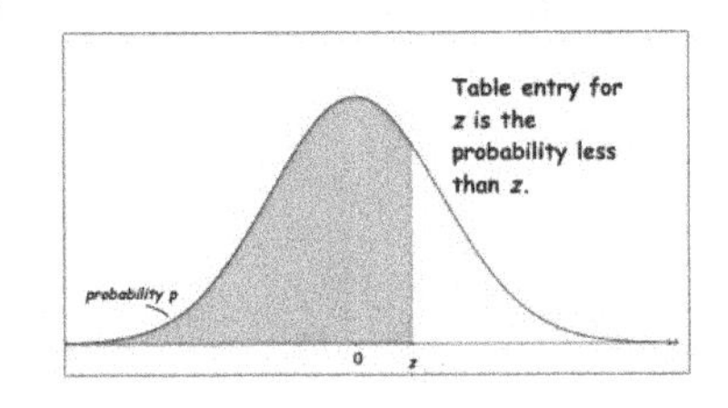
Table entry for z is the probability less than z.
probability p
0
z

z	.00	.01	.02	.03	.04	.05	.06	.07	.08	.09
0.0	.5000	.5040	.5080	.5120	.5160	.5199	.5239	.5279	.5319	.5359
0.1	.5398	.5438	.5478	.5517	.5557	.5596	.5636	.5675	.5714	.5753
0.2	.5793	.5832	.5871	.5910	.5948	.5987	.6026	.6064	.6103	.6141
0.3	.6179	.6217	.6255	.6293	.6331	.6368	.6406	.6443	.6480	.6517
0.4	.6554	.6591	.6628	.6664	.6700	.6736	.6772	.6808	.6844	.6879
0.5	.6915	.6950	.6985	.7019	.7054	.7088	.7123	.7157	.7190	.7224
0.6	.7257	.7291	.7324	.7357	.7389	.7422	.7454	.7486	.7517	.7549
0.7	.7580	.7611	.7642	.7673	.7704	.7734	.7764	.7794	.7823	.7852
0.8	.7881	.7910	.7939	.7967	.7995	.8023	.8051	.8078	.8106	.8133
0.9	.8159	.8186	.8212	.8238	.8264	.8289	.8315	.8340	.8365	.8389
1.0	.8413	.8438	.8461	.8485	.8508	.8531	.8554	.8577	.8599	.8621
1.1	.8643	.8665	.8686	.8708	.8729	.8749	.8770	.8790	.8810	.8830
1.2	.8849	.8869	.8888	.8907	.8925	.8944	.8962	.8980	.8997	.9015
1.3	.9032	.9049	.9066	.9082	.9099	.9115	.9131	.9147	.9162	.9177
1.4	.9192	.9207	.9222	.9236	.9251	.9265	.9279	.9292	.9306	.9319
1.5	.9332	.9345	.9357	.9370	.9382	.9394	.9406	.9418	.9429	.9441
1.6	.9452	.9463	.9474	.9484	.9495	.9505	.9515	.9525	.9535	.9545
1.7	.9554	.9564	.9573	.9582	.9591	.9599	.9608	.9616	.9625	.9633
1.8	.9641	.9649	.9656	.9664	.9671	.9678	.9686	.9693	.9699	.9706
1.9	.9713	.9719	.9726	.9732	.9738	.9744	.9750	.9756	.9761	.9767
2.0	.9772	.9778	.9783	.9788	.9793	.9798	.9803	.9808	.9812	.9817
2.1	.9821	.9826	.9830	.9834	.9838	.9842	.9846	.9850	.9854	.9857
2.2	.9861	.9864	.9868	.9871	.9875	.9878	.9881	.9884	.9887	.9890
2.3	.9893	.9896	.9898	.9901	.9904	.9906	.9909	.9911	.9913	.9916
2.4	.9918	.9920	.9922	.9925	.9927	.9929	.9931	.9932	.9934	.9936
2.5	.9938	.9940	.9941	.9943	.9945	.9946	.9948	.9949	.9951	.9952
2.6	.9953	.9955	.9956	.9957	.9959	.9960	.9961	.9962	.9963	.9964
2.7	.9965	.9966	.9967	.9968	.9969	.9970	.9971	.9972	.9973	.9974
2.8	.9974	.9975	.9976	.9977	.9977	.9978	.9979	.9979	.9980	.9981
2.9	.9981	.9982	.9982	.9983	.9984	.9984	.9985	.9985	.9986	.9986
3.0	.9987	.9987	.9987	.9988	.9988	.9989	.9989	.9989	.9990	.9990
3.1	.9990	.9991	.9991	.9991	.9992	.9992	.9992	.9992	.9993	.9993
3.2	.9993	.9993	.9994	.9994	.9994	.9994	.9994	.9995	.9995	.9995
3.3	.9995	.9995	.9995	.9996	.9996	.9996	.9996	.9996	.9996	.9997
3.4	.9997	.9997	.9997	.9997	.9997	.9997	.9997	.9997	.9997	.9998
3.5	.9998	.9998	.9998	.9998	.9998	.9998	.9998	.9998	.9998	.9998
3.6	.9998	.9998	.9999	.9999	.9999	.9999	.9999	.9999	.9999	.9999
3.7	.9999	.9999	.9999	.9999	.9999	.9999	.9999	.9999	.9999	.9999
3.8	.9999	.9999	.9999	.9999	.9999	.9999	.9999	.9999	.9999	.9999

Chi Square Distributions

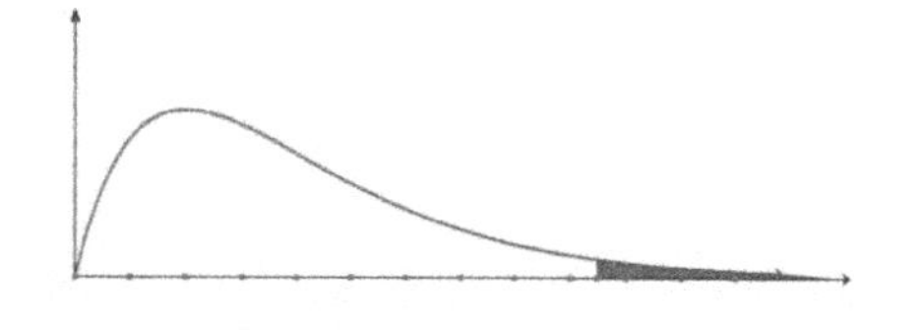

Probability α in the tail of the distribution

df	0.25	0.20	0.15	0.10	0.05	0.025	0.02	0.01	0.005	0.0025	0.001	0.0005
1	1.32	1.64	2.07	2.71	3.84	5.02	5.41	6.63	7.88	9.14	10.83	12.12
2	2.77	3.22	3.79	4.61	5.99	7.38	7.82	9.21	10.60	11.98	13.82	15.20
3	4.11	4.64	5.32	6.25	7.81	9.35	9.84	11.34	12.84	14.32	16.27	17.73
4	5.39	5.99	6.74	7.78	9.49	11.14	11.67	13.28	14.86	16.42	18.47	20.00
5	6.63	7.29	8.12	9.24	11.07	12.83	13.39	15.09	16.75	18.39	20.52	22.11
6	7.84	8.56	9.45	10.64	12.59	14.45	15.03	16.81	18.55	20.25	22.46	24.10
7	9.04	9.80	10.75	12.02	14.07	16.01	16.62	18.48	20.28	22.04	24.32	26.02
8	10.22	11.03	12.03	13.36	15.51	17.53	18.17	20.09	21.96	23.77	26.12	27.87
9	11.39	12.24	13.29	14.68	16.92	19.02	19.68	21.67	23.59	25.46	27.88	29.67
10	12.55	13.44	14.53	15.99	18.31	20.48	21.16	23.21	25.19	27.11	29.59	31.42
11	13.70	14.63	15.77	17.28	19.68	21.92	22.62	24.73	26.76	28.73	31.26	33.14
12	14.85	15.81	16.99	18.55	21.03	23.34	24.05	26.22	28.30	30.32	32.91	34.82
13	15.98	16.98	18.20	19.81	22.36	24.74	25.47	27.69	29.82	31.88	34.53	36.48
14	17.12	18.15	19.41	21.06	23.68	26.12	26.87	29.14	31.32	33.43	36.12	38.11
15	18.25	19.31	20.60	22.31	25.00	27.49	28.26	30.58	32.80	34.95	37.70	39.72
16	19.37	20.47	21.79	23.54	26.30	28.85	29.63	32.00	34.27	36.46	39.25	41.31
17	20.49	21.61	22.98	24.77	27.59	30.19	31.00	33.41	35.72	37.95	40.79	42.88
18	21.60	22.76	24.16	25.99	28.87	31.53	32.35	34.81	37.16	39.42	42.31	44.43
19	22.72	23.90	25.33	27.20	30.14	32.85	33.69	36.19	38.58	40.89	43.82	45.97
20	23.83	25.04	26.50	28.41	31.41	34.17	35.02	37.57	40.00	42.34	45.31	47.50
21	24.93	26.17	27.66	29.62	32.67	35.48	36.34	38.93	41.40	43.78	46.80	49.01
22	26.04	27.30	28.82	30.81	33.92	36.78	37.66	40.29	42.80	45.20	48.27	50.51
23	27.14	28.43	29.98	32.01	35.17	38.08	38.97	41.64	44.18	46.62	49.73	52.00
24	28.24	29.55	31.13	33.20	36.42	39.36	40.27	42.98	45.56	48.03	51.18	53.48
25	29.34	30.68	32.28	34.38	37.65	40.65	41.57	44.31	46.93	49.44	52.62	54.95
26	30.43	31.79	33.43	35.56	38.89	41.92	42.86	45.64	48.29	50.83	54.05	56.41
27	31.53	32.91	34.57	36.74	40.11	43.19	44.14	46.96	49.64	52.22	55.48	57.86
28	32.62	34.03	35.72	37.92	41.34	44.46	45.42	48.28	50.99	53.59	56.89	59.30
29	33.71	35.14	36.85	39.09	42.56	45.72	46.69	49.59	52.34	54.97	58.30	60.73
30	34.80	36.25	37.99	40.26	43.77	46.98	47.96	50.89	53.67	56.33	59.70	62.16
40	45.62	47.27	49.24	51.81	55.76	59.34	60.44	63.69	66.77	69.70	73.40	76.09
50	56.33	58.16	60.35	63.17	67.50	71.42	72.61	76.15	79.49	82.66	86.66	89.56
60	66.98	68.97	71.34	74.40	79.08	83.30	84.58	88.38	91.95	95.34	99.61	102.70
80	88.13	90.41	93.11	96.58	101.88	106.63	108.07	112.33	116.32	120.10	124.84	128.26
100	109.14	111.67	114.66	118.50	124.34	129.56	131.14	135.81	140.17	144.29	149.45	153.17

t-Distribution Critical Values

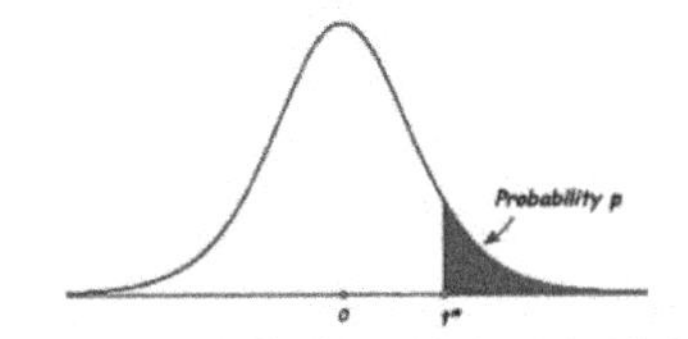

df	0.25	0.15	0.10	0.05	0.025	0.02	0.01	0.005	0.001	0.0005
	Probabilities in the right tail of the distribution									
1	1.000	1.963	3.078	6.314	12.706	15.895	31.821	63.657	318.309	636.619
2	0.816	1.386	1.886	2.920	4.303	4.849	6.965	9.925	22.327	31.599
3	0.765	1.250	1.638	2.353	3.182	3.482	4.541	5.841	10.215	12.924
4	0.741	1.190	1.533	2.132	2.776	2.999	3.747	4.604	7.173	8.610
5	0.727	1.156	1.476	2.015	2.571	2.757	3.365	4.032	5.893	6.869
6	0.718	1.134	1.440	1.943	2.447	2.612	3.143	3.707	5.208	5.959
7	0.711	1.119	1.415	1.895	2.365	2.517	2.998	3.499	4.785	5.408
8	0.706	1.108	1.397	1.860	2.306	2.449	2.896	3.355	4.501	5.041
9	0.703	1.100	1.383	1.833	2.262	2.398	2.821	3.250	4.297	4.781
10	0.700	1.093	1.372	1.812	2.228	2.359	2.764	3.169	4.144	4.587
11	0.697	1.088	1.363	1.796	2.201	2.328	2.718	3.106	4.025	4.437
12	0.695	1.083	1.356	1.782	2.179	2.303	2.681	3.055	3.930	4.318
13	0.694	1.079	1.350	1.771	2.160	2.282	2.650	3.012	3.852	4.221
14	0.692	1.076	1.345	1.761	2.145	2.264	2.624	2.977	3.787	4.140
15	0.691	1.074	1.341	1.753	2.131	2.249	2.602	2.947	3.733	4.073
16	0.690	1.071	1.337	1.746	2.120	2.235	2.583	2.921	3.686	4.015
17	0.689	1.069	1.333	1.740	2.110	2.224	2.567	2.898	3.646	3.965
18	0.688	1.067	1.330	1.734	2.101	2.214	2.552	2.878	3.610	3.922
19	0.688	1.066	1.328	1.729	2.093	2.205	2.539	2.861	3.579	3.883
20	0.687	1.064	1.325	1.725	2.086	2.197	2.528	2.845	3.552	3.850
21	0.686	1.063	1.323	1.721	2.080	2.189	2.518	2.831	3.527	3.819
22	0.686	1.061	1.321	1.717	2.074	2.183	2.508	2.819	3.505	3.792
23	0.685	1.060	1.319	1.714	2.069	2.177	2.500	2.807	3.485	3.768
24	0.685	1.059	1.318	1.711	2.064	2.172	2.492	2.797	3.467	3.745
25	0.684	1.058	1.316	1.708	2.060	2.167	2.485	2.787	3.450	3.725
26	0.684	1.058	1.315	1.706	2.056	2.162	2.479	2.779	3.435	3.707
27	0.684	1.057	1.314	1.703	2.052	2.158	2.473	2.771	3.421	3.690
28	0.683	1.056	1.313	1.701	2.048	2.154	2.467	2.763	3.408	3.674
29	0.683	1.055	1.311	1.699	2.045	2.150	2.462	2.756	3.396	3.659
30	0.683	1.055	1.310	1.697	2.042	2.147	2.457	2.750	3.385	3.646
35	0.682	1.052	1.306	1.690	2.030	2.133	2.438	2.724	3.340	3.591
40	0.681	1.050	1.303	1.684	2.021	2.123	2.423	2.704	3.307	3.551
45	0.680	1.049	1.301	1.679	2.014	2.115	2.412	2.690	3.281	3.520
50	0.679	1.047	1.299	1.676	2.009	2.109	2.403	2.678	3.261	3.496
60	0.679	1.045	1.296	1.671	2.000	2.099	2.390	2.660	3.232	3.460
70	0.678	1.044	1.294	1.667	1.994	2.093	2.381	2.648	3.211	3.435
80	0.678	1.043	1.292	1.664	1.990	2.088	2.374	2.639	3.195	3.416
90	0.677	1.042	1.291	1.662	1.987	2.084	2.368	2.632	3.183	3.402
100	0.677	1.042	1.290	1.660	1.984	2.081	2.364	2.626	3.174	3.390
125	0.676	1.041	1.288	1.657	1.979	2.075	2.357	2.616	3.157	3.370
150	0.676	1.040	1.287	1.655	1.976	2.072	2.351	2.609	3.145	3.357
175	0.676	1.040	1.286	1.654	1.974	2.069	2.348	2.604	3.137	3.347
200	0.676	1.039	1.286	1.653	1.972	2.067	2.345	2.601	3.131	3.340
250	0.675	1.039	1.285	1.651	1.969	2.065	2.341	2.596	3.123	3.330
400	0.675	1.038	1.284	1.649	1.966	2.060	2.336	2.588	3.111	3.315
1000	0.675	1.037	1.282	1.646	1.962	2.056	2.330	2.581	3.098	3.300
∞	0.674	1.036	1.282	1.645	1.960	2.054	2.326	2.576	3.091	3.291
	50%	70%	80%	90%	95%	96%	98%	99%	99.5%	99.9%
	Confidence Level C									